LECTURES ON THE PHYSICS OF STRONGLY CORRELATED SYSTEMS XII

Proceedings in the Series of
Lectures on the Physics of Highly Correlated Electron Systems

Year	Title	Publisher	ISBN
2007	Twelfth Training Course	AIP Conf. Proceedings Vol. 1014	978-0-7354-0535-6
2006	Eleventh Training Course	AIP Conf. Proceedings Vol. 918	978-0-7354-0428-1
2005	Tenth Training Course	AIP Conf. Proceedings Vol. 846	0-7354-0340-6
2004	Ninth Training Course	AIP Conf. Proceedings Vol. 789	0-7354-0279-5
2003	Eighth Training Course	AIP Conf. Proceedings Vol. 715	0-7354-0194-2
2002	Seventh Training Course	AIP Conf. Proceedings Vol. 678	0-7354-0147-0
2001	Sixth Training Course	AIP Conf. Proceedings Vol. 629	0-7354-0083-0
2000	Fifth Training Course	AIP Conf. Proceedings Vol. 580	0-7354-0019-9
1999	Fourth Training Course	AIP Conf. Proceedings Vol. 527	1-56396-950-5
1998	Third Training Course	*not published*	
1997	Second Training Course	AIP Conf. Proceedings Vol. 438	1-56396-789-8
1996	First Training Course	*not published*	

To learn more about these titles, or the AIP Conference Proceedings Series, please visit the webpage **http://proceedings.aip.org**

LECTURES ON THE PHYSICS OF STRONGLY CORRELATED SYSTEMS XII

Twelfth Training Course in the Physics of Strongly Correlated Systems

Salerno, Italy 1 – 12 October 2007

EDITORS
Adolfo Avella
Ferdinando Mancini
Università degli Studi di Salerno
Salerno, Italy

All papers have been peer reviewed

SPONSORING ORGANIZATIONS
Università degli Studi di Salerno, Italy
International Institute for Advanced Scientific Studies "E.R. Caianiello", Italy
Campania Region, Italy
COST P16

AMERICAN INSTITUTE OF PHYSICS

Melville, New York, 2008
AIP CONFERENCE PROCEEDINGS ■ VOLUME 1014

Editors:

Adolfo Avella
Ferdinando Mancini

Dipartimento di Fisica "E. R. Caianiello"
Università degli Studi di Salerno
Via S. Allende
I-84081 Baronissi (SA)
Italy

E-mail: avella@physics.unisa.it
mancini@physics.unisa.it

L.C. Catalog Card No. 2008926871
ISBN 978-0-7354-0535-6
ISSN 0094-243X

Printed in the United States of America

CONTENTS

Preface

The present volume contains the notes of the lectures delivered at the "Twelve Training Course in the Physics of Strongly Correlated Systems" held in Vietri sul Mare (Salerno, Italy) in October 2007. It also contains contributions by some of the participants, who delivered seminars in the afternoon sessions.

Following the tradition of previous years, the meeting was devoted to the training of young scientists in one of the most intriguing fields of condensed matter physics: strongly correlated systems. The intent was to bring together for two weeks four senior and about thirty junior researchers in a close location with an informal atmosphere, paying special attention to foster the active participation of the young researchers.

The course consisted of four lectures every morning, held by Professors Stephan Haas, Mark Jarrell, Hilbert von Loehneysen and Veljko Zlatic, and of afternoon activities (seminars delivered by the junior researchers, solving of specific problems, round table on hot topics, hands-on training on relevant numerical issues exploiting the well-equipped computer room of the IIASS...) aimed principally at promoting discussions between the attendees and the lecturers. The outcome of this type of course was a significant interchange of ideas among the participants thanks to both the enlightening morning lectures and the long afternoon sessions devoted to discussions.

It is hoped that both the meeting, which brought together leaders in the field as well as bright and eager beginners, and the present volume, which may be useful as an up-to-date book for researchers interested in this field, shall be considered as firm points by the scientific community.

We wish to acknowledge the support of those institutions that made the course possible. The main sponsor of the event has been the University of Salerno. Finally, we wish to thank Professor Maria Marinaro, President of the International Institute of Advanced Scientific Studies, who hosted the event in the wonderful and warm venue of Vietri sul Mare.

Adolfo Avella
Ferdinando Mancini
Salerno, April 8th, 2008

PART I

LECTURES

Quantum Magnetism, Nanomagnets and Entanglement

Stephan Haas

Department of Physics and Astronomy, University of Southern California, Los Angeles, CA 90089

Abstract. These lecture notes give an introduction to quantum magnetism. We start by a discussion of paramagnetism and the Weiss meanfield theory applied to the Ising model. In the following sections, we turn to the fully quantum-mechanical Heisenberg Hamiltonian. First, we consider the limit of small clusters whose properties we analyze via exact diagonalization. Bulk quantum magnets are then within the framework of the linear spinwave theory. A short discussion of Monte Carlo techniques follows. Finally, we discuss aspects of entanglement in quantum spin systems.

Keywords: Quantum Magnetism, Nanomagnets, Entanglement
PACS: 75.50.Ee

INTRODUCTION

The Heisenberg Hamiltonian,[1] describing the pairwise interactions between localized spins, is one of the most fundamental models of correlated quantum matter. In spite of its seemingly simple mathematical form, it has an almost unimaginable richness, arising from dimensionality and geometrical constraints, competing exchange interactions, the type of spin degrees of freedom, and additional interactions with external fields or other degrees of freedom, such as phonons.

Assuming that the pairwise interactions between the spins are isotropic, the basic Heisenberg Hamiltonian is commonly represented as

$$H = \sum_{i,j} J_{ij} \vec{S}_i \cdot \vec{S}_j - \vec{h} \sum_i \vec{S}_i \tag{1}$$

where J_{ij} are the exchange integrals between spins on sites i and j which typically decay rapidly with the distance between these sites. They can either be positive, favoring antiferromagnetic alignment, or negative, favoring ferromagnetism. In many cases only nearest-neighbor interactions are considered, and longer-range interactions are neglected. The spin operators $\vec{S}_i$ represent quantum spin degrees of freedom, i.e. with S=1/2 or 1, which arise from the ionic state of particular atoms, such as Fe, Cu or Ni. $\vec{h}$ represents an applied external magnetic field, measured in units of $\mu_B g$. Additional richness enters the Hamiltonian when anisotropies are considered. For example, a frequently studied variant of the isotropic Heisenberg model is the XXZ model with $\vec{S}_i \cdot \vec{S}_j \rightarrow \lambda S_i^z S_j^z + S_i^x S_j^x + S_i^y S_j^y$, where the anisotropy parameter λ interpolates between the XY limit ($\lambda = 0$), the Heisenberg point ($\lambda = 1$) and the Ising limit ($\lambda = \infty$).

Even in its simplest, most studied form, $H = J\sum_{<ij>} \vec{S}_i \cdot \vec{S}_j$, i.e. the spin-1/2 Heisenberg antiferromagnet with nearest-neighbor interactions only, the zero-temperature

CP1014, *Lectures on the Physics of Strongly Correlated Systems XII: Twelfth Training Course*,
edited by A. Avella and F. Mancini

ground state of this system can assume radically different forms, depending on dimensionality and geometry. For hypercubic lattices of dimensionality $D \geq 2$, the ground state is a long-ranged antiferromagnet, characterized by a finite alternating sublattice magnetization which is the order parameter. In a one-dimensional chain, however, quantum fluctuations are sufficiently strong to suppress long-range antiferromagnetic order, and instead the system is in a quasi-long-range-ordered state, characterized by a vanishing sublattice magnetization and algebraically decaying spin-spin correlation functions, $\langle \vec{S}_i \cdot \vec{S}_j \rangle \propto \sqrt{\ln r}/r$. Another quasi-one-dimensional configuration that has been extensively studied is the ladder geometry, i.e. systems of coupled chains.[2] Surprisingly, ladders can have different types of ground states, depending on whether they involve even or odd numbers of chains. Odd-leg ladders are quasi-long-ranged with properties similar to the single chain. In contrast, even-leg ladders have short-ranged ground states with exponentially decaying correlation functions, $\langle \vec{S}_i \cdot \vec{S}_j \rangle \propto \exp(-|i-j|/\xi)$, where the antiferromagnetic correlation length $\xi = v/\Delta$ is inversely proportional to the energy gap between the resonating valence bond ground state and the lowest-lying triplet excitation.

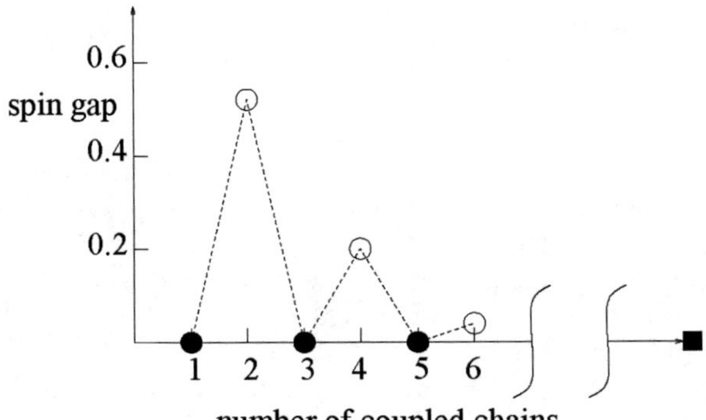

FIGURE 1. Sketch of the spin gap, i.e. the energy gap between the ground state and the first excited triplet state, in quasi-one-dimensional ladder systems. Odd-leg ladders (filled circles) are gapless, whereas even-leg ladders (open circles) have a finite spin gap. The two-dimensional limit (filled square) is long-range ordered.

Frustration effects due to competing antiferromagnetic exchange interactions are known to have equally profound effects on the ground state of the Heisenberg model. These can arise either through frustrated lattice structures, e.g. in a triangular lattice, or in bipartite lattices via longer-range interactions. A prototypical example is the spin-1/2 Heisenberg antiferromagnet on a square lattice with competing nearest-neighbor J_1 and next-nearest-neighbor J_2 interactions. As a function of the ratio J_2/J_1, the system undergoes a series of quantum phase transitions from an ordered Heisenberg antiferromagnet at sufficiently small J_2/J_1 to a quantum-disordered regime at intermediate values of J_2/J_1, and ultimately to a phase with collinear order at sufficiently large J_2/J_1. The precise nature of the intermediate quantum-disordered phase has not been settled, in spite of more than 20 years of extensive scrutiny. However, there is some numerical evidence for it to be in some type of resonating valence bond liquid state.

Continued interest in the Heisenberg model arises because of various reasons. First, there are many magnetic materials whose properties can be well explained based on variants of the Heisenberg Hamiltonian. Second, because of its simplicity this model is often used as a generic starting point to test new ideas and concepts in correlated quantum matter systems. Third, quantum information theory relies on models of interacting quantum bits which are often modeled by Heisenberg clusters.

INDEPENDENT SPINS AND WEISS MEANFIELD THEORY

Paramagnetism: Let us start by considering the response of a single quantum spin, or of an ensemble of independent spins, to an applied magnetic field.[3, 4] For now, we will focus on the field and temperature dependence of two basic observables: the magnetization $m(h,T)$ and the uniform susceptibility $\chi(T) = \partial m(h,T)/\partial h|_{h\to 0}$. The most basic example assumes one spin-1/2 degree of freedom with projections $S_z = \pm 1/2$. The corresponding Hamiltonian can be represented by a 2*2 matrix in the Ising basis,

$$H = -h\hat{S}_z = \frac{h}{2}\begin{pmatrix} 1 & 0 \\ 0 & 1 \end{pmatrix}, \tag{2}$$

which is already diagonal with eigenvalues $E_S = \pm h/2$. The magnetization of this system depends on the applied field and the temperature,

$$m(h,T) = \frac{\sum_{S_z} S_z \exp(-\beta E_{S_z})}{\sum_{S_z} \exp(-\beta E_{S_z})} = \frac{1}{2}\tanh\left(\frac{\beta h}{2}\right). \tag{3}$$

The corresponding magnetic susceptibility is obtained by taking the derivative with respect to the applied magnetic field,

$$\chi(T) = \frac{\beta}{4}\mathrm{sech}^2\left(\frac{\beta h}{2}\right)_{h\to 0} = \frac{\beta}{4}. \tag{4}$$

This result is known as the Curie law for isolated spins, i.e. an inverse dependence on temperature with a prefactor that contains information about the number and type of spins that respond to the applied field.

It is useful to visualize the physical content of $m(h,T)$ by plotting it and considering various limiting cases. For a given temperature, increases with increasing applied magnetic field, saturating at its maximum value of 1/2. The effects of temperature are inverse, i.e. increasing temperature decreases $m(h,T)$. Note that for any infinitesimal applied field, the zero-temperature magnetization is 1/2, whereas in the absence of an applied field it is 0. Taking the limits $x \to 0$ and $x \to \infty$ of $(1/2)\tanh(x)$ with $x = \beta h/2$, one finds $m(h,T) \to \beta h/4$ for small arguments, i.e. large T or small h, and $m(h,T) \to (1 - \exp(-\beta h))/2$ for large arguments.

Note that there is an alternative approach to obtain the above results. Start by evaluating the partition function of the system, $Z(h,T) = \sum_{S_z} exp(-\beta E_{S_z}) = 2\cosh(\beta h/2)$.

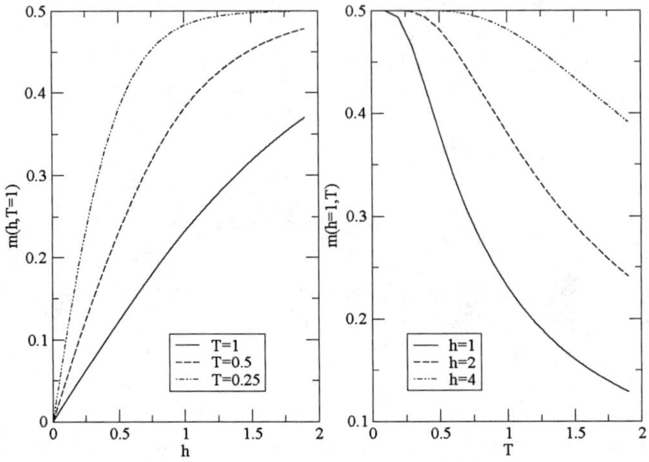

FIGURE 2. The uniform magnetization of a spin-1/2 paramagnet as a function of temperature and applied magnetic field. $m(h,T)$ increases with decreasing temperature and with increasing applied field.

The free energy is then given by $F(h,T) = -\beta^{-1}(\ln 2 + \ln\cosh(\beta h/2))$, and the magnetization is obtained via a derivative, $m(h,T) = \partial F(h,T)\partial h = (1/2)\tanh(\beta h/2)$.

As a formal exercise, let us now consider the case where the applied field is along a direction other than the z-axis. This results in an off-diagonal Hamiltonian matrix,

$$H = -h\hat{S}_x = \frac{h}{2}\begin{pmatrix} 0 & 1 \\ 1 & 0 \end{pmatrix}, \tag{5}$$

with eigenvalues $E_{1,2} = \pm h/2$ and corresponding eigenfunctions $|\psi\rangle_{1,2} = (1/\sqrt{2})(|1\rangle \mp |2\rangle)$. These are the ingredients needed to build up the density matrix of the system,

$$\rho(h,T) = \sum_{i=1}^{2} \exp(-\beta E_i)\psi_i\rangle\langle\psi_i = \begin{pmatrix} \cosh(\beta h/2) & \sinh(\beta h/2) \\ \sinh(\beta h/2) & \cosh(\beta h/2) \end{pmatrix} \tag{6}$$

with a trace of $2\cosh(\beta h/2)$. In addition, $tr(\rho(h,T)\cdot S_x) = \sinh(\beta h/2)$, which leads to a uniform magnetization $m(h,T) = tr(\rho(h,T)\cdot S_x)/tr(\rho(h,T)) = (1/2)\tanh(\beta h/2)$. Naturally, the final result does not depend on the initial choice of quantization axis. However, it is helpful for later purposes to remember this derivation for cases involving non-diagonal Hamiltonian matrices.

Moving on towards the case of arbitrary spin, let us next consider spin-1 degrees of freedom with projections $S_z = 1, 0, -1$. Following the same steps as before, we arrive at $m(h,T) = 2\sinh(\beta h)/(1 + 2\cosh(\beta h))$ and $\chi(T) = 2\beta/3$. From this we see that the prefactor of the Curie law indeed depends on the value of the spin in a non-trivial fashion.

Let us next consider the general spin-S case. In order to evaluate the partition function, we will make use of the finite geometric series $\sum_{k=0}^{n} r^k = (1-r^{n+1})/(1-r)$. The partition function of a spin-S system has 2S+1 terms which can be factorized and summed using the above series:

$$\begin{aligned} Z(h,T) &= \exp(\beta hS) + \exp(\beta h(S-1)) + \cdots + \exp(-\beta h(S-1)) + \exp(-\beta hS) \\ &= \exp(\beta hS)\frac{1-\exp(-\beta h(S+1))}{1-\exp(-\beta h)} + \exp(-\beta hS)\frac{1-\exp(\beta hS)}{1-\exp(\beta h)} \\ &= \frac{\sinh((\beta h/2)(2S+1))}{\sinh(\beta h/2)}. \end{aligned} \tag{7}$$

The corresponding free energy hence becomes $F(h,T) = -\beta^{-1}[\ln \sinh((\beta h/2)(2S+1) - n \sinh(\beta h/2)]$, and the uniform magnetization is

$$m(h,T) = -\frac{\partial F(h,T)}{\partial h} = \left(\frac{2S+1}{2}\right)\coth\left(\frac{(\beta h/2)(2S+1)}{2}\right) - \frac{1}{2}\coth(\beta h/2). \tag{8}$$

This expression is useful in visualizing the quantum-to-classical crossover that occurs with increasing S. Basically, as S increases more quantization steps occur in the spin projection S_z, which eventually merge into a quasi-continuum.

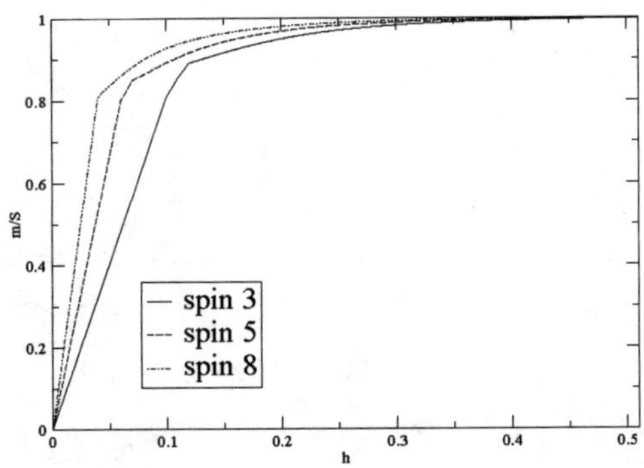

FIGURE 3. Quantum-classical crossover in the magnetization of a paramagnet as a function of the spin S at a given temperature (here $T = 10$). Because of greater quantum fluctuations, the smaller spins require a higher applied magnetic field to achieve saturation. Note that a numerical evaluation of the expression for $m(h,T)$ is tricky because it requires exact cancellations of divergencies at low fields. This is the reason for the "spikes" at $h \leq 0.15$ in the figure.

As $s \to \infty$, one can replace the sum by an integral, and thus obtain

$$m(h,T) \to \frac{\int_{-s}^{s} ds' s' \exp(\beta hs')}{\int_{-s}^{s} ds' \exp(\beta hs')} = s \cdot \coth(\beta hs) - \frac{1}{\beta h}, \tag{9}$$

which turns out to be a good approximation to the sum for $s \geq 3$.

Molecular Field Theory for the Ising Model: Following this brief review of paramagnetism, it is useful to consider the molecular field theory for the Ising model. The idea underlying this approximation is that the spin-spin interactions in the model give rise to an *effective* magnetic field, which then can be treated in a similar manner as the magnetic field in the above treatment of paramagnets.

In the presence of am applied magnetic field, the quantum ferromagnetic Ising model is given by

$$H = -J \sum_{\langle i,j, \rangle} S_i^z \cdot S_j^z - h \sum_i S_i^z, \tag{10}$$

where we have assumed the simplest case in which the external magnetic field points along the quantization direction (z-axis). Let us for now consider ferromagnetic interactions, and discuss the antiferromagnetic case later on. The spin-spin interaction term can be represented as $S_i^z \cdot S_j^z = (S_i^z - m) \cdot (S_j^z - m) - m^2 + m \cdot (S_i^z + S_j^z)$, where m is the mean value of the spin projection along the z-axis. Therefore, the first term in this representation can be assumed to be small, and is neglected in the molecular field approximation. The remaining two terms yield a constant off-set in energy and an effective field $h_{eff} = -Jzm/2$. Here z is the number of nearest neighbors (note that the sum in the Ising interaction is assumed to run over all nearest-neighbor pairs), and the factor 1/2 avoids double-counting of nearest-neighbor pairs. The resulting Ising meanfield Hamiltonian is hence given by

$$H_{meanfield} = \frac{Jm^2Nz}{2} - \left(\frac{Jmz}{2} - h\right) \sum_{i=1}^{N} S_i^z, \tag{11}$$

which is of the same form as the paramagnetic Hamiltonian with a zero-energy offset. A self-consistent transcendental equation for the meanfield magnetization is obtained by minimizing $H_{meanfield}$ with respect to m, and by borrowing the results for paramagnets from the previous section:

$$m(h,T) = \frac{N}{2} \tanh\left[\frac{\beta}{2}\left(h + Jzm(h,T)\right)\right]. \tag{12}$$

This equation can be solved numerically for $m(h,T)$, using standard techniques such as bisection, Newton's method or simple iteration. Moreover, analytical solutions are available in limits such as $T \to 0$ and $T \to T_c$, where T_c is the system's critical temperature which we will discuss in the following.

A graphic solution of the transcendental equation for $m(h,T)$ is shown in Fig. 4 below. Below a critical temperature T_c, there are three solutions, two of which are magnetic with $m(h,T) > 0$ and $m(h,T) < 0$, and one is paramagnetic with $m(h,T) = 0$. These solutions reflect the rotational symmetry of the problem: in the absence of an external magnetic field, perfect ferromagnetic alignment can occur in two directions, and these two solutions are energetically degenerate. They are also degenerate with the paramagnetic solution, but the free energy of the ferromagnetic solutions is lower. Above T_c, only the paramagnetic solution remains.

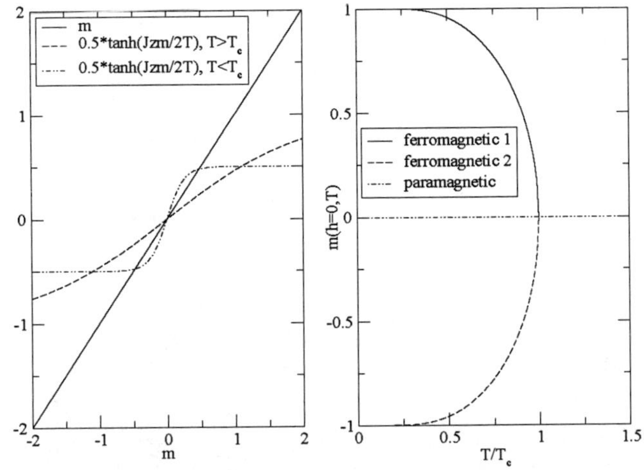

FIGURE 4. Graphic solution of the transcendental equation arising from the mean field solution of the Ising model. Below the transition temperature, there are three points of intersection, indicating 2 magnetic and 1 paramagnetic solutions. Above the transition temperature, only the paramagnetic solution remains.

Let us quickly sketch the free energy argument. At zero temperature, we can simply compare the ground state energies of the paramagnet $U_{para} = 0$ with that of the ferromagnet $U_{ferro} = -JS^2Nz/2$. At finite temperatures, we need to minimize the free energy, defined by $F = U - TS$ for the three competing solutions. The free energies can be calculated directly from the partition function, and since $F = -\ln(Z)/\beta$, we are looking for the solution(s) with the maximum partition function. For the paramagnetic case, the partition function is given by $Z_{para} = 2^N$, whereas for the magnetic solutions the partition function is $Z_{mag} = (2\cosh(\beta Jzm/2))^N$ which for $T < T_c$ is larger than Z_{para}.

In the absence of an external magnetic field, the critical temperature of the system can be obtained by minimizing the transcendental equation with respect to the magnetization m, which yields

$$1 = \frac{\partial}{\partial m}\left[\frac{1}{2}\tanh\left(\frac{\beta Jzm}{2}\right)\right] = \frac{\beta Jz}{4}\operatorname{sech}^2\left(\frac{\beta Jzm}{2}\right)_{m=0}, \tag{13}$$

and hence

$$T_c = Jz/4. \tag{14}$$

This simple result is quite remarkable. It states that the critical temperature, below which the system orders magnetically, depends on the energy scale J, the lattice geometry via z, and on the spin, which here gives rise to the factor 1/4.

The transcendental equation can be expanded in the limits $T \to 0$ and $T \to T_c$, yielding $m(h=0, T\to 0) \approx 1-2\cdot\exp(-2T_c/T)$ and $m(h=0, T\to T_c) \approx \sqrt{3(1-(T/T_c))}$, which close to the transition temperature gives the typical meanfield exponent $m \propto t^{1/2}$ (where $t \equiv 1-(T/T_c)$). This result is, in fact, not correct. A more thorough examination of the Ising model close to the critical point, e.g. using renormalization group and Monte Carlo techniques, gives $m \propto t^{1/3}$ in three spatial dimensions.

The above formalism can be extended to finite applied fields, and used to calculate the magnetic susceptibility. One finds that

$$m = \frac{1}{2}\tanh\left[\frac{h+Jzm}{2T}\right] \to_{h\to 0, T>T_c} \frac{h+Jzm}{4T}. \tag{15}$$

This gives approximately $m \approx \frac{h}{4T} + \frac{mT_c}{T}$, which implies $m \approx \frac{h}{4(T-T_c)}$ and $\chi \approx \frac{1}{4(T-T_c)}$, i.e. the well-known Curie-Weiss law.

At finite applied magnetic fields, only one of the three solutions of the transcendental equation survives at all temperatures. In this case, the system is always in a paramagnetic state. One can visualize this by considering what happens to the graphic solution in Fig. 4, when the x-axis is off-set for the straight line, corresponding to the effect of a finite applied field. Then there will be only one intercept between the straight line and the tanh(x), corresponding to the paramagnetic solution.

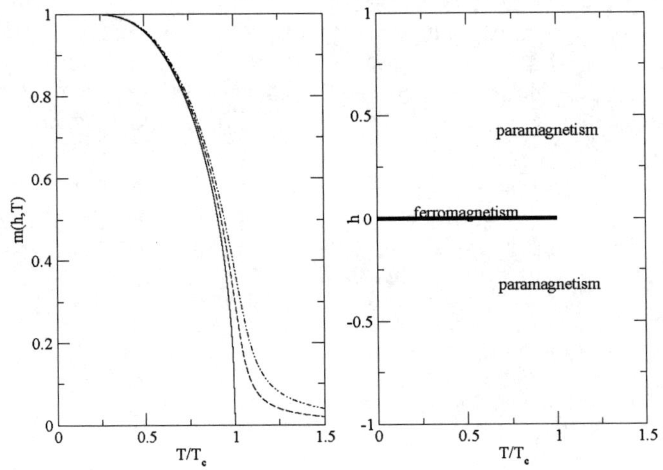

FIGURE 5. Meanfield solution of the Ising ferromagnet in a magnetic field. Left: the black line corresponds to $h = 0$, whereas the dashed and dot-dashed lines are for finite applied fields, in which case there is no phase transition as a function of temperature. The resulting phase diagram is shown in the right figure.

It is interesting to use the meanfield equation in the presence of a finite external field to extract the critical exponent of the system as a function of the applied field.

Obviously, the critical field for this system is $h = 0$, and the question of interest is how the magnetization of the system scales as we approach $h \to h_c = 0$. Let us start by rescaling the self-consistency equation in the following form: $\sigma = \tanh[((h/2) + \sigma T_c)/(T_c)] \equiv \tanh(\tilde{h} + \sigma)$, where we set $\tilde{h} \equiv h/(2T_c) \ll \sigma$ since we are investigating the small field critical behavior. Then we can expand the right-hand-side of the transcendental equation, $\sigma \approx (\tilde{h} + \sigma) - (1/3)(\tilde{h} + \sigma)^3$, from which follows that $\sigma \propto h^{1/3}$, because the $\tilde{h}$ in the $(\tilde{h} + \sigma)^3$ term can be neglected.

Hence, based on the meanfield analysis we obtain the meanfield critical exponents for the Ising model, $\beta = 1/2, \delta = 3$, which should be contrasted with the scaling exponents which are obtained numerically using Monte Carlo. For example, in the 3D case, one finds $\beta = 0.325, \delta = 4.82$.

The molecular field approximation can be extended to the case of antiferromagnets with not too many changes. The ground state energy of the classical antiferromagnet is $E_{antiferro} = -JS^2Nz/2$, i.e. the energy off-set is the same as for the ferromagnet. An important difference, however, is that because of the alternating spin order in an antiferromagnet, one now needs to consider sublattice magnetizations m_A and m_B instead of the uniform magnetization in the ferromagnetic case. In particular, one finds a set of coupled self-consistent equations,

$$|m_A| = \frac{1}{2}\tanh\left(\frac{\beta(h - Jz|m_B|)}{2}\right) \quad (16)$$

$$|m_B| = \frac{1}{2}\tanh\left(\frac{\beta(h - Jz|m_A|)}{2}\right), \quad (17)$$

which a phase diagram analogous to the ferromagnet, simply with the magnetization now replaced by the sublattice magnetization, and with a transition temperature, i.e. the Neel temperature $T_N = Jz/4$. For finite but small applied magnetic fields, these equations can be expanded as done above for the ferromagnetic case, and the susceptibility can be calculated. For temperatures $T > T_N$ one finds $\chi_{antiferro} = 1/(4(T + T_N))$, which is to be contrasted with $\chi_{ferro} = 1/(4(T - T_c))$.

Below the Neel temperature, one needs to distinguish between different directions of the applied field. If the applied field is parallel to the quantization axis of the antiferromagnet, i.e. here the z-axis, one finds for the uniform susceptibility

$$\chi_{\parallel} = \frac{\partial}{\partial h}\frac{m_A + m_B}{2}|_{h=0} \to \exp(-\beta h). \quad (18)$$

On the other hand, if the applied field is perpendicular to the quantization axis, one considers the projection onto the the original direction at $h = 0$, $|m_A| = |m_B| = \sin(\theta/2) \approx h/2$, which yields $\chi_{\perp} = const$.

We finish this section with a summary of the uniform susceptibilities in a paramagnet, a ferromagnet and an antiferromagnet, shown in Fig. 6. The uniform susceptibility is one of the most straightforward experimental observables to distinguish the magnetic properties of magnetic materials. It can be measured using nuclear magnetic resonance techniques.

The paramagnet follows the simple Curie law. This means that the uniform susceptibility diverges at zero temperature, and the system shows no "resistance" to change in

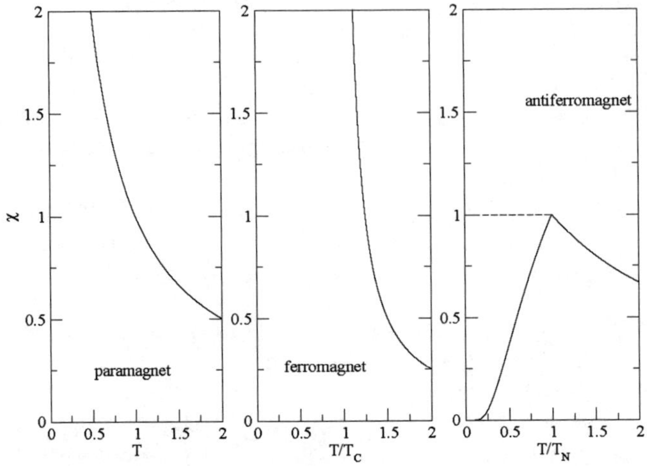

FIGURE 6. Comparison of the uniform susceptibilities in a paramagnet, a ferromagnet and an antiferromagnet, all in the absence of an applied magnetic field. For the antiferromagnet, the susceptibility below the Neel temperature depends on the direction of the external field (solid: parallel to the quantization axis, dashed: perpendicular).

the applied field, i.e. the spins align even along infinitesimal external magnetic fields. The ferromagnet follows the Curie-Weiss law. Above the Curie-Weiss temperature, it behaves like an effective paramagnet. However, below the transition temperature it diverges. Since the critical magnetic field is zero, all spins align along externally applied fields, even if they are of infinitesimal strength. In an antiferromagnet, the uniform susceptibility is finite for all temperatures. At high temperatures, it behaves like an effective paramagnet. Below, the Neel temperature the susceptibility depends on the direction of the applied field, as discussed above. Note since the antiferromagnet's susceptibility is finite, it does "resist" external magnetic fields, at least to a certain extend.

FINITE HEISENBERG CLUSTERS

In order to gain a better understanding of the quantum properties of magnetic systems, it is useful to consider small Heisenberg clusters whose eigenstates are accessible to direct diagonalization. It turns out that many of the most prominent quantum effects are most visible in the smallest clusters. This in turn is of relevance to nanoscience and quantum information theory. Moreover, it is a first step towards extrapolating the properties of bulk systems.

TABLE 1. 2-site spin-1/2 Heisenberg cluster

eigenvalue	eigenvector	spin-rotation	S^z_{tot}
$-3J/4$	$(\vert\uparrow\downarrow\rangle - \vert\downarrow\uparrow\rangle)/\sqrt{2}$	antisymmetric	0
$J/4$	$(\vert\uparrow\downarrow\rangle + \vert\downarrow\uparrow\rangle)/\sqrt{2}$	symmetric	0
$J/4$	$\vert\uparrow\uparrow\rangle$	symmetric	1
$J/4$	$\vert\downarrow\downarrow\rangle$	symmetric	-1

2-site cluster: Let us start by examining the simplest possible Heisenberg cluster, i.e. the spin-1/2 two-site dimer whose Hamiltonian is given by

$$H = J\vec{S}_1 \cdot \vec{S}_2 = J\left(S^z_1 S^z_2 + \frac{1}{2}(S^+_1 S^-_2 + S^-_1 S^+_2)\right). \tag{19}$$

The Hilbert space of this system is of size 2^N , and a commonly used set of basis states is given by $|11\rangle, |10\rangle, |01\rangle, |00\rangle$, where 1 denotes an up-spin, and 0 denotes a down-spin. We can construct a matrix representation of the Hamiltonian by acting on each of these basis states:

$$\begin{aligned} H|11\rangle &= \frac{J}{4}|11\rangle \\ H|10\rangle &= -\frac{J}{4}|10\rangle + \frac{J}{2}|01\rangle \\ H|01\rangle &= -\frac{J}{4}|01\rangle + \frac{J}{2}|10\rangle \\ H|00\rangle &= \frac{J}{4}|00\rangle \end{aligned}$$

which leads to the matrix representation:

$$H = \frac{J}{4}\begin{pmatrix} 1 & 0 & 0 & 0 \\ 0 & -1 & 2 & 0 \\ 0 & 2 & -1 & 0 \\ 0 & 0 & 0 & 1 \end{pmatrix}. \tag{20}$$

This Hamiltonian is block-diagonal with respect to subspaces of definite S^z_{tot}, i.e. for subspaces $S^z_{tot} = 1$ and $S^z_{tot} = -1$ there is only one entry, corresponding to eigenvalues $J/4$, and for the subspace $S^z_{tot} = 0$, the block matrix is

$$H_{S^z_{tot}=0} = \frac{J}{4}\begin{pmatrix} -1 & 2 \\ 2 & -1 \end{pmatrix}, \tag{21}$$

with eigenvalues $-3J/4$ and $J/4$. Hence, the ground state is a singlet, and the first excited state is a triplet. We can summarize the result of the matrix diagonalization in Table 1.

When adding a Zeeman term with an external magnetic field to the Hamiltonian, i.e.

$$H = J\vec{S}_1 \cdot \vec{S}_2 - h(S^z_1 + S^Z_2), \tag{22}$$

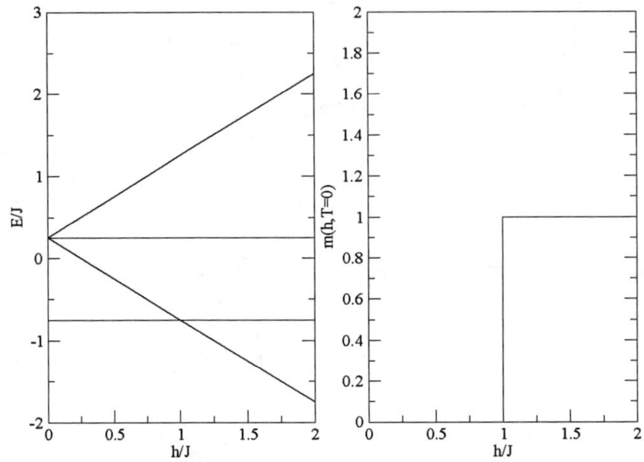

FIGURE 7. Energy levels and zero-temperature magnetization of a two-site Heisenberg cluster in an applied magnetic field. At a critical field strength $h_c = J$, there is a crossover in energy eigenstates, at which the system magnetization jumps from 0 to 1.

the Hamiltonian matrix changes to

$$H = \begin{pmatrix} J/4-h & 0 & 0 & 0 \\ 0 & -J/4 & J/2 & 0 \\ 0 & J/2 & -J/4 & 0 \\ 0 & 0 & 0 & J/4+h \end{pmatrix}, \tag{23}$$

and the eigenvalues become $J/4+h, J/4-h, J/4, -3J/4$. This means that the threefold degenerate triplet state at zero magnetic field splits into three distinct states as the magnetic field is applied. Moreover, there is a crossover at $h_c = J$, where the $S^z_{tot} = 1$ state becomes the ground state, and therefore the magnetization jumps from 0 to 1.

This approach can be generalized to finite temperatures, i.e. the uniform magnetization of the system becomes

$$m(h,T) = \frac{\sum_i S^z_i \exp(-\beta E_i)}{\sum_i \exp(-\beta E_i)} = \frac{2\sinh(\beta h)}{2\cosh(\beta h) + 1 + \exp(\beta h)}, \tag{24}$$

whose temperature and field dependence is depicted in Fig. 9. At finite temperatures and with increasing applied magnetic field, the jump in the magnetization, observed at zero temperature, is smoothed out. This is very typical in such systems, and from an experimental point of view it is important to know at which temperatures quantum jumps, such as this one, are not observable anymore because they are wiped out by thermal fluctuations. The temperature dependence of the uniform magnetization depends on the strength of the applied magnetic field. If the applied field is smaller than the critical field strength, the zero-temperature magnetization vanishes, and $m(h,T)$ is thermally activated, i.e. $m(h,T) \propto \exp(-\beta h)$. On the other hand, if the applied field exceeds the critical field strength at zero temperature, the uniform magnetization is activated as $m(h,T) \propto 1-\exp(-\beta h)$. In both cases, however, at sufficiently large temperatures, $m(h,T)$ shows a Curie-like decay, common to all Heisenberg clusters, as we will see later on.

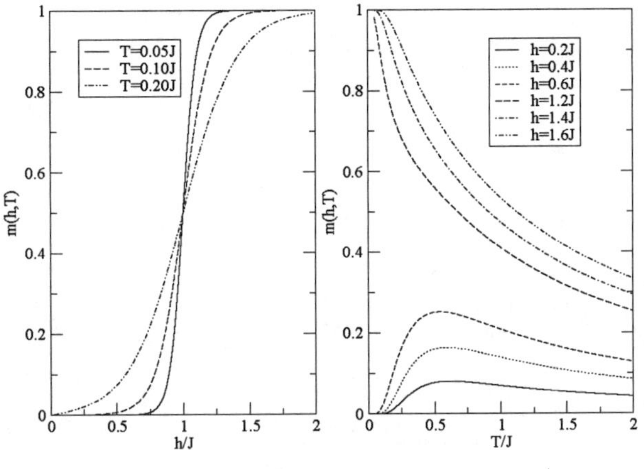

FIGURE 8. Temperature and field dependence of the uniform magnetization of a 2-site Heisenberg cluster. Thermal fluctuations smooth out the step function seen in $m(h,T=0)$. At high temperatures $m(h,T)$ follows a Curie law. At low temperatures, its behavior depends on the field, starting at 1 for $h > H_c$, and at 0 for $h < h_c$.

3-site clusters: Next, let us consider three-site spin-1/2 Heisenberg clusters. For this case, there are two possible distinct geometries: a closed triangle or a three-site open chain. Let us discuss these two cases in parallel. The Hilbert space does not depend on the geometry. It has $2^3 = 8$ states. In the same Ising basis used for the previous two-site example, the basis states are $|111\rangle, |110\rangle, |101\rangle, |011\rangle, |001\rangle, |010\rangle, |100\rangle, |000\rangle$. The first basis state has $S^z_{tot} = 3/2$, the next three basis states are in the $S^z_{tot} = 1/2$ subspace, the next three span the $S^z_{tot} = -1/2$ subspace, and $|000\rangle$ is the only element of the $S^z_{tot} = -3/2$ subspace, and therefore automatically an eigenstate of Hamiltonians that preserve S^z_{tot} as a good quantum number.

The antiferromagnetic Heisenberg Hamiltonian indeed commutes with S^z_{tot}. For the linear three-site chain it is $H_{linear} = J(\vec{S}_1 \cdot \vec{S}_2 + \vec{S}_2 \cdot \vec{S}_3)$, and for the triangle it is $H_{linear} = J(\vec{S}_1 \cdot \vec{S}_2 + \vec{S}_2 \cdot \vec{S}_3 + \vec{S}_3 \cdot \vec{S}_1)$. The eigenvalues and eigenstates of these systems can be obtained by acting with the Hamiltonian on the basis states, and subsequent diagonalization of the block matrices in the various S^z_{tot} subspaces. Since the subspaces for these cases contain at most three spins, this can still be achieved by analytical methods, or with the help of an analytic diagonalization routines, such as provided e.g. by Math-

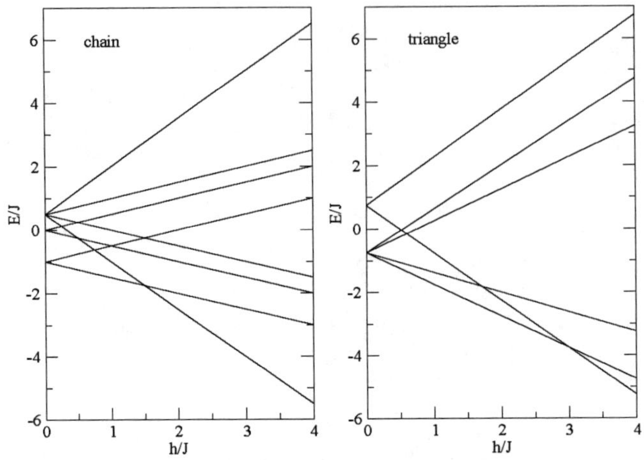

FIGURE 9. Energy levels in three-site spin-1/2 antiferromagnetic Heisenberg clusters. In both cases, there is a critical field, at which the system ground state jumps from $S^z_{tot} = 1/2$ to $S^z_{tot} = 3/2$, resulting in a jump in the system's uniform magnetization, similar to the 2-site cluster.

TABLE 2. Eigenvalues of 3-site spin-1/2 Heisenberg clusters

linear	triangle	S^z_{tot}
$(J-3h)/2$	$(3J-6h)/4$	3/2
$(J+3h)/2$	$(3J+6h)/4$	-3/2
$(-2J+h)/2$	$(-3J+4h)/4$	1/2
$(J+h)/2$	$(-3J+4h)/4$	1/2
h	$(3J+4h)/4$	1/2
$(-2J-h)/2$	$(3J+4h)/4$	-1/2
$(J-h)/2$	$(3J+4h)/4$	-1/2
$-h$	$(3J-4h)/4$	-1/2

ematica. The resulting eigenenergy levels, including a Zeeman field, are listed in Table 2.

The triangle has multiple degeneracies which are not split by the external magnetic field, whereas in the chain case all degeneracies at zero field are split at finite fields. Furthermore, the critical at which the ground state crosses over from $S^z_{tot} = 1/2$ to $S^z_{tot} = 3/2$ is different for the two cases: $h_c = 3J/2$ for the chain and $h_c = 3J$ for the triangle, indicating that the triangular geometry protects the zero-field $S^z_{tot} = 1/2$ ground state more efficiently. Note also, that at zero magnetic field, there is a degenerate ground state, similar to an isolated spin-1/2 particle which has degenerate up-spin and down-spin configurations. This degeneracy is removed by infinitesimal applied fields. For the three-site clusters this means that strictly at h=0 the magnetization is 0, then it jumps to 1/2 at infinitesimal fields, and then it jumps to 3/2 at the critical magnetic field.

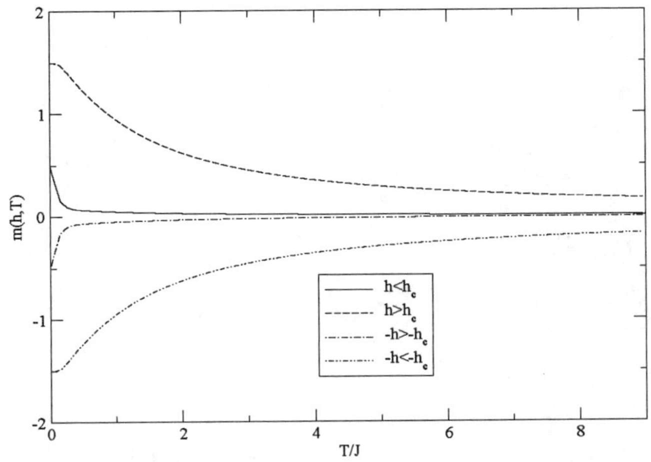

FIGURE 10. Temperature dependence of the uniform magnetization in a 3-site spin-1/2 antiferromagnetic Heisenberg chain.

The finite-temperature properties of the three-site clusters are obtained in analogy to the cases discussed above, using $m(h,T) = (\sum_i S_i^z \exp(-\beta E_i))/(\sum_i \exp(-\beta E_i))$. The resulting temperature dependence is plotted in Fig. 10. Similar to the two-site cluster, the low-temperature behavior depends on the specific magnetic field that it is applied, whereas the high-temperature behavior is Curie-like as usual.

4-site clusters: Moving on to the 4-site case, we will try to identify patterns that depend on the fact whether the clusters have an even or odd number of spins. For the 4-site Heisenberg model, there are 3 topologically distinct configurations: a square, a linear 4-site chain, and a linear 3-site chain with an additional spin site linked to the central spin. The Hilbert space for these systems is of size $2^4 = 16$ and contains states with $S_{tot}^z = \pm 2, \pm 1, 0$. As for the above cases, the Hamiltonian matrix is therefore block-diagonal with respect to the quantum number S_{tot}^z. However, even the blocks can be too large to be handled by straightforward analytical diagonalization, e.g. the largest block with $S_{tot}^z = 0$ has 6 states.

We can, however, further reduce these blocks by utilizing another symmetry, i.e. translational invariance. Note that not every cluster is translationally invariant, so let's here concentrate on the square geometry which can also be interpreted as a 4-site chain with periodic boundary conditions. Periodicity implies that the eigenstates of this cluster are eigenstates of the translation operator, i.e. $\hat{T}_r|\psi\rangle = \exp(i\vec{k}\cdot\vec{r}|\psi\rangle$, and this gives us a recipe to construct translationally invariant basis states with an additional fixed quantum number $k = \pi n/N$, where N is the lattice size and $n = 0,1,2,...,N$. As an example for this procedure, let us examine the $S^z_{tot} = 0$ subspace of the 4-site square. Initially, it contains the 6 basis states $|1100\rangle, |1001\rangle, |0110\rangle, |0011\rangle, |1010\rangle, |0101\rangle$. Applying the translation operator to this basis states, these can be further sorted with respect to the available k-values:

$$
\begin{aligned}
|k=0\rangle_1 &= \frac{1}{2}(|1100\rangle + |0110\rangle + |0011\rangle + |1001\rangle) \\
|k=0\rangle_2 &= \frac{1}{\sqrt{2}}(|1010\rangle + |0101\rangle) \\
|k=\pi\rangle_1 &= \frac{1}{2}(|1100\rangle - |0110\rangle + |0011\rangle - |1001\rangle) \\
|k=\pi\rangle_2 &= \frac{1}{\sqrt{2}}(|1010\rangle - |0101\rangle) \\
|k=\pi/2\rangle_1 &= \frac{1}{2}(|1100\rangle + i|0110\rangle - |0011\rangle - i|1001\rangle) \\
|k=\pi/2\rangle_2 &= \frac{1}{2}(|1100\rangle - i|0110\rangle - |0011\rangle + i|1001\rangle)
\end{aligned}
$$

The advantage gained from rewriting the $S^z_{tot} = 0$ basis states in these combinations is that the $S^z_{tot} = 0$ sub-block of the Hamiltonian matrix is split in three further 2 by 2 sub-blocks, sorted by their quantum numbers k:

$$
H(S^z_{tot} = 0 = \begin{pmatrix} 0 & \sqrt{2}J & 0 & 0 & 0 & 0 \\ \sqrt{2}J & -J & 0 & 0 & 0 & 0 \\ 0 & 0 & 0 & 0 & 0 & 0 \\ 0 & 0 & 0 & -J & 0 & 0 \\ 0 & 0 & 0 & 0 & 0 & 0 \\ 0 & 0 & 0 & 0 & 0 & 0 \end{pmatrix}, \tag{25}
$$

using the order of basis states as listed above. Hence, the largest diagonalization problem to be solved is that for a $2*2$ matrix, yielding the zero-field ground state energy $E_0 = -2J$ and $|\psi_0\rangle = \frac{1}{\sqrt{12}}[2|1010\rangle + 2|0101\rangle - |1001\rangle - |0110\rangle - |1100\rangle - |0011\rangle]$. It is easy to show that this state is in fact a d-wave resonant valence bond state (RVB), i.e. a superposition of singlet coverings of the 4-site square. It turns out that many finite-cluster ground states for even-site clusters are RVB states.

Why such an extended discussion of such a particular problem? Because it elucidates the procedure that is taken when performing large-matrix numerical diagonalizations. The matrix is reduced into block-diagonal form using all known quantum numbers of the system. This has the additional advantage that the particular eigenstates that emerge from the diagonalization are already partially characterized with respect to the quantum numbers that have been used to build up the Hamiltonian sub-matrix. For Heisenberg clusters with an original Hilbert space of 2^N, this implies at least three reductions due to (i) S^z_{tot}, (ii) k, and (iii) for the case of $S^z_{tot} = 0$ a $\pm$ spin-inversion symmetry, at least for clusters with an even number of spins. Carrying through the full matrix diagonalization for the $N = 4$ square cluster, one finds a level spectrum reminiscent of the previously discussed cases. In the absence of a magnetic field, the ground state, like for the $N = 2$ case, is a singlet. In fact, this holds true for all even-site clusters, whereas the odd-site clusters have doublet ground states. Furthermore, the 4-site cluster has two critical fields, $h_{c1} = J$ and $h_{c2} = 2J$, corresponding to jumps in the cluster's uniform magnetization form 0 to 1 and from 1 to 2.

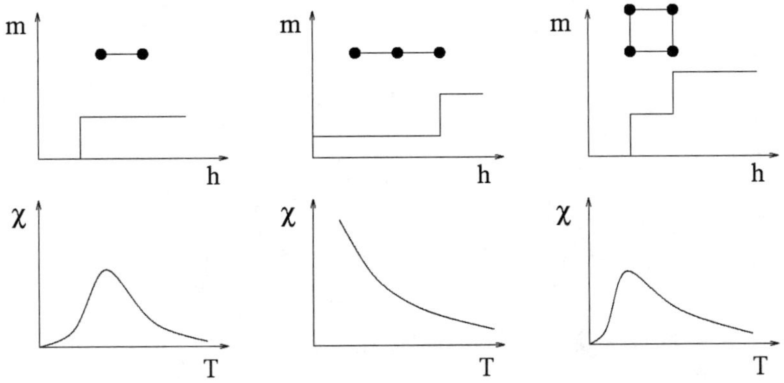

FIGURE 11. Comparison of uniform magnetization and susceptibility in various small clusters.

In Fig. 11, we compare a few results from the exact diagonalization of ultra-small Heisenberg clusters. One observes that

1. There is an even-odd alternation (singlet vs. doublet ground states) depending on the number of spins in the cluster. This results in very different low-temperature properties, i.e. for the uniform susceptibility exponential activation in even-site clusters, vs. Curie-type divergence for the odd-site clusters.
2. On the other hand, the high-temperature properties of all clusters are similar: in this regime, thermal fluctuations wipe out the interesting low-temperature quantum behavior.
3. For even-site clusters we find at low temperatures $\chi \propto \exp(-\Delta/T)$ with $\Delta(N)$ decreasing as $N \to \infty$ (see also Fig. 1). Upon further analysis, one finds that the functional dependence of $\Delta(N)$ depends on the geometry of structures that are studied. For example, for chains $\Delta(N) = \Delta(\infty) - \alpha/N$, where α is a fit parameter. For ladder structures, one finds instead $\Delta(N) = 1 - \alpha \cdot \exp(N/N_0)$.

In bulk systems, applied magnetic fields can trigger quantum phase transitions, i.e. phase transitions at zero temperature, the precursors of which we already witness in the smallest clusters.

LINEAR SPINWAVE THEORY

Spinwave theory is a traditional approach to Heisenberg-like Hamiltonians, which is based on an expansion about classical ground states, taking into account quantum fluctuations which lead to a renormalization of the classical state.[5] However, spin-wave theory breaks down when the quantum fluctuations become large, e.g. in the one-dimensional case and in strongly frustrated higher-dimensional systems.

Ferromagnet: Let us start by considering spinwave theory for the Heisenberg ferromagnet with nearest-neighbor interactions. Its classical zero-temperature ground state is known to be simple alignment of all spins. As we will see, the effect of quantum fluctuations in this case is to admix wave-like precessions about this state, i.e. spinwaves, into the ground state.

The first step is to represent the spin operators as bosons, using the Holstein-Primakoff transformation:

$$\begin{aligned} S_i^z &= S_i - a_i^\dagger a_i \\ S_i^+ &= a_i^\dagger \sqrt{2S}\sqrt{1-\frac{a_i^\dagger a_i}{2S}} \approx a_i^\dagger \sqrt{2S} \\ S_i^- &= \sqrt{2S}\sqrt{1-\frac{a_i^\dagger a_i}{2S}}\, a_i \approx \sqrt{2S}\, a_i \end{aligned} \tag{26}$$

where a and a^+ are boson annihilation and creation operators. Note that the original mapping preserves spin statistics, i.e. $[S^x, S^y] = i\varepsilon_{ijk}S^z$, whereas the approximation "$\approx$" does not. Also note, that in the approximation above we have only kept the leading (linear) terms, i.e. "linear spinwave approximation". In some cases it may be necessary to go to higher order in this expansion.

Assuming that we are dealing with a hyper-cubic lattice, we can substitute the Holstein-Primakoff transformation into the Hamiltonian:

$$\begin{aligned} H &= J\sum_{\langle ij\rangle}\left[S_i^z S_j^z + \frac{1}{2}\left(S_i^+ S_j^- + S_i^- S_j^+\right)\right] \\ &\simeq J\sum_{\langle ij\rangle}\left[S^2 - S(a_i^\dagger a_i + a_j^\dagger a_j) + S(a_i^\dagger a_j + a_i a_j^\dagger)\right] \\ &\approx \frac{JNzS^2}{2} + SJ\sum_{\langle lm\rangle}\left[a_l^\dagger a_m + a_m^\dagger a_l - 1 - a_l^\dagger a_l - a_m^\dagger a_m\right]. \end{aligned}$$

This bilinear form can be diagonalized using simple Fourier transforms, i.e. $a_i = N^{-1/2}\sum_k a_k \exp(ikl)$ and $a_i^\dagger = N^{-1/2}\sum_{k'} a_{k'}^\dagger \exp(-ik'l)$, resulting in

$$\begin{aligned} H \approx\ & \frac{JNzS^2}{2} - \frac{SJNz}{2} \\ +\ & \frac{SJ}{2N}\sum_{l,\delta,k,k'}\left[a_{k'}^\dagger a_k \exp(-ik'l)\exp(ik(l+\delta)) + a_{k'}^\dagger a_k \exp(-ik'(l-\delta))\exp(ikl)\right] \\ -\ & 2a_{k'}^\dagger a_k \exp(-ik'l)\exp(ikl), \end{aligned}$$

where δ enumerates the neighbors of site l, i.e. $\vec{\delta} = \pm\hat{x}, \pm\hat{y}$ on a square lattice. The sum now simplifies via the identity $\delta_{k,k'} = N^{-1}\sum_k \exp(-i(k-k'))$:

$$H \approx \frac{JNzS^2}{2} - \frac{SJNz}{2} + JSz\sum_{\delta,k}\left[a_k^\dagger a_k(\exp(i\vec{k}\cdot\vec{\delta}) + \exp(-i\vec{k}\cdot\vec{\delta})\right] - \frac{SJ}{N}\sum_k a_k^\dagger a_k, \quad (27)$$

which yields the dispersion relation for ferromagnetic magnons, $E_k = E_0 - JSz(1-\gamma_k)$, where $E_0 = JNSz(S-1)/2$ and $\gamma_k = z^{-1}\sum_\delta \cos(\vec{k}\cdot\vec{\delta})$. For long wavelengths (small k), the ferromagnetic dispersion can be expanded, using $\gamma_k \approx 1-(k\delta)^2$: $E_k \approx JSz(k\delta)^2$, which means that ferromagnetic spinwaves disperse quadratically.

This quadratic dispersion law has some immediate consequences for thermodynamic response functions which can be measured experimentally. For example, it can be shown from this expansion of the low-energy dispersion that the low-temperature magnetization of the ferromagnet depends algebraically on the temperature with a characteristic power law of 3/2:

$$\begin{aligned} m &= \langle S^z_{tot}\rangle = \langle\sum_{i=1}^N S^z_i\rangle = NS - \sum_k a_k^\dagger a_k \approx NS\sum_k \frac{1}{\exp(\beta ck^2)-1} \\ &\approx NS - \frac{4\pi}{V}\int dkk^2 \frac{1}{\exp(\beta ck^2)-1} \approx NS - T^{3/2}\left[\frac{2\pi}{Vc^{5/2}}\frac{1}{\exp(x)-1}\right], \end{aligned} \quad (28)$$

where we have set $c = JSz$, and have used the scaling substitution $x \equiv ck/T$, which renders the last integral independent of temperature, leading to $m \approx NS - const \cdot T^{3/2}$ at sufficiently low temperatures.

These quantized spinwaves are also called “magnons". They are collective modes, i.e. they are global excitations involving a macroscopic number of spins in the system. At sufficiently low energies (temperatures), they determine the thermodynamic response of the system. Of course, at higher energies the system contains other types of excitations, e.g. local excitations, as well.

Antiferromagnet: A similar treatment of the antiferromagnetic case is possible, but as we will see now it involves a further transformation. The starting point for antiferromagnetic spinwave theory is the classical Neel state, i.e. perfect up-down alignment of neighboring spins. For simplicity, let us again consider the case of a two-dimensional square lattice. In this case, there are two sublattices, which we denote as the A- and B-sublattice. The Holstein-Primakoff transformation of the spin operators then depends on the sublattice.

A-sublattice:

$$\begin{aligned} S^z_i &= S_i - a_i^\dagger a_i \\ S^+_i &= a_i^\dagger\sqrt{2S}\sqrt{1-\frac{a_i^\dagger a_i}{2S}} \approx a_i^\dagger\sqrt{2S} \\ S^-_i &= \sqrt{2S}\sqrt{1-\frac{a_i^\dagger a_i}{2S}}a_i \approx \sqrt{2S}a_i \end{aligned} \quad (29)$$

B-sublattice:

$$\begin{aligned} S_j^z &= -S + b_j^\dagger b_j \\ S_j^- &= b_j^\dagger \sqrt{2S}\sqrt{1-\frac{b_j^\dagger b_j}{2S}} \approx b_j^\dagger\sqrt{2S} \\ S_j^+ &= \sqrt{2S}\sqrt{1-\frac{b_j^\dagger b_j}{2S}}\, b_j \approx \sqrt{2S}b_j \end{aligned} \tag{30}$$

and the substitution into the Heisenberg Hamiltonian, followed by the Fourier transform now yields

$$\begin{aligned} H &= \frac{J}{2}\sum_{l,\delta}\left[S_l^z S_{l+\delta}^z + \frac{1}{2}\left(S_l^+ S_{l+\delta}^- + S_l^- S_{l+\delta}^+\right)\right] \\ &\approx \frac{J}{2}\sum_{l,\delta}\left[-S^2 - S\left(a_l^\dagger a_l + b_{l+\delta}^\dagger b_{l+\delta}\right) + S\left(a_l^\dagger b_{l+\delta}^\dagger + a_l b_{l+\delta}\right)\right] \\ &\approx -\frac{JNS^2 z}{2} - \frac{JSz}{2}\sum_k\left(a_k^\dagger a_k + b_k^\dagger b_k\right) + \frac{JSz}{2}\sum_k\left(a_k^\dagger b_k^\dagger + a_k b_k\right). \end{aligned} \tag{31}$$

Hence, in the antiferromagnetic case the Fourier transform is not sufficient to diagonalize the Hamiltonian. An additional transformation is needed. This is the Boguliubov transformation,

$$\begin{aligned} a &= u\alpha + v\beta^+ \\ a^+ &= u\alpha^+ + v\beta \\ b &= u\beta + v\alpha^+ \\ b^+ &= u\beta^+ v\alpha. \end{aligned}$$

For convenience of notation, we will suppress the site and momentum indices here. The new operators α^+ and β^+ create bosonic Bogoliubov quasiparticles. The coefficients $u = \cosh\theta$, $v = \sinh\theta$ ensure that the new variables satisfy bosonic commutation relations. They are determined by the constraint that enforces the transformed Hamiltonian to be diagonal in these new operators:

$$\begin{aligned} H &\propto -(a^+a + b^+b) + \gamma(a^+b^+ + ab) \\ &\propto -(u^2\alpha^+\alpha + uv\alpha^+\beta^+ + uv\beta\alpha + v^2\beta\beta^+ + u^2\beta^+\beta + UV\alpha\beta + v^2\alpha\alpha^+) \\ &+ \gamma(u^2\alpha\beta + uv\alpha^+\alpha + uv\beta^+\beta + v^2\beta^+\alpha^+ + u^2\alpha^+\beta^+ + uv\alpha^+\alpha + uv\beta\beta^+ + v^2\beta\alpha) \\ &\propto (\alpha^+\alpha + \beta^+\beta)(-u^2 - v^2 + 2\gamma uv) - 2v^2 + 2\gamma uv + (\alpha^+\beta^+ + \alpha\beta)(-2uv + \gamma(u^2+v^2)) \\ &\propto (\alpha^+\alpha + \beta^+\beta + 1)(-u^2 - v^2 + 2\gamma uv) + u^2 - v^2. \end{aligned}$$

In order to render the Hamiltonian diagonal, we set the last term in the penultimate line to zero, i.e. $0 = -2uv + \gamma(u^2+v^2)$, which is the constraint that helps us determine the values of the coefficients u and v. Using this constraint together with $u^2 - v^2 = 1$, one finds that $\gamma = -\tanh(2\theta) = -2uv/(u^2+v^2)$, and hence

$$u^2 = \frac{1}{2}\left[\frac{1}{\sqrt{1-\gamma^2}}+1\right], \quad (32)$$

$$v^2 = \frac{1}{2}\left[\frac{1}{\sqrt{1-\gamma^2}}-1\right]. \quad (33)$$

As desired, the Hamiltonian is now diagonal in the new operators,

$$H \approx -\frac{JNzS(S+1)}{2} + \sum_k \omega_k(\alpha_k^+\alpha_k + \beta_k^+\beta_k + 1), \quad (34)$$

with an energy dispersion $\omega_k = JSz\sqrt{1-\gamma_k^2}$, which at small energies be expanded about $\vec{k} = (\pi,\pi)$ as $\omega_k \approx JSzk\delta$. In contrast to the ferromagnetic case, this is a linear dispersion law.

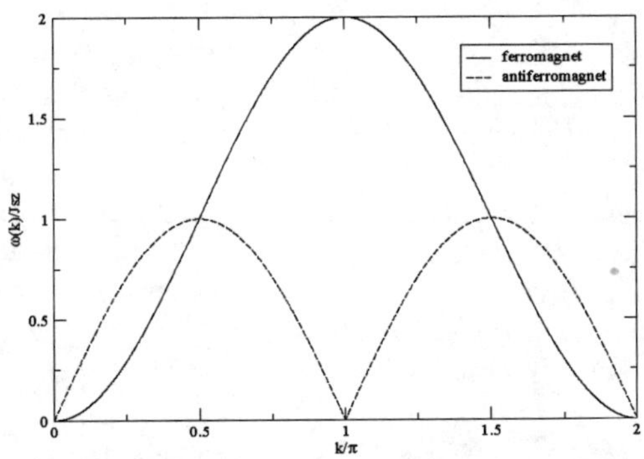

FIGURE 12. Comparison of magnon dispersions in the ferromagnetic and antiferromagnetic Heisenberg models, obtained from linear spinwave theory. Note that the dominant low-energy contribution is quadratic in the ferromagnet and linear in the antiferromagnet.

The thermodynamic activation of the antiferromagnetic order parameter is therefore expected to be different from the ferromagnetic case. Following the same steps outlined above for the ferromagnet, one finds

$$m_A = NS - \sum_k \langle a_k^+ a_k \rangle \approx NS - \frac{4\pi}{V}\int dk k^2 \frac{1}{\exp(\beta ck)-1}\cdot\frac{1}{\sqrt{1-\gamma_k^2}} \approx NS - const\cdot T^2, \quad (35)$$

which implies a slightly stronger temperature dependence than for the ferromagnetic case.

TABLE 3. Sublattice magnetization per site for the spin-1/2 Heisenberg antiferromagnet on hypercubic lattices, obtained by the linear spinwave approximation.

chain (1D)	square (2D)	cube (3D)
0	$\frac{1}{2}-\frac{0.294}{2}$	$\frac{1}{2}-\frac{0.156}{2}$

The temperature dependence of the uniform susceptibility can be obtained in a similar fashion, starting with the linear dispersion including a Zeeman field along the z-axis $E_k \approx ck - dh$, where we have absorbed all the physical coefficients in the constants c and d, and only the dependences on the momentum k and the field h are shown explicitly. Using similar scaling arguments, we find for the resulting uniform susceptibility along the z-axis

$$\chi_{\|} = \frac{\partial m(h)}{\partial h}|_{h\to 0} \approx -\frac{4\pi}{V}\int dkk \frac{\beta d \exp(\beta(ck-dh))}{(\exp(\beta(ck-dh))-1)^2} \propto T. \tag{36}$$

Next, we consider the effect of an applied magnetic field along the perpendicular direction, i.e. the x-axis: $H = H_0 - h\sum_{i=1}^{N} S_i^x$. Using the above transformations, we can substitute $S^x = (S^+ + S^-)/2 \approx (a^+ + a)\sqrt{S/2} = (u\alpha + u\alpha^+ + v\beta + v\beta^+)\sqrt{S/2}$. The ground state expectation values of these linear Boguliubov operators vanish, and hence the uniform susceptibility is constant in the plane perpendicular to the z-axis.

Zero-point fluctuations: Even at zero temperature, there is a renormalization of the antiferromagnetic order parameter, i.e. the sublattice magnetization, due to quantum fluctuations. This reduction can be determined quantitatively within the spinwave approximation. First note that at zero temperature $\langle\alpha_k^+\alpha\rangle = \langle\beta_k^+\beta_k\rangle = 1$, whereas the off-diagonal correlations vanish. Hence,

$$\begin{aligned}\frac{m_A}{N} &= S - \frac{1}{N}\sum_k a_k^+ a_k = S - \frac{1}{N}\sum_k \left[u_k^2\langle\alpha_k^+\alpha_k\rangle + u_k v_k\langle\alpha_k^+\beta_k^+\rangle + u_k v_k\langle\beta_k\alpha_k\rangle + v_k^2\langle\beta_k\beta_k^+\rangle\right] \\ & S - \frac{1}{N}\sum_k (u_k^2 + v_k^2 - 1) = S - \frac{1}{2N}\sum_k \left(\frac{1}{\sqrt{1-\gamma_k^2}} - 1\right).\end{aligned} \tag{37}$$

This sum can be evaluated numerically. Let us compare the cases of a linear chain with $\gamma_k = \cos(ka)$, the two-dimensional square lattice with $\gamma_k = \cos(k_x a) + \cos(k_y a)$, and the three-dimensional cube with $\gamma_k = \cos(k_x a) + \cos(k_y a) + \cos(k_z a)$. The numerical values for m_A/N are listed in Table 3.

They indicate that the renormalization of the 0th order result of the sublattice magnetization, i.e. $m_A/N = 1/2$, is weaker in higher dimensions, and stronger in lower dimensions. In fact, in the one-dimensional case the integral expression containing the correction due to quantum fluctuations diverges logarithmically, which at face value would give the unphysical result of a negatively divergent sublattice magnetization, but this is of course just a consequence of the breakdown of the spinwave approximation in this case. Therefore, we set $m_A/N = 0$ in the above comparison.

Since the spinwave theory is an expansion about a classical state, it is not applicable to quantum disordered systems which do not have a classical counterpart. For example, the frustrated antiferromagnetic Heisenberg model with competing nearest-neighbor and neaxt-nearest-neighbor interactions J and J' falls into this category. This model is known to have three phases: (i) an antiferromagnetically ordered phase with at small J'/J, (ii) a collinear phase with at large J'/J, and (iii) a quantum disordered phase at intermediate J'/J. There is no known classical state in this intermediate which could serve as the starting point of a spinwave expansion. However, in the two ordered phases one can at least treat the frustration effect perturbatively, using spinwave theory. For example, in the $\vec{k} = (\pi, \pi)$ regime the sublattice magnetization is reduced due to the neaxt-nearest-neighbor interactions, yielding,

$$\frac{m_A}{N} = S - \frac{1}{2N}\sum_k \left[\frac{1-(J'/J)(1-\eta_k)^2}{\sqrt{1-(J'/J)(1-\eta_k)^2 - \gamma_k^2}} - 1 \right], \tag{38}$$

where $\eta_k = \cos(k_x a)\cos(k_y a)$. For $J' = 0$, this results correctly reduces to the expression we obtained above for the nearest-neighbor Heisenberg antiferromagnet.

CLASSICAL AND QUANTUM MONTE CARLO

Monte Carlo is a widely used numerical technique in which a random number generator is used to evaluate multi-dimensional integrals.[6] Especially in the context of statistical physics, this stochastic method is much more efficient than deterministic quadratures, such as those based on Taylor expansion (e.g. Simpson method) whose error can be estimated to be of the order $O(n^{-l/d})$, where n is the number of partitions in a given direction, d is the dimensionality of the integral, and l is the order of the quadrature, e.g. $l = 4$ for the Simpson method, using

$$\int_a^b dx f(x) \approx \frac{h}{3} \sum_{k=0}^{(h/2)-1} [f_{2k} + 4 f_{2k+1} + f_{2k+2}] + O(h^4). \tag{39}$$

In contrast to deterministic methods, the error in Monte Carlo does not depend on dimensionality, but only on the number of points that are used to approximate the integral, i.e. the error is $O(n^{-1/2})$, and therefore an upper "critical dimension" can be determined by comparing the to error estimates, $l/d = 1/2$, above which the Monte Carlo method becomes more precise. For example, comparing the efficiency of Monte Carlo with the Simpson method, one finds for the upper critical dimension $d = 8$.

As a first introduction how to use a random number generator to evaluate an integral, consider the example of determining the area of a pond A_{pond} inside a rectangular area $A_{rect} = a \cdot b$ whose size is known. You can achieve this by randomly tossing stones into the area A_{rect} and counting how many times the water is hit. The area of the pond is then simply given by $A_{pond} = A_{rect} \cdot N_{pond}/N_{total}$, where N_{pond} counts the number of hits

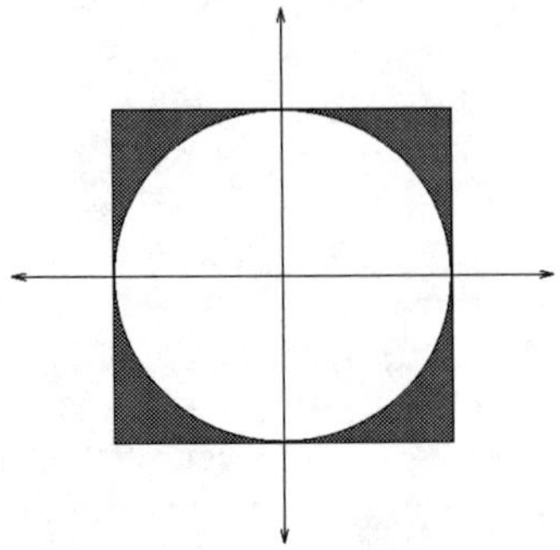

FIGURE 13. Illustration of the simple Monte Carlo evaluation of the area of a circle with radius 1, discussed in the text. The number π is obtained by randomly generating pairs of numbers that fit into the square box, and counting the number of "hits" inside the white circle.

of N_{tot} throws. A simple computer example would be to determine the number π by embedding a circle of radius $R = 1$ into a square of area $2 \cdot 2$. Using a random number generator, pairs of numbers $[x_i, y_i]$ are randomly created in the interval within which the square is positioned. The area of the circle should then be $A_{circle} = \pi \pm O(N^{-1/2})$ where N is the number of randomly generated coordinates.

More generally, Monte Carlo can be described as solving integrals by randomly sampling their kernel, using

$$\int_a^b dx f(x) \approx \frac{b-a}{N} \sum_{k=1}^{N} f(x_k) + O(N^{-1/2}), \tag{40}$$

which of course can be easily generalized to multiple dimensions. The fact that the Monte Carlo error decays with the inverse square root of the number of trials an be shown rigorously by evaluating the variance for M measurements of the quantity m_k, $\sigma_m^2 = \frac{1}{M}\sum_{k=1}^{M}[m_k - \bar{m}]$, where $\bar{m}$ is the mean, $m_k = \frac{1}{N}\sum_{i=1}^{N} f_{ki}$, and f_{ki} is the ith random evaluation of the function $f(x)$. The result is $\sigma_m^2 = \sigma^2/N$, which gives rise to the inverse square root error.

Many integral kernels have regions of strong variation, and are thus more efficiently evaluated by an adaptive mesh that tracks these regions. In Monte Carlo, this is called "importance sampling". It means that instead of randomly sampling in the interval of integration, a weight function is defined that mimics the kernel function $f(x)$ over which the integral is performed, i.e.

$$\int_a^b dx f(x) = \int_a^b dx w(x) \frac{f(x)}{w(x)} = \int_a^b dx' \frac{f(x')}{w(x')}, \tag{41}$$

where by setting $dx' = dx w(x)$ the integral is now a restricted integral, sampling phase space according to the weight function. The weight function must be chosen such that $\int_a^b dx w(x) = 1$. Since the weight function mimics $f(x)$, the new kernel $f(x')/w(x')$

should be roughly constant, and therefore

$$\int_a^b dx f(x) \approx \frac{b-a}{N} (\sum_{i=1}^{N})' \frac{f(x_i)}{w(x_i)} \tag{42}$$

should be a good approximation.

Note that $(\sum)'$ denotes a restricted sum that preferably samples points where is large. In order to achieve this specific weighted sampling, the weight function must satisfy the detailed balance condition:

$$w(x_{old})T(x_{old} \rightarrow x_{new}) = w(x_{new})T(x_{new} \rightarrow x_{old}), \tag{43}$$

where T is the transition rate between the sample points x_{old} and x_{new}. In practice, this can be achieved very easily. Imagine that $\eta_i \in [0,1]$ is a random number drawn from a uniform distribution, then the new sampling point is created from the old sampling point via $x_{new} = x_{old} + \Delta(2\eta_i - 1)$, where Δ is an adjustable step size parameter. Metropolis showed that this step should be accepted when $w(x_{new})/w(x_{old}) \geq \eta_i'$, where η_i' is another random number from the uniform distribution.[7] This updating scheme fulfills the detailed balance condition, and thus correctly samples according to the restricted sum $\sum'$.

The steps of the Metropolis algorithm can be summarized as follows:

1. Generate a random x_{old}, and evaluate $w(x_{old})$.
2. Generate $x_{new} = x_{old} + \Delta(2\eta_i - 1)$, and evaluate $w(x_{new})$.
3. Accept this trial step if $w(x_{new})/w(x_{old}) \geq \eta_i' \in [0,1]$.
4. Periodically evaluate $\int_a^b dx f(x) \approx \frac{1}{N} \sum_{k=n}^{n+Mn} f(x_k)$, where M ensures that the sampling points are statistically uncorrelated.

This last step is necessary because x_{old} and x_{new} are intrinsically correlated, and to ensure the statistically independent integrity of the sampling process, several steps (2) should be taken between evaluations (4).

In statistical mechanics, e.g. for the spin-1/2 Ising model, we often want to evaluate sums of the type $m(h,T) = \sum_S S \exp(-\beta E_S) / \sum_S \exp(-\beta E_S)$, and hence a natural choice for the weight function for the Ising model and related systems is $w(S) = \exp(\beta E_S)$, which naturally weighs the low-energy states higher than the high-energy states. The power of Monte Carlo is that we do not need to completely evaluate this sum, but just use a random number generator combined with the Metropolis algorithm to sample $(1/N) \sum_S' S$ efficiently. For this case, the transition rate satisfying the detailed balance condition is given by $T_{old \rightarrow new} = \min(1, \exp(-\beta(E_{new} - E_{old})))$. The Metropolis algorithm for this case is:

1. Generate a random initial spin configuration $\{S_1, ..., S_N\}_{old}$, and evaluate $E(\{S_1, ..., S_N\}_{old})$.
2. Create a new spin configuration by flipping one random spin, i.e. $S_i(new) = -S_i(old)$, and evaluate $E(\{S_1, ..., S_N\}_{new})$
3. Accept this local spin update if $w(x_{new})/w(x_{old}) = \exp(-\beta(E_{new} - E_{old})) \geq \eta_i' \in [0,1]$.

4. Periodically evaluate $|m| = \frac{1}{N}(\sum_{i=1}^{N})'|S_i|$.

The magnitude sign in the last step is necessary because of the spontaneous symmetry breaking in the Ising model. Without taking the magnitude of the average spin, the statistical average would otherwise lead to $m = 0$.

The local spin updates used in step (2) become increasingly inefficient in lowering the system's energy as $T \to T_c$. This phenomenon, known as "critical slowing down" is due to the fact that close to the transition temperature, extended subclusters of magnetically aligned (in the ferromagnetic case) or anti-aligned (in the antiferromagnetic case) spins form, which ultimately grow to the size of the entire system at T_c. Local spin flips, such as performed in step (3), are unlikely to improve the system energy in this case, but flips of entire subclusters may. These are called global updates. Improvements of the Monte Carlo algorithm, such as the technique by Swendsen and Wang, therefore include smart algorithms that identify and update (flip) entire subclusters to overcome the critical slowing down.

The outcome of a Monte Carlo simulation of the Ising model would be e.g. curves of $|m(T)|$ for various linear system sizes L. In the thermodynamic limit $L \to \infty$ one expects this quantity to be finite for $T < T_c$ and to vanish for $T > T_c$. For finite systems, instead, $|m(T)|$ shows precursor behavior of a phase transition, i.e. an inflection in $|m(T)|$. A finite-size equivalent of the critical temperature can thus be found by identifying $\partial m/\partial T|_{max}$, or alternatively by evaluating the midpoint $(m(0) - m(\infty))/2$. The thermodynamic critical temperature can then be extracted by fitting the finite-size Monte Carlo data by $T_c(L) = T_c(\infty) - aL^{-1/\nu}$. The result of this procedure for the two-dimensional spin-1/2 Ising ferromagnet on a square lattice is $T_c = 2.26$ and $\nu = 1$. Since the correlation length, i.e. the characteristic size of the aligned clusters, diverges as $\xi \approx L \propto (T_c - T)^{-\nu}$, and at the same time the magnetization vanishes as $m \propto (T_c - T)^{\beta}$, we find for the finite-size dependence of the magnetization at the critical temperature $m(L) \propto L^{-\beta/\nu}$ which can be used to determine the scaling exponent $\beta = 1/8$ by fitting the Monte Carlo results for the magnetization at the inflection point to $m(L) = aL^{-b}$. These Monte Carlo results $(\nu = 1, \beta = 1/8)$ should be contrasted with the meanfield critical exponents $(\nu = 1/2, \beta = 1/2)$.

Finally, for true quantum Hamiltonians, the partition function of a many-body system becomes $Z = tr(\exp(-\beta H))$, and unless it is abundantly easy to diagonalize the Hamiltonian which is mostly not the case some trick needs to be applied to numerically sample this trace. A very popular algorithm that has been used extensively in the recent years is the "stochastic series expansion" technique.[8] It is based on what formally looks like a high-temperature expansion of the partition function, i.e.

$$Z = tr(\exp(-\beta H)) = \sum_{\{S\}} \sum_{n=0}^{\infty} \frac{(-\beta)^n}{n!} \langle S|H^n|S\rangle, \tag{44}$$

where $\{S\}$ is a complete basis set over which the trace is performed. As usual, it comes down to how to stochastically evaluate and where terminate the other sum. In practice, the Hamiltonian is rewritten as a sum of bond operators acting on two neighboring sites, $H = \sum_{i=1}^{N_b} H_{bi}$, and the power expansion of the partition function is then recast in terms of all possible strings, i.e. products, of bond operators,

$$Z \approx \sum_{\{S\}} \sum_{n=0}^{L} \frac{\beta^n (L-n)!}{L!} \langle S | \prod_{i=1}^{L} H_{bi} | S \rangle, \tag{45}$$

which is terminated at an appropriate maximum string length L. In practice, identity operators can be inserted such that all string lengths in the simulation are equal. There are two types of contributions arising from the bond operators: diagonal contributions due to the $S_i^z S_j^z$ interaction and the Zeeman terms in the Hamiltonian, and off-diagonal contributions due to the spin-flip $S_i^+ S_j^-$ terms.

ENTANGLEMENT IN QUANTUM SPIN SYSTEMS

Classical bits, used for computation, are often denoted as "1" and "0". They can be represented by an effective spin-1/2 degree of freedom, where "1" corresponds to the spin pointing up, and "0" represents a spin pointing down. To solve a complex many-body problem, such as determining the ground state of a Heisenberg Hamiltonian, requires 2^N classical bits for a system with N sites. Therefore, using classical computers we have been restricted to numerically diagonalize Heisenberg which are not larger than 36 sites, because for larger system sizes the Hilbert space is prohibitive.

Quantum bits encode information differently: they use superpositions of "1" and "0", such as which offer a much more efficient basis for quantum many-body problems. There are further advantages of qubits, e.g. information encoded in qubits can be protected from eavesdropping.

Entanglement is a measure of the quantum character of many-body states. Consider for example a simple two-body state,

$$|\psi\rangle = \frac{\alpha|11\rangle + (1-\alpha)|00\rangle}{\sqrt{\alpha^2 + (1-\alpha)^2}}, \tag{46}$$

where we assume that α is real. This state extrapolates between two classical two-body states, i.e. $|11\rangle$ and $|00\rangle$. Hence an appropriate entanglement measure should vanish in these two limits, whereas it should be at its maximum for $\alpha = 1/2$. The von-Neumann entropy offers such a measure.[9] In this section we will discuss the von-Neumann entropy for various quantum states in details, and we will look for generalization of it to account for global entanglement. In the language of quantum magnetism, the question to be answered is how much a subset of spins in a many-body system is entangled with the remaining spins. Hence, we need to start by defining an "inside" subset whose bipartite entanglement with the remaining "outside" spins needs to be evaluated for a given state $|\psi\rangle$, which could be but does not have to be the ground state of a many-body Hamiltonian, such as the Heisenberg model.

The recipe to construct the von-Neumann entropy is straightforward:

1. Start with a quantum many-body state $|\psi\rangle$ which you may obtain for example from exact diagonalization of a Heisenberg cluster.

2. Use this state to construct the corresponding density matrix, i.e. $\rho = |\psi\rangle\langle\psi|$.
3. Define the "inside" area A, and trace out the outside environment B out of the density matrix: $\rho_A = tr\rho_B$.
4. Use this reduced density matrix to calculate the von-Neumann entropy defined by $S = -\rho_A tr\rho_A$.
5. The initial state is maximally entangled if $S = \ln 2$ and minimally entangled if $S = 0$.

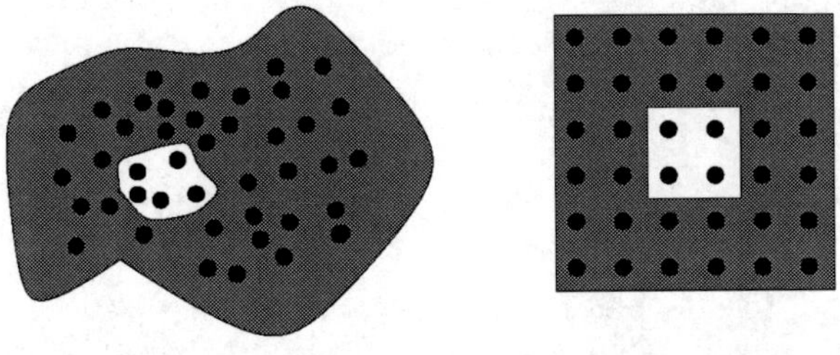

FIGURE 14. Illustrations of "inside" areas (white) and "outside" environment (shaded) in ensembles of qubits. The qubits are represented as black circles. Left: inhomogenous ensemble, Right: square lattice.

To illustrate this concept, let us start by evaluating the von-Neumann entropy for two cases, i.e. a non-entangled classical state $|F\rangle = |1111\rangle$ and a fully entangled quantum state $W\rangle = \frac{1}{\sqrt{2}}(|1111\rangle + |0000\rangle)$. Defining the "inside" area, we will consider the entanglement of the central (inside) 2 spins with the 2 spins on the outside of the system. For the ground state of the classical ferromagnet $|F\rangle = |1111\rangle$, we find a density matrix $\rho = |1111\rangle\langle 1111|$. Tracing out the outer two spins, we arrive at a reduced density matrix $\rho_A = \langle 0|_1\langle 0|_4\rho|0\rangle_1|0\rangle_4 + \langle 1|_1\langle 1|_4\rho|1\rangle_1|1\rangle_4 = |11\rangle_{23}\langle 11|_{23}$, where the subscripts refer to the positions of the spins in the cluster. The reduced density matrix for this state only has one element. Hence its von-Neumann entropy is $S = -tr(|11\rangle\langle 11|\ln(|11\rangle\langle 11|)) = -tr 1\ln(1) = 0$, since the logarithm of 1 vanishes. Therefore, the classical ferromagnetic state is not entangled.

In contrast the state $W\rangle = \frac{1}{\sqrt{2}}(|1111\rangle + |0000\rangle)$ is maximally entangled. After tracing out the outer two spins, its reduced density matrix is

$$\rho_A = tr_B\rho = \frac{1}{2}(|11\rangle_{23}\langle 11|_{23} + |00\rangle_{23}\langle 00|_{23}) = \frac{1}{2}\begin{pmatrix} 1 & 0 \\ 0 & 1 \end{pmatrix}.$$

Therefore the corresponding von-Neumann entropy is

$$S = -tr\frac{1}{2}\begin{pmatrix} 1 & 0 \\ 0 & 1 \end{pmatrix}\ln\frac{1}{2}\begin{pmatrix} 1 & 0 \\ 0 & 1 \end{pmatrix} = -tr\frac{1}{2}\begin{pmatrix} 1 & 0 \\ 0 & 1 \end{pmatrix}\begin{pmatrix} \ln(1/2) & 0 \\ 0 & \ln(1/2) \end{pmatrix} = \ln(2).$$

Entanglement is a property of states, regardless whether they are eigenstates or ground states of Hamiltonians. Therefore we can pose problems that otherwise may seem quite unconventional in the context of solid-state physics. For example, we can ask how a state with a variational parameter can be optimized to yield maximum entanglement. As a toy example, let us consider the two-bit state $|\psi\rangle = (|10\rangle + \alpha|01\rangle)/\sqrt{1+\alpha^2}$. Constructing the density matrix, and tracing out one of the two spins, we arrive at the reduced density matrix,

$$\rho_A = \frac{1}{1+\alpha^2}\begin{pmatrix}1 & 0\\ 0 & \alpha^2\end{pmatrix}.$$

This leads to the von-Neumann entropy $S = \ln(1+\alpha^2) - (\alpha^2 \ln\alpha^2)/(1+\alpha^2)$. Extrema of this entropy can be found by setting the derivative $\partial S/\partial\alpha$ equal to zero. There are three minimally entangled solutions at $\alpha = 0 \pm \infty$, corresponding to classical states, and two maximally entangled solutions at $\alpha \pm 1$, which are also (but generally do not have to be) eigenstates of a 2-site Heisenberg Hamiltonian.

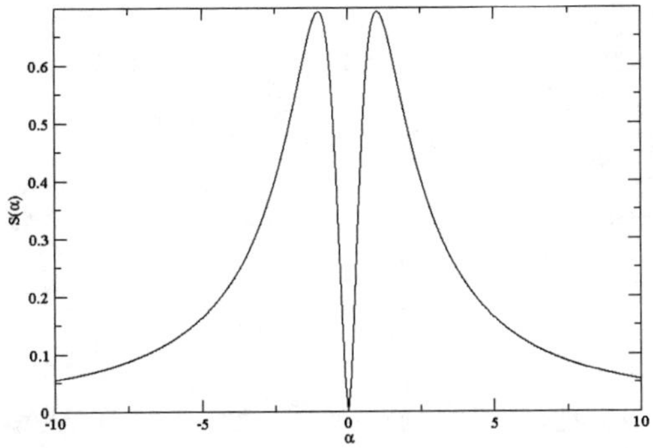

FIGURE 15. Von-Neumann entropy of a two-bit state. The parameter α encodes the admixture of the two states. Extrema of $S(\alpha)$ correspond to maximally and minimally entangled combinations, discussed in the text.

Next, let us verify that if we exchange the definition of the "inside" and the "outside" area in a system, the von-Neumann entropy should be the same. Let us study the example of a 3-site state $|\psi\rangle = [|100\rangle + |010\rangle + |001\rangle]/\sqrt{3}$. Tracing out the 2nd and 3rd site, we arrive at the reduced density matrix

$$\rho_1 = \begin{pmatrix}1/3 & 0\\ 0 & 2/3\end{pmatrix},$$

which yields $S = \ln(3) - 2\ln(2)/3$. Tracing out the 1st site, the reduced density matrix is

$$\rho_{23} = \frac{1}{3}\begin{pmatrix}1 & 1 & 0\\ 1 & 1 & 0\\ 0 & 0 & 1\end{pmatrix}.$$

In order to determine the entropy corresponding to this density matrix, we need to diagonalize it first. If the diagonalized reduced density matrix is $\rho_D = A^+\rho A$, then $S = -tr\rho\ln\rho = -tr(A\rho_D A^+ \ln(A\rho_D A^+)) = -tr(A\rho_D A^+ \ln(\rho_D)A^+) = -tr\rho_D \ln(\rho_D)$.

For the given three-site state

$$\rho_D = \begin{pmatrix} 0 & 0 & 0 \\ 0 & 2/3 & 0 \\ 0 & 0 & 1/3 \end{pmatrix},$$

and hence the resulting von-Neumann entropy is the same as determined from ρ_1.

Next, let us consider a state for which the von-Neumann entropy depends on how the inside and the outside areas are defined. We can construct such a state on a three-site chain e.g. by multiplying an up-spin at site 1 with a singlet between the spins on sites 2 and 3. This yields $|\psi\rangle = (|110\rangle - |101\rangle)/\sqrt{2}$. You can verify that the entropy resulting from tracing out spins 2 and 3 gives $S = 0$, i.e. spin 1 is not entangled with spins 2 and 3. On the other hand, tracing sites 1 and 2, one finds $S = \ln(2)$, since site 3 is maximally entangled with site 2. This means that there is local bipartite entanglement, although there is no global entanglement in this state. This opens up the question how to determine global entanglement in a many-body state.

Several proposed measures for global entropy are currently discussed. One of the starting points of this discussion is the first proposed Meyer-Wallach measure,[10]

$$Q = \frac{1}{n} \sum_{k=1}^{n} 2(1 - tr\rho_k^2), \tag{47}$$

which is an algebraic average over single-site entropies. However, as we will see now, this measure can miss the lack of entanglment in factorized states. Let us first apply the Meyer-Wallach measure to a state which we know to be maximally entangled by construction, i.e. the 4-site state $|W\rangle = (|1111\rangle + |0000\rangle)/\sqrt{2}$, which we have already examined at the beginning of this section. Its full density matrix is given by $\rho = (|1111\rangle\langle 1111| + |0000\rangle\langle 0000| + |1111\rangle\langle 0000| + |0000\rangle\langle 1111|)/2$, and when tracing out an arbitrary 3 spins, we arrive at he reduced single-site density matrix

$$\frac{1}{2}\begin{pmatrix} 1 & 0 \\ 0 & 1 \end{pmatrix}.$$

The Meyer-Wallach measure thus becomes, confirming maximum global entanglement.

However, since the Meyer-Wallach measure is solely based on the entanglment of single sites with the remaining system, it can miss lack of entanglement in factorized states. Let us construct such a state by multiplying two two-site Bell states with each other, i.e.

$$|\psi\rangle = \frac{1}{\sqrt{2}}(|11\rangle_{12} + |00\rangle_{12}) \frac{1}{\sqrt{2}}(|11\rangle_{34} + |00\rangle_{34}) = \frac{1}{2}(|1111\rangle + |1100\rangle + |0011\rangle + |0000\rangle).$$

We know that if we trace out the first two or the second two spins, the resulting von-Neumann entropy must vanish. However, since the single-site reduced density matrix for this system is $\rho_k = (|1\rangle_k\langle 1|_k + |0\rangle_k\langle 0|_k)/2$ the Meyer-Wallach measure gives $Q = 1$, and thus obviously does not reflect true global entanglement.

A more recently proposed global entanglement measure that improves on the Mayer-Wallach measure takes into account all possible partitions, not just single-site traces, and is thus a generalization of the former.[11] This Scott measure is given by

$$Q_m = \binom{n}{m}^{-1} \sum_{|S|=m}^{n} \frac{2^m}{2^m - 1}(1 - tr\rho_S^2). \tag{48}$$

Let us again compare the states $|W\rangle = (|1111\rangle + |0000\rangle)/\sqrt{2}$ and $|\psi\rangle = (|1111\rangle + |1100\rangle + |0011\rangle + |0000\rangle)/2$ on a 4-site chain. For both cases, we first have to evaluate the binomial prefactor $\binom{n}{m}$ which counts the number of partitions. For single sites, this prefactor is $\binom{4}{1} = \frac{4!}{3!1!} = 4$, for 2-site partitions it is 6, and for 3-site partitions it is 4 again. For the state $|W\rangle = (|1111\rangle + |0000\rangle)/\sqrt{2}$, the corresponding reduced density matrices are $\rho_1 = [|1\rangle\langle 1| + |0\rangle\langle 0|]/2, \rho_2[|11\rangle\langle 11| + |00\rangle\langle 00|]/2$, and $\rho_3 = [|111\rangle\langle 111| + |000\rangle\langle 000|]/2$, leading to $Q_1 = 1$, $Q_2 = 2/3$, and $Q_3 = 4/7$. On the other hand, for the factorized state $|\psi\rangle = (|1111\rangle + |1100\rangle + |0011\rangle + |0000\rangle)/2$ one finds $Q_1 = 1$ and $Q_2 = 4/9$. This is a small improvement compared to the Meyer-Wallach measure, since this measure at least distinguishes between an optimally entanglement and a factorized state, since $4/9 < 2/3$. However, one would hope that measures of global entanglement should vanish altogether if there exists even a single type of partition that leads to a factorized state. This could be achieved by recently proposed geometrical averages,[12] e.g.

$$Q = \prod \frac{2^m}{2^m - 1}(1 - tr\rho_S^2). \tag{49}$$

ACKNOWLEDGMENTS

Many thanks to the organizers and participants of the XII Training Course in the Physics of Strongly Correlated Systems at Vietri. In particular, I would like to acknowledge Adolfo Avella, who has been most helpful.

REFERENCES

1. For a review see E. Manousakis, Rev. Mod. Phys. **63**, 1 (1991).
2. See e.g. E. Dagotto and T.M. Rice, Science **271**, 618 (1996).
3. D.C. Mattis, "The theory of magnetism made simple", World Scintific Publisher (2006).
4. S. Blundell, "Magnetism in condensed matter", Oxford University Press (2001).
5. P.W. Anderson, Phys. Rev. **86** 694 (1952).
6. T. Pan, "An introduction to computational physics", Cambridge University Press (1997).
7. N. Metropolis and S. Ulam, J. Am. Stat. Assoc. **44**, 335 (1949).
8. A.W. Sandvik, Phys. Rev. B **59**, R14160 (1999).
9. J. von Neumann, "Mathematische Grundlagen der Quantenmechanik", Springer (1955).
10. D.A. Meyer and N.R. Wallach, quant-ph/0108104.
11. A.J. Scott, Phys. Rev. A **69**, 052330 (2004).
12. P.J. Love et al., Quant. Inf. Proc. **6**, 187 (2007).

The Dynamical Cluster Approximation with Quantum Monte Carlo Cluster Solvers

M. Jarrell*, A. Macridin*, K. Mikelsons*, D.G.S.P. Doluweera* and J.E. Gubernatis†

*Department of Physics, University of Cincinnati, Cincinnati, Ohio 45221
†Theoretical Division, Los Alamos National Laboratory

Abstract. We present a pedagogical discussions of the dynamical mean field (DMFA) and dynamical cluster (DCA) approximations and associated Monte Carlo and entropy-based methods of Bayesian data analysis. The DMFA and DCA methods are developed as coarse-graining approximations and the relationship between the cluster and lattice problems are detailed. The Hirsch-Fye and continuous time Quantum Monte Carlo (QMC) algorithms are used to solve the cluster problem. The algorithms are discussed, together with methods for efficient measurements and the modifications required by the self-consistency of the DMFA/DCA. Then, several principles of Bayesian data analysis are presented. When coupled with information theory, this analysis produces a precise and systematic way to analytically continue Matsubara-time QMC results to real frequencies. We show how to use Bayesian inference to qualify the solution of the continuation and optimize the inputs. Besides developing the Bayesian formalism, we also present a detailed description of the data qualification, sketch an efficient algorithm to solve for the optimal spectra, give cautionary notes where appropriate, and present two detailed case studies to demonstrate the method.

Keywords: Dynamical mean field, Cluster, Quantum Monte Carlo, Maximum entropy
PACS: 02.60.Pn, 71.27.+a, 89.70.Cf, 02.70.Hm, 02.70.Ss, 02.70.Uu, 07.05.P

CONTENTS

CP1014, *Lectures on the Physics of Strongly Correlated Systems XII: Twelfth Training Course*, edited by A. Avella and F. Mancini

1. INTRODUCTION

Some of the most exotic properties of materials, including high-temperature superconductivity, magnetism, and heavy Fermion and non-Fermi liquid behaviors, are due to strong electronic correlations. The materials which display these properties are characterized by either narrow electronic bands or compact orbitals with large angular momentum in the valence shell. In either case, the potential energy associated with some of these electronic degrees of freedom is of similar magnitude or larger than their electronic kinetic energy (bandwidth), which invalidates conventional perturbative approaches. Thus, we resort to the construction of simplified models to study these systems.

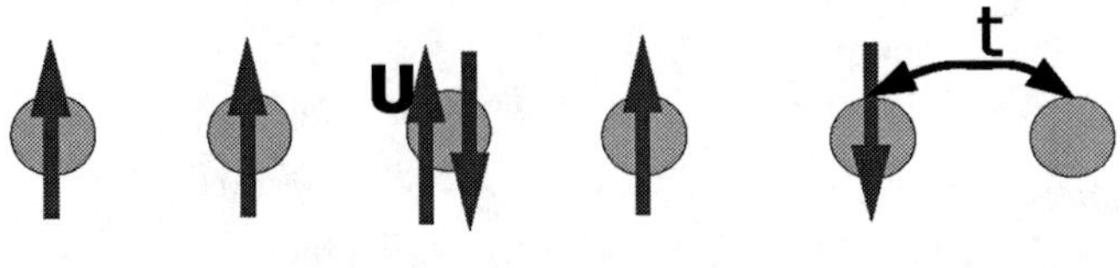

FIGURE 1. Cartoon of the Hubbard model, characterized by a single band with near-neighbor hopping t, and local repulsion U.

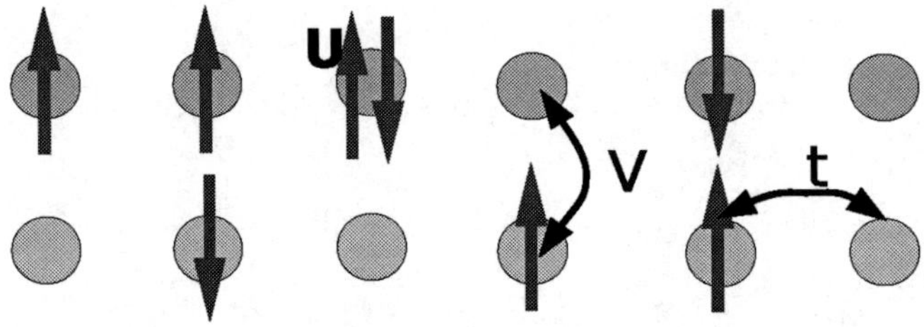

FIGURE 2. Cartoon of the periodic Anderson model, characterized by two bands one with near-neighbor hopping t with a local hybridization V to a second band with local repulsion U.

For example, the Hubbard model [1] is the simplest model of a correlated electronic lattice system. Both it and the $t-J$ model are thought to at least qualitatively describe some of the properties of transition metal oxides, and high temperature superconductors [2]. The periodic Anderson model along with various Kondo lattice models have been proposed to describe both the actinide and lanthanide heavy fermion systems and the Anderson insulators. The Holstein model incorporates the essential physics of strongly interacting electrons and phonons. All of these model Hamiltonians contain at least two major ingredients: a local interaction term and a non-local hopping term. For example, the Hubbard model Hamiltonian is

$$H = -t \sum_{\langle j,k\rangle\sigma} (c^{\dagger}_{j\sigma}c_{k\sigma} + c^{\dagger}_{k\sigma}c_{j\sigma}) + \varepsilon\sum_{j}(n_{j\uparrow}+n_{j\downarrow}) + U\sum_{j}(n_{j\uparrow}-1/2)(n_{j\downarrow}-1/2), \quad (1)$$

where $c^{\dagger}_{j\sigma}$ ($c_{j\sigma}$) creates (destroys) an electron at site j with spin σ, $n_{i\sigma} = c^{\dagger}_{i\sigma}c_{i\sigma}$, t is the nearest neighbor hopping which sets the unit of energy and U is the on-site Coulomb repulsion between the electrons. The periodic Anderson model (PAM) Hamiltonian is

$$H = -t \sum_{<ij>\sigma} (d^{\dagger}_{i\sigma}d_{j\sigma} + d^{\dagger}_{j\sigma}d_{i\sigma}) + V\sum_{i\sigma}(d^{\dagger}_{i\sigma}f_{i\sigma} + f^{\dagger}_{i\sigma}d_{i\sigma}) + \frac{U}{2}\sum_{i\sigma}(n^{f}_{i,\sigma}-\frac{1}{2})(n^{f}_{i,-\sigma}-\frac{1}{2}) \quad (2)$$

where $d_{i\sigma}$ and $f_{i\sigma}$ ($d^{\dagger}_{i\sigma}$ and $f^{\dagger}_{i\sigma}$) destroy (create) a d- and f-electron on site i with spin σ, U is the screened Coulomb-matrix element for the localized f-states and V characterizes the mixing between the two subsystems, f and respectively d orbitals.

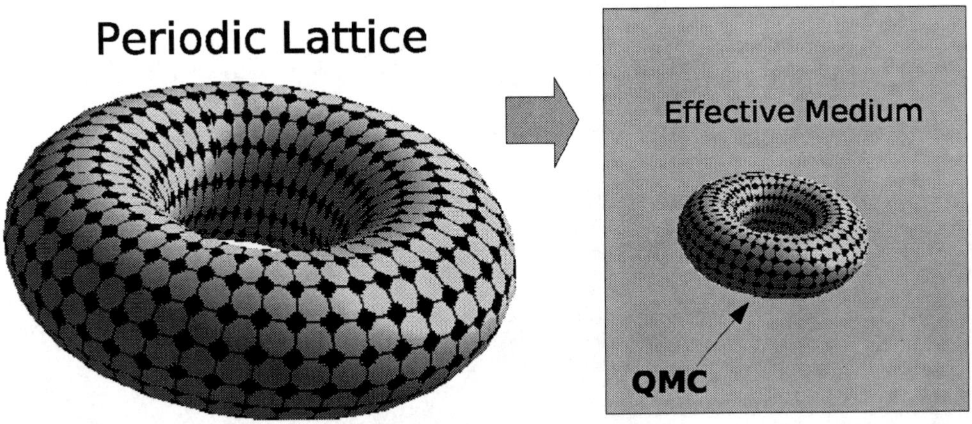

FIGURE 3. Quantum cluster approaches, like the DMFA and DCA, map the infinite lattice problem onto a self-consistently embedded cluster problem.

However, except for special limits, even such simplified models like Eq. 1 cannot be solved exactly. For example, for the Hubbard model, no exact solutions exist except in one dimension, where the knowledge is in fact rather complete [3]. The periodic Anderson model is only solvable in the limit where the orbital degeneracy diverges [4], and the Holstein model is only solvable in the Eliashberg-Migdal limit where vertex corrections may be neglected. Clearly a new approach to these models is needed if nontrivial exact solutions are desired.

Metzner and Vollhardt suggested such a new approach [5, 6, 7] based on an expansion in $1/d$ about the point $d = \infty$ to study these strongly correlated lattice models. The resulting formalism neglects dynamical intersite correlations while retaining the important local dynamical correlations. The resulting formalism is called the Dynamical Mean Field Approximation (DMFA) since it may be employed in any dimension, but is only exact on infinite dimensional lattices. In finite dimensions, the Dynamical Cluster Approximation (DCA) is used to study systematic non-local corrections to the DMFA [8, 9]. Quantum cluster approaches such as the DMFA and DCA work by mapping an infinite periodic lattice onto a self-consistently embedded cluster problem, as illustrated in Fig. 3. Correlations up to the cluster size are treated explicitly, while those at longer length scales are treated in a mean field. The DMFA/DCA cluster problem may be solved by a variety of methods; however, Quantum Monte Carlo (QMC) is the first numerically exact method employed [10] and remains the most powerful and adaptable method.

In this article, we will present a pedagogical discussion of a complete suite of QMC-based formalisms and algorithms for the DMFA and DCA. In Sec. 2 we will first rederive the DMFA as a course-graining approximation, extend this logic to derive the DCA, and then describe how physical quantities are calculated in this formalism. In Sec. 3 we will discuss two powerful QMC algorithms used to solve the embedded cluster problem of the DMFA/DCA. Finally, in Sec. 4 we will discuss entropy-based Bayesian data analysis and its use to analytically continue the Matsubara time or frequency QMC data to real

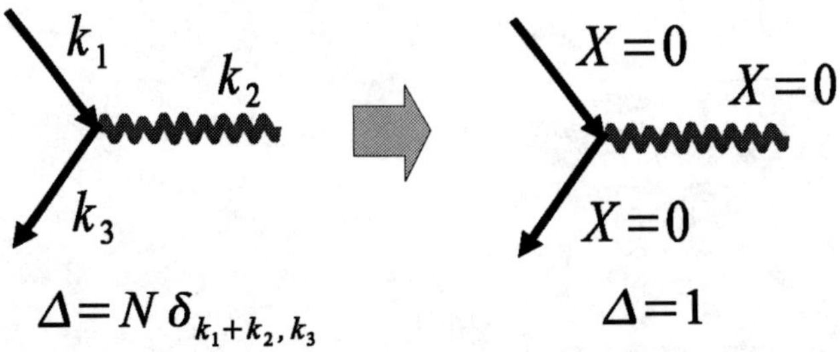

FIGURE 4. The Laue function Δ, which described momentum conservation at a vertex (left). In the DMFA, $\Delta = 1$, so momentum conservation is neglected for compact graphs (right) so that we may freely sum over the momentum labels leaving only local propagators and interactions.

frequencies.

2. THE DYNAMICAL MEAN FIELD AND CLUSTER APPROXIMATIONS

2.1. The Dynamical mean-field approximation

The DMFA is a local approximation which was used by Kuramoto in perturbative calculations as a simplification of the k-summations which render the problem intractable [11]. But it was after the work of Metzner and Vollhardt [5] and Müller-Hartmann [6] who showed that this approximation becomes exact in the limit of infinite dimension that it received extensive attention. In this limit, the spatial dependence of the self-energy disappears, retaining only its variation with time. Please see the reviews by Pruschke *et al* [12] and Georges *et al* [13] for a more extensive treatment.

In this section, we will show that it is possible to re-interpret the DMFA as a course graining approximation. For a two-dimensional lattice, this is equivalent to averaging, or coarse-graining, the Green's functions used to calculate the irreducible diagrammatic insertions over the Brillouin zone.

Müller-Hartmann [6] showed that this coarse-graining becomes exact in the limit of infinite-dimensions. For Hubbard-like models, the properties of the bare vertex are completely characterized by the Laue function Δ which expresses the momentum conservation at each vertex. In a conventional diagrammatic approach

$$\Delta(\mathbf{k}_1,\mathbf{k}_2,\mathbf{k}_3,\mathbf{k}_4) = \sum_{\mathbf{r}} \exp\left[i\mathbf{r}\cdot(\mathbf{k}_1+\mathbf{k}_2-\mathbf{k}_3-\mathbf{k}_4)\right] \qquad (3)$$
$$= N\delta_{\mathbf{k}_1+\mathbf{k}_2,\mathbf{k}_3+\mathbf{k}_4}$$

where $\mathbf{k}_1$ and $\mathbf{k}_2$ ($\mathbf{k}_3$ and $\mathbf{k}_4$) are the momenta entering (leaving) each vertex through its legs of G. However as the dimensionality $D \to \infty$ Müller-Hartmann showed that the Laue function reduces to [6]

$$\Delta_{D\to\infty}(\mathbf{k}_1,\mathbf{k}_2,\mathbf{k}_3,\mathbf{k}_4) = 1 + \mathcal{O}(1/D) \quad . \qquad (4)$$

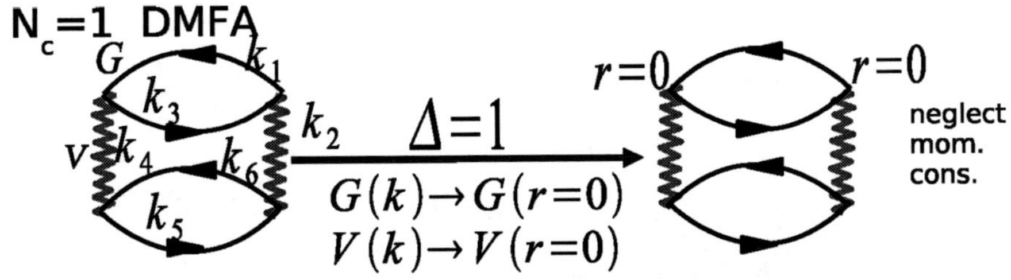

FIGURE 5. The second order contribution the generating functional Φ. As we apply the DMFA coarse-graining approximation, Eq. 4, Φ becomes a functional of the local Green's function and interaction.

The DMFA assumes the same Laue function, $\Delta_{DMFA}(\mathbf{k}_1, \mathbf{k}_2, \mathbf{k}_3, \mathbf{k}_4) = 1$, even in the context of finite dimensions. More generally, for an electron scatting from an interaction (boson) pictured in Fig. 4, $\Delta_{DMFA}(\mathbf{k}_1, \mathbf{k}_2, \mathbf{k}_3) = 1$. Thus, the conservation of momentum at internal vertices is neglected. We may freely sum over the internal momentum labels of each Green's function leg and interaction leading to a collapse of the momentum dependent contributions leaving only local terms.

This argument may then be applied to the generating functional Φ. It is the sum over all closed connected compact graphs constructed from the dressed Green's function G and the bare interaction. The self energy Σ_σ is obtained from a functional derivative of Φ, $\Sigma_\sigma = \delta\Phi/\delta G_\sigma$, and the irreducible vertices $\Gamma_{\sigma\sigma'} = \delta\Sigma_\sigma/\delta G'_\sigma$. The second order contribution to Φ for a Hubbard-like model is illustrated in Fig. 5, becomes a functional of the local interaction and Green's function. The self energy Σ may be obtained from a functional derivative of Φ with respect to the Green's function G, which effectively breaks one of the Green's function lines.

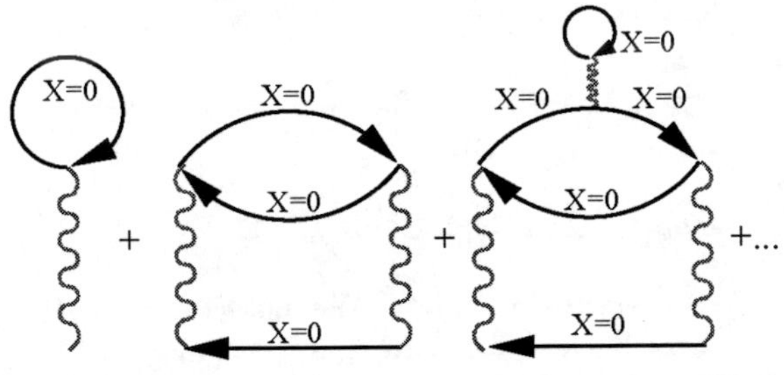

FIGURE 6. The DMFA self energy. Note that it contains local self energy corrections See, e.g., the third graph. To prevent overcounting these contributions, the local self energy must be excluded, c.f., Eq. 5 from the Green's function line used in most cluster solvers.

The perturbative series for Φ, Σ and the irreducible vertices Γ in the DMFA are identical to those of the corresponding impurity model, so that conventional impurity solvers may be used. However, since most impurity solvers can be viewed as methods that sum all the graphs, not just the skeletal ones, it is necessary to exclude $\Sigma(i\omega_n)$ from the local propagator $\mathcal{G}$ input to the impurity solver in order to avoid overcounting the

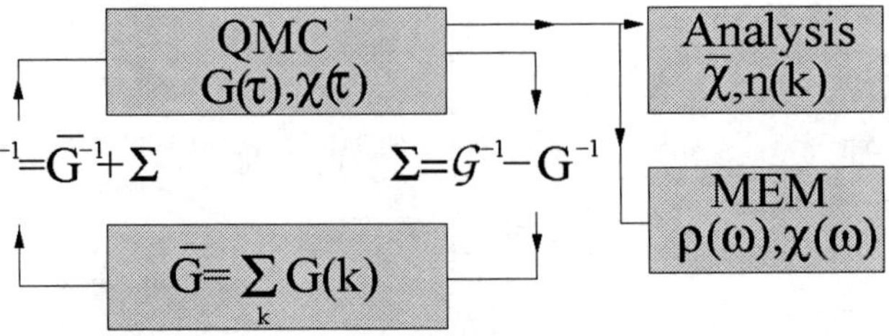

FIGURE 7. The DMFA algorithm. QMC is used as a cluster solver. Once convergence is reached, $G = \bar{G}$, and the irreducible quantities are used in the analysis and Maximum Entropy Method (MEM) codes to calculate the phase diagram and spectra, respectively.

local self-energy $\Sigma(i\omega_n)$ ($i\omega_n = (2n+1)\pi T$ is the Matsubara frequency).

$$\mathscr{G}(i\omega_n)^{-1} = G(i\omega_n)^{-1} + \Sigma(i\omega_n) \tag{5}$$

where $G(i\omega_n)$ is the full local Green's function. Hence, in the local approximation, the Hubbard model has the same diagrammatic expansion as an Anderson impurity with a bare local propagator $\mathscr{G}(i\omega_n;\Sigma)$ which is determined self-consistently.

An algorithm constructed from this approximation is the following: (i) An initial guess for $\Sigma(i\omega_n)$ is chosen (usually from perturbation theory). (ii) $\Sigma(i\omega_n)$ is used to calculate the corresponding local Green's function

$$G(i\omega_n) = \int d\eta \frac{\rho^0(\eta)}{i\omega_n - \eta - \varepsilon - \Sigma(i\omega_n)}, \tag{6}$$

where ρ^0 is the non-interacting density of states. (iii) Starting from $G(i\omega_n)$ and $\Sigma(i\omega_n)$ used in the second step, the host Green's function $\mathscr{G}(i\omega_n)^{-1} = G(i\omega_n)^{-1} + \Sigma(i\omega_n)$ is calculated which serves as bare Green's function of the impurity model. (iv) Starting with $\mathscr{G}(i\omega_n)$, the local Green's function $G(i\omega_n)$ is obtained using the Quantum Monte Carlo method (or another technique). (v) Using the QMC output for the cluster Green's function $G(i\omega_n)$ and the host Green's function $\mathscr{G}(i\omega_n)$ from the third step, a new $\Sigma(i\omega_n) = \mathscr{G}(i\omega_n)^{-1} - G(i\omega_n)^{-1}$ is calculated, which is then used in step (ii) to reinitialize the process. Steps (ii) - (v) are repeated until convergence is reached. In step (iv) the QMC algorithm of Hirsch and Fye [14, 15] may be used to compute the local Green's function $G(\tau)$ or other physical quantities in imaginary time. Local dynamical quantities are then calculated by analytically continuing the corresponding imaginary-time quantities using the Maximum-Entropy Method (MEM) [16].

2.2. The Dynamical Cluster Approximation

In this section, we will review the formalism which leads to the dynamical cluster approximation. Here, we first motivate the fundamental idea of the DCA which is coarse-graining, we then describe the mapping to an effective cluster problem and discuss the relationship between the cluster and lattice at the one and two-particle level.

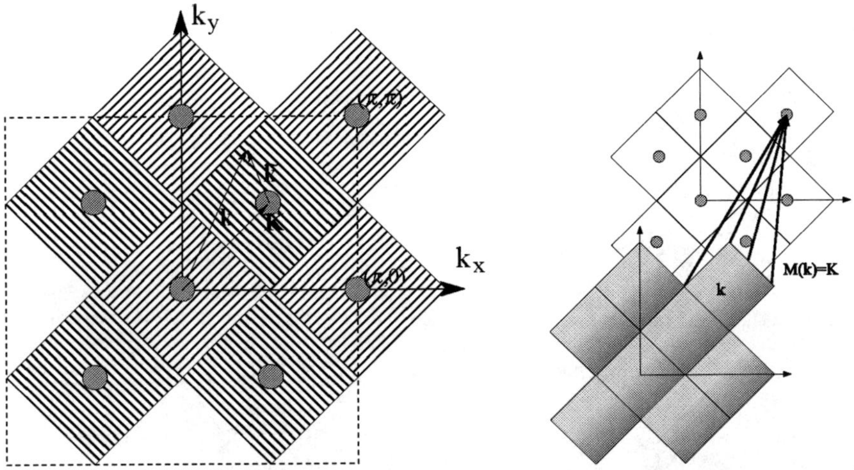

FIGURE 8. (left) Coarse-graining cells for $N_c = 8$ (differentiated by alternating fill patterns) that partition the first Brillouin Zone (dashed line). Each cell is centered on a cluster momentum $\mathbf{K}$ (filled circles). (right) To construct the DCA cluster (e.g. for $N_c = 8$) we map a generic $\mathbf{k}$ to the nearest cluster point $\mathbf{K} = \mathbf{M}(\mathbf{k})$ so that $\tilde{\mathbf{k}} = \mathbf{k} - \mathbf{K}$ remains in the cell around $\mathbf{K}$.

2.2.1. Coarse-Graining

Like the DMFA, the DCA may be intuitively motivated with a coarse-graining transformation. In the DMFA, the propagators used to calculate Φ and its functional derivatives were coarse-grained over the entire Brillouin zone, leading to local (momentum independent) irreducible quantities. In the DCA, we wish to relax this condition and systematically restore momentum conservation and non-local corrections. Thus, in the DCA, the reciprocal space of the lattice (Fig. 8) which contains N points is divided into N_c cells of identical linear size Δk. The coarse-graining transformation is set by averaging the Green's function within each cell. If $N_c = 1$ the original lattice problem is mapped to an impurity problem, and we recover the DMFA. If N_c is larger than one, then non-local corrections of length $\approx \pi/\Delta k$ to the DMFA are introduced. Provided that the propagators are sufficiently weakly momentum dependent, this is a good approximation. If N_c is chosen to be small, the cluster problem can be solved using conventional techniques such as QMC. This averaging process also establishes a relationship between the systems of size N and N_c. A simple and unique choice which will be discussed in Sec. 2.2.2 is to equate the irreducible quantities (self energy, irreducible vertices) of the cluster to those in the lattice.

2.2.2. A diagrammatic derivation

This coarse graining procedure and the relationship of the DCA to the DMFA is illustrated by a microscopic diagrammatic derivation of the DCA starting again from the

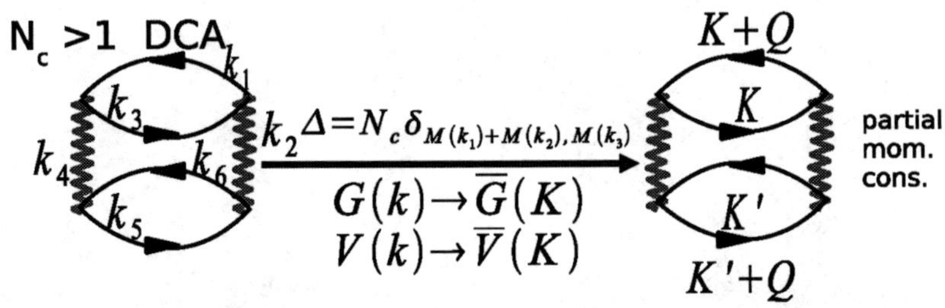

FIGURE 9. A second-order term in the generating functional of the Hubbard model. Here the undulating line represents the interaction U, and on the LHS (RHS) the solid line, the lattice (coarse-grained) single-particle Green's functions. When the DCA Laue function is used to describe momentum conservation at the internal vertices, the momenta collapse onto the cluster momenta and each lattice Green's function is replaced by the coarse-grained result.

Baym generating functional Φ [19]. The DCA systematically restores the momentum conservation at internal vertices of Φ relinquished by the DMFA. The Brillouin-zone is divided into $N_c = L^D$ cells of size $\Delta k = 2\pi/L$ (c.f. Fig. 8 for $N_c = 8$). Each cell is represented by a cluster momentum $\mathbf{K}$ in the center of the cell. We require that momentum conservation is (partially) observed for momentum transfers between cells, i.e., for momentum transfers larger than Δk, but neglected for momentum transfers within a cell, i.e., less than Δk. This requirement can be established by using the Laue function [9]

$$\Delta_{DCA}(\mathbf{k}_1, \mathbf{k}_2, \mathbf{k}_3, \mathbf{k}_4) = N_c \delta_{\mathbf{M}(\mathbf{k}_1)+\mathbf{M}(\mathbf{k}_2),\mathbf{M}(\mathbf{k}_3)+\mathbf{M}(\mathbf{k}_4)} \quad , \tag{7}$$

where $\mathbf{M}(\mathbf{k})$ is a function which maps $\mathbf{k}$ onto the momentum label $\mathbf{K}$ of the cell containing $\mathbf{k}$ (see, Fig. 8). This choice for the Laue function systematically interpolates between the exact result, Eq. 4, which it recovers when $N_c \to N$ and the DMFA result, Eq. 4, which it recovers when $N_c = 1$. With this choice of the Laue function the momenta of each internal leg may be freely summed over the cell.

This is illustrated for the second-order term in the generating functional in Fig. 9. Each internal leg $G(\mathbf{k})$ in a diagram is replaced by the coarse–grained Green's function $\bar{G}(\mathbf{M}(\mathbf{k}))$, defined by

$$\bar{G}(\mathbf{K}) \equiv \frac{N_c}{N} \sum_{\tilde{\mathbf{k}}} G(\mathbf{K} + \tilde{\mathbf{k}}) \quad , \tag{8}$$

where N is the number of points of the lattice, N_c is the number of cluster $\mathbf{K}$ points, and the $\tilde{\mathbf{k}}$ summation runs over the momenta of the cell about the cluster momentum $\mathbf{K}$ (see, Fig. 8). The diagrammatic sequences for the generating functional and its functional derivatives are unchanged; however, the complexity of the problem is greatly reduced since $N_c \ll N$.

As with the DMFA, the coarse-graining approximation will be applied to only the compact part of the free energy, Φ, and its functional derivatives. Physically, this is justified by the fact that irreducible terms like the self energy are short ranged, while reducible quantities like G must be able to capture long length and time scale physics. This

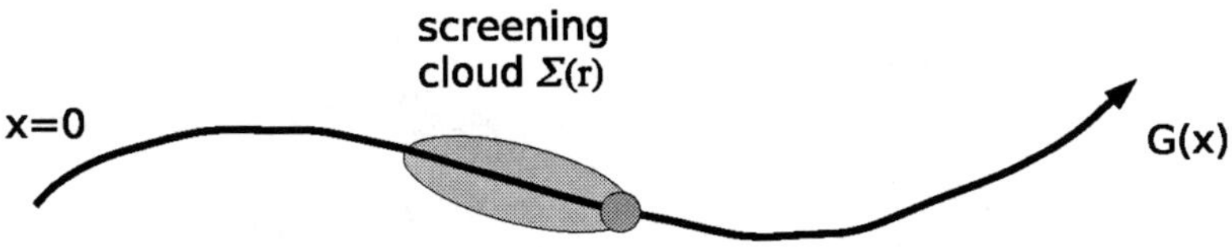

FIGURE 10. Screening of a propagating particle. The single particle Green's function, which describes the quantum phase and amplitude the particle accumulates, is poorly approximated by a small cluster calculation. Its self energy, which describes generally short ranged screening processes, can be well approximated by a small cluster calculation.

is motivated in Fig. 10. As the particle propagates from the origin to space-time location x, the quantum phase and amplitude it accumulates is described by the single-particle Green's function $G(x)$. Consequently if x is larger than the size of the DCA cluster, then $G(x)$ is poorly approximated by the cluster Green's function. However, the self energy Σ describes the many-body processes that produce the screening cloud surrounding the particle. These processes are generally short ranged in a strongly correlated many-body system, so the self energy is often well approximated by the cluster quantity. Formally, we have justified this elsewhere by exploring the Δk-dependence of the compact and non-compact parts of the free energy [17]. The generating functional is the sum over all of the closed connected compact diagrams, such as the one shown in Fig. 9.

The corresponding DCA estimate for the free energy is

$$F_{DCA} = -k_B T \left(\Phi_c - \mathrm{Tr}\left[\Sigma_\sigma G_\sigma\right] - \mathrm{Tr}\ln\left[-G_\sigma\right]\right), \tag{9}$$

where Φ_c is the cluster generating functional. The trace indicates summation over frequency, momentum and spin. F_{DCA} is stationary with respect to $\mathbf{G}_\sigma$ when

$$\frac{-1}{k_B T}\frac{\delta F_{DCA}}{\delta G_\sigma(\mathbf{k})} = \Sigma_{c\sigma}(M(\mathbf{k})) - \Sigma_\sigma(\mathbf{k}) = 0, \tag{10}$$

which means that $\Sigma(\mathbf{k}) = \Sigma_{c\sigma}(M(\mathbf{k}))$ is the proper approximation for the lattice self energy corresponding to Φ_c. The corresponding lattice single-particle propagator is then given by

$$G(\mathbf{k},z) = \frac{1}{z - \varepsilon_{\mathbf{k}} - \varepsilon - \Sigma_c(M(\mathbf{k}),z)}. \tag{11}$$

A similar procedure is used to construct the two-particle quantities needed to determine the phase diagram or the nature of the dominant fluctuations that can eventually destroy the quasi-particle. This procedure is a generalization of the method of calculating response functions in the DMFA [10, 18]. In the DCA, the introduction of the momentum dependence in the self-energy will allow one to detect some precursor to transitions which are absent in the DMFA; but for the actual determination of the nature of the instability, one needs to compute the response functions. These susceptibilities are thermodynamically defined as second derivatives of the free energy with respect to external fields. $\Phi_c(G)$ and $\Sigma_{c\sigma}$, and hence F_{DCA} depend on these fields only through G_σ

and the bare G^0_σ. Following Baym [19] it is easy to verify that, the approximation

$$\Gamma_{\sigma,\sigma'} \approx \Gamma_{c\sigma,\sigma'} \equiv \delta\Sigma_{c\sigma}/\delta G_{\sigma'} \tag{12}$$

yields the same estimate that would be obtained from the second derivative of F_{DCA} with respect to the applied field. For example, the first derivative of the free energy with respect to a spatially homogeneous external magnetic field h is the magnetization,

$$m = \mathrm{Tr}\left[\sigma G_\sigma\right]. \tag{13}$$

The susceptibility is given by the second derivative,

$$\frac{\partial m}{\partial h} = \mathrm{Tr}\left[\sigma \frac{\partial G_\sigma}{\partial h}\right]. \tag{14}$$

We substitute $G_\sigma = \left(G^{0-1}_\sigma - \Sigma_{c\sigma}\right)^{-1}$, and evaluate the derivative,

$$\frac{\partial m}{\partial h} = \mathrm{Tr}\left[\sigma \frac{\partial G_\sigma}{\partial h}\right] = \mathrm{Tr}\left[G^2_\sigma \left(1 + \sigma \frac{\partial \Sigma_{c\sigma}}{\partial G_{\sigma'}} \frac{\partial G_{\sigma'}}{\partial h}\right)\right]. \tag{15}$$

If we identify $\chi_{\sigma,\sigma'} = \sigma \frac{\partial G_{\sigma'}}{\partial h}$, and $\chi^0_\sigma = G^2_\sigma$, collect all of the terms within both traces, and sum over the cell momenta $\tilde{\mathbf{k}}$, we obtain the two–particle Dyson's equation

$$\begin{aligned} 2(\bar{\chi}_{\sigma,\sigma} \;&-\; \bar{\chi}_{\sigma,-\sigma}) \\ &= \; 2\bar{\chi}^0_\sigma + 2\bar{\chi}^0_\sigma \left(\Gamma_{c\sigma,\sigma} - \Gamma_{c\sigma,-\sigma}\right)\left(\bar{\chi}_{\sigma,\sigma} - \bar{\chi}_{\sigma,-\sigma}\right). \end{aligned} \tag{16}$$

We see that again it is the irreducible quantity, i.e., the vertex function, for which cluster and lattice correspond.

2.2.3. Cluster Selection

The geometry of the DCA cluster is relevant for systematic studies of phase transitions. All cluster geometries which can be used to tile the lattice without gaps are valid (c.f. Fig. 11). In this approximation, the lattice has long range order once the correlation length ξ of the order reaches the linear cluster size, since then the lattice tiled with such clusters would be ordered. A scaling ansatz for T_c which captures these ideas is [20]

$$\xi\left(T_c(N_c)\right) = N_c^{1/D}. \tag{17}$$

So for a typical transition, with $\xi(T) \propto |T - T_c^*|^{-\nu}$ (where T_c^* is the transition temperature of the thermodynamic lattice), the scaling formula is $T_c^* = T_c(N_c) - A(N_c)^{-1/D\nu}$, where A, ν, and T_c^* are fit to the data $T_c(N_c)$. However, clearly this approach will not apply to clusters with strange geometries, like e.g., rods in a 2D system.

Ideal clusters should properly represent all length scales up to $N_c^{1/D}$, and no others. Consider the 16-site clusters shown in Fig. 12, the number of neighbors in each neighbor

FIGURE 11. A variety of cluster geometries which may be used to tile a two-dimensional square lattice.

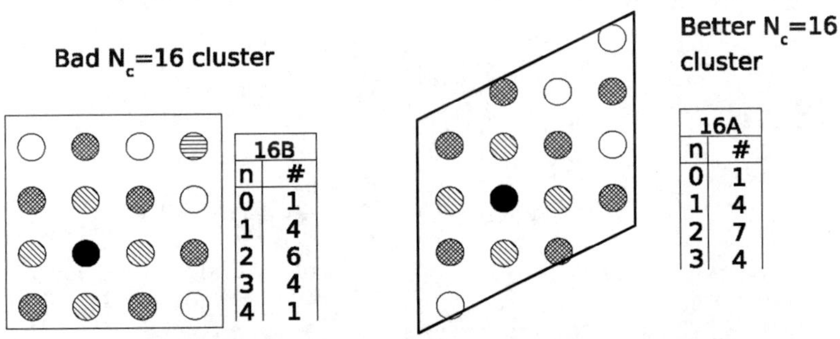

FIGURE 12. Two sixteen ($N_c = 16$) site periodic clusters. Here, the neighbors to the black site are shown different patterns.

shell are shown in the adjacent tables [21]. On the lattice, the nth shell has 4n neighbors. Both 16A and 16B have complete near neighbor (n=1) shells with four near neighbors. However, the conventional square cluster 16B, has 6 (not 8) sites in the n=2 shell, has 4 (not 12) in the n=3 shell, and even on site in the n=4 shell. The cluster 16A is far better, with no site in the n=4 shell and a nearly complete n=2 shell.

Betts [21] explored different cluster geometries, and realized that there are few clusters with the same geometry (point group) as the lattice, but far more with lower symmetry, that are often superior. He selected clusters based on neighbors in a given shell, symmetry, and squareness. Since for a given cluster size, one incomplete shell is usually inevitable, he classified the imperfection of each cluster by the number of sites missing in nearer shells, rather than the number in an incomplete one and the number of sites in farther shells. Thus, 16B has an imperfection of three while 16A has an imperfection of one. The imperfection number is the most important selection criteria.

2.3. Calculation of Physical Properties

Most experiments measure quantities which we can express theoretically as reducible one or two-particle Green's functions. As discussed above, the appropriate way to calculate these quantities is to first extract the corresponding irreducible quantity from the cluster calculation, and then use it to calculate the reducible quantity. For example, to calculate the single-particle Green's function (relevant for angle-resolved photoemission spectroscopy) we first extract the cluster self energy and use the Dyson equation to construct the lattice Green's function. To calculate the phase diagram, we calculate the irreducible vertices in the different scattering channels Γ, and insert them into the Bethe-Salpeter equations for the lattice. In this subsection we will provide more details about the relationship between the lattice and cluster two-particle Green's functions and describe how a lattice susceptibility may be calculated efficiently.

Particle-hole channel. As a specific example, we will describe the calculation of the two-particle Green's function

$$\begin{aligned}\chi_{\sigma,\sigma'}(q,k,k') &= \int_0^\beta\int_0^\beta\int_0^\beta\int_0^\beta d\tau_1 d\tau_2 d\tau_3 d\tau_4 \\ &\times\ e^{i((\omega_n+\nu)\tau_1-\omega_n\tau_2+\omega_{n'}\tau_3-(\omega_{n'}+\nu)\tau_4)} \\ &\times\ \langle T_\tau c^\dagger_{\mathbf{k}+\mathbf{q}\sigma}(\tau_1)c_{\mathbf{k}\sigma}(\tau_2)c^\dagger_{\mathbf{k}'\sigma'}(\tau_3)c_{\mathbf{k}'+\mathbf{q}\sigma'}(\tau_4)\rangle\quad ,\end{aligned}$$

where we adopt the conventional notation [22] $k=(\mathbf{k},i\omega_n)$, $k'=(\mathbf{k},\omega_n')$, $q=(\mathbf{q},\nu_n)$ and T_τ is the time ordering operator.

$\chi_{\sigma,\sigma'}(q,k,k')$ and $\Gamma_{\sigma,\sigma'}(q,k,k')$ are related to each other through the Bethe-Salpeter equation (Fig. 13):

$$\begin{aligned}\chi_{\sigma,\sigma'}(q,k,k') &= \chi^0_{\sigma,\sigma'}(q,k,k')+\chi^0_{\sigma,\sigma''}(q,k,k'') \\ &\times\ \Gamma_{\sigma'',\sigma'''}(q,k'',k''')\chi_{\sigma''',\sigma'}(q,k''',k')\end{aligned}\qquad (18)$$

where $\Gamma_{\sigma,\sigma'}(q,k,k')$ is the two-particle irreducible vertex which is the analogue of the self-energy, $\chi^0_{\sigma,\sigma'}(q,k,k'')$ is the non-interacting susceptibility constructed from a pair of fully-dressed single-particle Green's functions. As usual, a summation is to be made for repeated indices.

We now make the DCA substitution $\Gamma_{\sigma,\sigma'}(\mathbf{q},\mathbf{k},\mathbf{k}')\rightarrow\Gamma_{c\sigma,\sigma'}(\mathbf{q},\mathbf{M}(\mathbf{k}),\mathbf{M}(\mathbf{k}'))$ in Eq. 18 (where frequency labels have been suppressed). Note that only the bare and dressed two-particle Green's functions χ depend upon the momenta $\tilde{\mathbf{k}}$ within a cell.

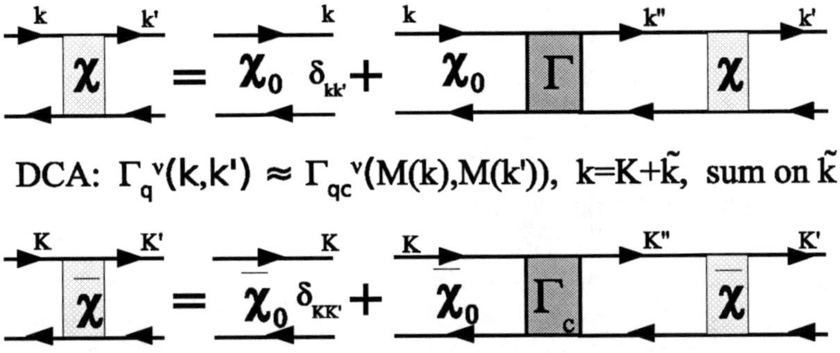

FIGURE 13. The Bethe-Salpeter equation in the DCA. We approximate the lattice irreducible vertex Γ^ν by the Γ_c^ν from the DCA cluster and coarse-grain over the $\tilde{\mathbf{k}}$. The remaining equation is a function of the cluster **K** only and may be solved by inversion.

Since χ and χ^0 in the product on the RHS of Eq. 18 share no common momentum labels, we may freely sum over the momenta $\tilde{\mathbf{k}}$ within a cell, yielding

$$\bar{\chi}_{\sigma,\sigma'}(q,K,K') = \bar{\chi}^0_{\sigma,\sigma'}(q,K,K') + \bar{\chi}^0_{\sigma,\sigma''}(q,K,K'') \times \Gamma_{c\sigma'',\sigma'''}(q,K'',K''')\bar{\chi}_{\sigma''',\sigma'}(q,K''',K'). \tag{19}$$

By coarse-graining the Bethe-Salpeter equation, we have greatly reduced its complexity; each of the matrices above is sufficiently small that they may be easily manipulated using standard techniques.

In contrast with the single-particle case where the coarse-grained quantities are identical to those of the cluster, $\chi_{c\sigma,\sigma'}(q,K,K')$ is not equal to $\bar{\chi}_{\sigma,\sigma'}(q,K,K')$. This is because the self-consistency is made only at the single-particle level. Unlike the single particle case where both $\Sigma(K)$ and $\bar{G}(K)$ are directly calculated, neither $\Gamma_{\sigma,\sigma'}(q,K,K')$ nor the coarse-grained susceptibility $\bar{\chi}_{\sigma,\sigma'}(q,K,K')$ are calculated during the self-consistency. Instead, the coarse-grained non-interacting susceptibility $\bar{\chi}^0_{\sigma,\sigma'}(q,K,K')$ is calculated in a separate program after the DCA converges using the following relation

$$\bar{\chi}^0_{\sigma,\sigma'}[(\mathbf{q},i\nu_n);(\mathbf{K},i\omega_n);(\mathbf{K}',i\omega_n')] = \delta_{\sigma,\sigma'}\delta_{\mathbf{K},\mathbf{K}'}\delta_{\omega_n,\omega_n'} \times \frac{N_c}{N}\sum_{\tilde{\mathbf{k}}} G_\sigma(\mathbf{K}+\tilde{\mathbf{k}},i\omega_n)G_\sigma(\mathbf{K}+\tilde{\mathbf{k}}+\mathbf{q},i\omega_n+\nu_n) \quad . \tag{20}$$

The corresponding cluster susceptibility is calculated by the cluster solver and the vertex function is extracted by inverting the cluster two-particle Bethe-Salpeter equation

$$\chi_{c\sigma,\sigma'}(q,K,K') = \chi_{c\sigma,\sigma'}^{0}(q,K,K') + \chi_{c\sigma,\sigma''}^{0}(q,K,K'') \times\Gamma_{c\sigma'',\sigma'''}(q,K'',K''')\chi_{c\sigma''',\sigma'}(q,K''',K'). \tag{21}$$

If we combine Eqs. 21 and 19, then the coarse-grained susceptibility may be obtained after elimination of $\Gamma(q,K,K')$ between the two equations. It reads

$$\bar{\chi}^{-1} = \chi_c^{-1} - \chi_c^{0^{-1}} + \bar{\chi}^{0^{-1}}, \tag{22}$$

where, for example, $\bar{\chi}$ is the matrix formed from $\bar{\chi}_{\sigma,\sigma'}(q,K,K')$ for fixed q. The charge (ch) and spin (sp) susceptibilities $\chi_{ch,sp}(q,T)$ are deduced from $\bar{\chi}$

$$\chi_{ch,sp}(q,T) = \frac{(k_B T)^2}{N_c^2} \sum_{KK'\sigma\sigma'} \lambda_{\sigma\sigma'} \bar{\chi}_{\sigma,\sigma'}(q,K,K') \quad , \tag{23}$$

where $\lambda_{\sigma\sigma'} = 1$ for the charge channel and $\lambda_{\sigma\sigma'} = \sigma\sigma'$ for the spin channel.

Particle-particle channel. The calculation of susceptibilities in the particle-particle channel is essentially identical to the above. The exception to this rule occurs when we calculate susceptibilities for transitions to states of lower symmetry than the lattice symmetry. For example, in order to obtain the pair function of the desired symmetry (s,p,d), the two-particle Green's function must be multiplied by the corresponding form factors $g(\mathbf{k})$ and $g(\mathbf{k}')$. In the study of the Hubbard model below, we will be particularly interested in $g(\mathbf{k}) = 1$ (s wave), $g(\mathbf{k}) = cos(k_x) + cos(k_y)$ (extended s wave) and $g(\mathbf{k}) = cos(k_x) - cos(k_y)$ ($d_{x^2-y^2}$ wave). These symmetries have been evoked as possible candidates for the superconducting ground state.

These factors modify the Bethe-Salpeter equations

$$\begin{aligned} g(\mathbf{k})\chi(q,k,k')g(\mathbf{k}') &= g(\mathbf{k})\chi^0(q,k,k')g(\mathbf{k}') \\ &+ \; g(\mathbf{k})\chi^0(q,k,k'') \times \Gamma(q,k'',k''') \times \chi(q,k''',k')g(\mathbf{k}') . \end{aligned} \tag{24}$$

where

$$\begin{aligned} \chi(q,k,k') &= \int_0^\beta \int_0^\beta \int_0^\beta \int_0^\beta d\tau_1 d\tau_2 d\tau_3 d\tau_4 \\ &\times \; e^{i((\omega_n+\nu)\tau_1 - \omega_n\tau_2 + \omega_{n'}\tau_3 - (\omega_{n'}+\nu)\tau_4)} \\ &\times \; \langle T_\tau c^\dagger_{\mathbf{k}+\mathbf{q}\sigma}(\tau_1) c^\dagger_{-\mathbf{k}-\sigma}(\tau_2) c_{-\mathbf{k}'-\sigma}(\tau_3) c_{\mathbf{k}'+\mathbf{q}\sigma}(\tau_4)\rangle \quad . \end{aligned} \tag{25}$$

On the LHS, we have dropped the spin indices since we will consider only opposite-spin pairing. Eq. 24 cannot be easily solved if it is coarse-grained, since this will partially convolve $\chi(q,k,k')$ with *two* factors of g on the LHS and *one* factor on the RHS. Hence for the pairing susceptibilities, or for any situation where non-trivial form factors must be used, we use the equivalent equation involving the reducible vertex F (instead of the irreducible vertex Γ)

$$\begin{aligned} g(\mathbf{k})\chi(q,k,k')g(\mathbf{k}') &= g(\mathbf{k})\chi^0(q,k,k')g(\mathbf{k}') \\ &+ \; g(\mathbf{k})\chi^0(q,k,k'') \\ &\times \; F(q,k'',k''')\chi^0(q,k''',k')g(\mathbf{k}'), \end{aligned} \tag{26}$$

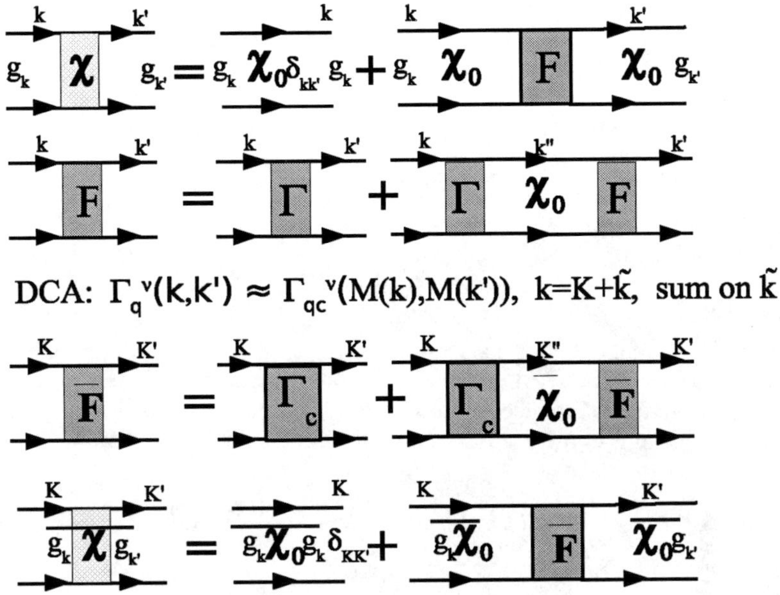

FIGURE 14. Calculation of particle-particle projected susceptibilities. Often we want to calculate a projected particle-particle susceptibility (e.g. d-wave, with $g_{\mathbf{k}} = \cos(k_x) - \cos(k_y)$). Here the Bethe-Salpeter equation is rewritten in terms of the irreducible vertex F. We approximate the lattice irreducible vertex Γ^ν by the Γ^ν_c from the DCA cluster and coarse-grain over the $\tilde{\mathbf{k}}$. Then the projected bare bubbles are calculated, and the remaining equation is a function of the cluster $\mathbf{K}$ only and may be solved by inversion.

where

$$\begin{aligned} F(q,k,k') &= \Gamma(q,k,k') \\ &+ \chi^0(q,k,k'')\Gamma(q,k'',k''')\chi^0(q,k''',k') + \cdots \end{aligned} \tag{27}$$

We define

$$\Pi_{g,g}(q,k,k') = g(\mathbf{k})\chi(q,k,k')g(\mathbf{k}') \tag{28}$$

$$\Pi^0_{g,g}(q,k,k') = g(\mathbf{k})\chi^0(q,k,k')g(\mathbf{k}') \tag{29}$$

$$\Pi^0_{g}(q,k,k') = g(\mathbf{k})\chi^0(q,k,k')\,. \tag{30}$$

The remaining steps of the calculation are similar to the particle-hole case. We invert the cluster particle-particle Bethe-Salpeter equation with $g = 1$ for the cluster, in order to extract Γ_c. We then coarse-grain Eq. 27, and use Γ_c to calculate the coarse-grained $\bar{F} = \Gamma_c\left(1 - \bar{\chi}^0\Gamma_c\right)^{-1}$. We then coarse-grain Eq. 26, and use the coarse-grained $\bar{F}$ to calculate the coarse-grained $\bar{\Pi}_{g,g}$

$$\bar{\Pi}_{g,g}(q,K,K') = \bar{\Pi}^0_{g,g}(q,K,K') \tag{31}$$

$$+ \quad \bar{\Pi}^0_g(q,K,K'')\bar{T}_2(q,K'',K''')\bar{\Pi}^0_g(q,K''',K') .$$

The pairing susceptibility of a desired symmetry is given by

$$P_g(q,T) = \frac{(k_B T)^2}{N_c^2} \sum_{K,K'} \bar{\Pi}_{gg}(q,K,K') \quad . \tag{32}$$

2.4. Summary

Coarse-graining methods are used to derive the DMFA, and DCA which map the lattice onto a self-consistently embedded cluster problem. The DMFA is a local approximation, while the DCA incorporates systematic non-local corrections. Irreducible quantities from the cluster are used to calculate reducible lattice quantities.

3. QUANTUM MONTE CARLO ALGORITHMS FOR THE QUANTUM CLUSTER PROBLEM

3.1. Introduction

The Hirsch-Fye and Continuous time Quantum Monte Carlo (QMC) algorithms are powerful and adaptable methods which may be used to study models of impurity clusters embedded in a host. As such, they are ideal cluster solvers for the embedded impurity problem at the heart of the Dynamical Mean Field and Dynamical Cluster Approximations.

We will sketch the Hirsch-Fye QMC in Sec. 3.2 and the weak coupling continuous time QMC algorithm in Sec. 3.3. Since both methods can be considered as expansions about the free electron limit, Wick's theorem applies to both, so similar methods for making and conditioning measurements described in Sec. 3.4 can be used for both HFQMC and CTQMC.

3.2. Hirsch-Fye QMC

To derive the Hirsch-Fye algorithm, we start with the Hubbard Hamiltonian (1)

$$H = H_0 + H_1, \tag{33}$$

where H_0 is the noninteracting part of the Hamiltonian

$$H_0 = -t \sum_{\langle j,k\rangle\sigma} (c^\dagger_{j\sigma}c_{k\sigma} + c^\dagger_{k\sigma}c_{j\sigma}) + \varepsilon \sum_j (n_{j\uparrow} + n_{j\downarrow}) \tag{34}$$

and H_1 describes the local interaction on a cluster $\mathscr{C}$ (or impurity).

$$H_1 = U \sum_{i \in \mathscr{C}} (n_{i\uparrow} - \frac{1}{2})(n_{i\downarrow} - \frac{1}{2}) . \tag{35}$$

The summation in Eq.34 is taken over the entire lattice whereas in Eq.35 it is taken only over the cluster $\mathscr{C}$ sites (see Fig.15). Our derivation of the equations follows the one presented by Hirsch and Fye for the impurity problem [14, 23], but we extend the derivation to treat the embedded cluster in the DMFA and DCA [15].

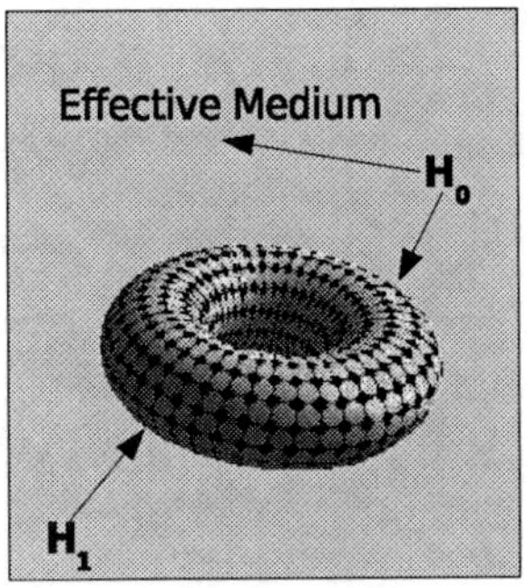

FIGURE 15. Hamiltonian breakup for the Hirsch-Fye derivation. H_0 describes both the host and non-interacting degrees of freedom of the cluster (impurity), while H_1 describes the interactions on the cluster.

By dividing the imaginary time (the inverse of the temperature) $\beta = \frac{1}{T}$ into L slices,

$$\Delta\tau = \frac{\beta}{L} , \tag{36}$$

the partition function can be written as

$$Z = Tr(e^{-\beta H}) = Tr\left(\prod_{l=1}^{L} e^{-\Delta\tau H}\right) . \tag{37}$$

From the Suzuki-Trotter formula

$$e^{-\Delta\tau H} = e^{-\Delta\tau H_0/2} e^{-\Delta\tau H_1} e^{-\Delta\tau H_0/2} + \mathscr{O}(\Delta\tau^3) , \tag{38}$$

one derives

$$Z = Tr(e^{-\beta H}) \approx Tr\left(\prod_{l=1}^{L} e^{-\Delta\tau H_0/2} e^{-\Delta\tau H_1} e^{-\Delta\tau H_0/2}\right) , \tag{39}$$

which has leading errors proportional to $\Delta\tau^2$ (since the Suzuki-Trotter formula was applied L times and $L \sim 1/\Delta\tau$). Then due to the periodic property of the trace, it is easy to see that this is the same as the Trotter decomposition

$$Z = Tr(e^{-\beta H}) \approx Tr\left(\prod_{l=1}^{L} e^{-\Delta\tau H_0} e^{-\Delta\tau H_1}\right) , \tag{40}$$

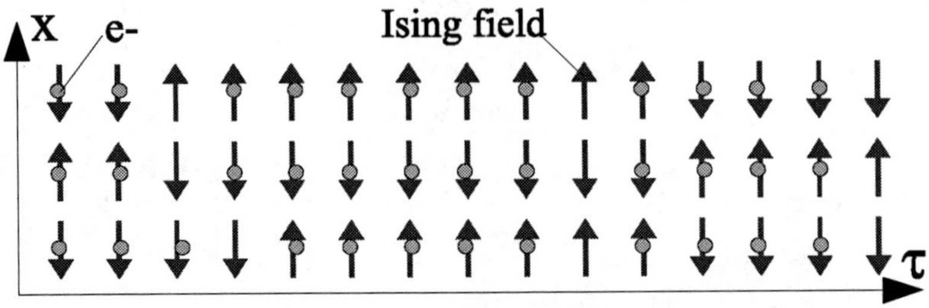

FIGURE 16. The Hirsch-Hubbard-Stratonovich transformation Eq. 43 maps an interacting systems of electrons onto a system of non-interacting electrons interacting with a time and space dependent Ising-like fields which coupling to the z-component of the electron spin.

with leading errors still proportional to $\Delta\tau^2$.

We introduce the identity operator in the occupation number basis

$$I = \sum_m |m\rangle\langle m| \tag{41}$$

between exponents of operators at adjacent imaginary time slices. The partition function becomes

$$\begin{aligned} Z = \ & \sum\nolimits_{m_L, m_{L-1}, .. m_1} \langle m_L | e^{-\Delta\tau H_0} e^{-\Delta\tau H_1} | m_{L-1}\rangle \langle m_{L-1} | e^{-\Delta\tau H_0} e^{-\Delta\tau H_1} | m_{L-2}\rangle \\ & \ldots \langle m_1 | e^{-\Delta\tau H_0} e^{-\Delta\tau H_1} | m_L\rangle \end{aligned} \tag{42}$$

With the identity due to Hirsch [24]

$$e^{-\Delta\tau U\left(n_{i\uparrow}n_{i\downarrow} - \frac{1}{2}(n_{i\uparrow}+n_{i\downarrow})\right)} = \frac{1}{2}\sum_{s_i=\pm 1} e^{\alpha s_i (n_{i\uparrow}-n_{i\downarrow})}\,, \tag{43}$$

with

$$\cosh\alpha = e^{\Delta\tau U/2}\,. \tag{44}$$

It is possible to introduce an auxiliary binary (± 1) field (called Hirsch-Hubbard-Stratonovich field, or HHS field) at every cluster site and at every time point. Thereby, the interacting problem described by Eq. 42 is replaced by a summation over all possible auxiliary field configurations of noninteracting terms. Eq. 43 can be proved by applying both the left and right hand sides on the four possible vectors (empty site, one electron up, one electron down and double occupied site) which span the local Hilbert space.

For each HHS field configuration $\{s\}$, we have a noninteracting problem composed of electrons scattering off a space and time dependent Ising-like field (c.f. Fig. 16).

Therefore, by defining

$$O_\sigma(s) = \begin{pmatrix} I & 0 & 0 & \dots & e^{-\Delta\tau K} e^{V_\sigma^L(s)} \\ e^{-\Delta\tau K} e^{V_\sigma^1(s)} & I & 0 & \dots & 0 \\ 0 & e^{-\Delta\tau K} e^{V_\sigma^2(s)} & I & 0 & \dots \\ \dots & & & & \\ \dots & \dots & \dots & e^{-\Delta\tau K} e^{V_\sigma^{L-1}(s)} & I \end{pmatrix}, \tag{45}$$

with

$$H_0 = \sum_{i,j,\sigma} c_{i\sigma}^\dagger K_{ij} c_{j\sigma} \tag{46}$$

and

$$V_\sigma^l(s)_{ij} = \begin{cases} \alpha\sigma s_{il}\delta_{i,j} & i \in \mathscr{C} \\ 0 & i \notin \mathscr{C} \end{cases}, \tag{47}$$

where s_{il} represents the value of HHS field at site i and time l, it is possible to write Eq. 42 as

$$Z = Tr_{\{s\}}\left[\det O_\uparrow(s)\ \det O_\downarrow(s)\right] \tag{48}$$

. Note that every matrix element shown in Eq. 45 is in fact a $N \times N$ sub-block, N being the dimension of the K matrix defined in Eq. 46. Thus O is a $(N \times L) \times (N \times L)$ matrix.

The matrix O is the inverse of the one-particle Green's function

$$G_\sigma^{-1}(s) = O_\sigma(s)\ . \tag{49}$$

By defining

$$T_{il;jl'} = \delta_{l-1,l'}\delta_{i,j} \tag{50}$$

and

$$V_\sigma^l(s)_{il;jl'} = \begin{cases} \alpha\sigma s_{il}\delta_{l,l'}\delta_{i,j} & i \in \mathscr{C} \\ 0 & i \notin \mathscr{C}\ or\ j \notin \mathscr{C} \end{cases}, \tag{51}$$

Eqs. 45 and 49 can be written as

$$G_\sigma^{-1} = I - Te^{-\Delta\tau K} e^{V_\sigma}\ . \tag{52}$$

In Eq. 52 and in the following equations we omit writing the explicit HHS dependence for simplicity reasons. By multiplying Eq. 52 at right with e^{-V_σ} (which is diagonal) the following equation is obtained

$$G_\sigma^{-1} e^{-V_\sigma} = e^{-V_\sigma} - Te^{-\Delta\tau K}. \tag{53}$$

Eq. 53 is used to establish a relation between the Green's functions G' and G which correspond to two different field configurations $\{s\}$ and respectively $\{s'\}$

$$G_\sigma'^{-1} e^{-V_\sigma'} - G_\sigma^{-1} e^{-V_\sigma} = e^{-V_\sigma'} - e^{-V_\sigma}\ . \tag{54}$$

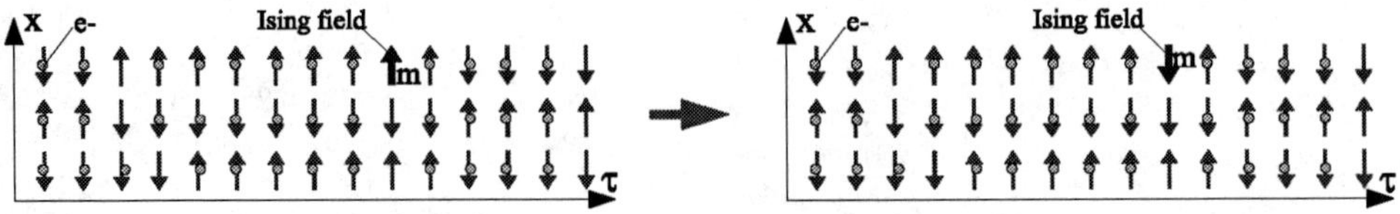

FIGURE 17. The original Hirsch-Fye algorithm involved local flips of the HHS fields.

Using

$$A^{-1} - B^{-1} = C \Longleftrightarrow A = B - BCA \ , \tag{55}$$

the following equation is obtained

$$e^{V'_\sigma} G'_\sigma = e^{V_\sigma} G_\sigma + e^{V_\sigma} G_\sigma (e^{-V_\sigma} - e^{-V'_\sigma}) e^{V'_\sigma} G'_\sigma \tag{56}$$

. After some easy manipulations, Eq. 56 can be written as

$$G'_\sigma = G_\sigma + (G_\sigma - I)(e^{V'_\sigma - V_\sigma} - I) G'_\sigma \ . \tag{57}$$

Another useful equation is

$$G_\sigma G'^{-1}_\sigma = I - (G_\sigma - I)(e^{V'_\sigma - V_\sigma} - I) \ , \tag{58}$$

obtained by multiplying Eq. 57 at right with G'^{-1}_σ. Eq. 58 will be used in the QMC process for calculating the transition probability form one configuration to another and Eq. 57 for updating to the new configuration when the transition is accepted.

The QMC algorithm implies generating different field configurations with a probability proportional to their weight, given by (see Eq. 48)

$$W(s) = \det G_\uparrow^{-1}(s) \det G_\downarrow^{-1}(s). \tag{59}$$

In the QMC process, the HHS field configuration is updated by proposing local (in both the cluster site and time index) flips of the HHS field (Fig. 17). We consider such two configurations (s and s') that differ only by a flip at point m, where m denotes both a cluster site and a time (i.e. $m \equiv il, i \in \mathscr{C}$). The ratio between the weights of these configurations is

$$R = \frac{W(s')}{W(s)} = \frac{\det G_\uparrow^{-1}(s') \det G_\downarrow^{-1}(s')}{\det G_\uparrow^{-1}(s) \det G_\downarrow^{-1}(s)}. \tag{60}$$

The ratio R determines the acceptance probability of the new configuration, according to either the Metropolis [25] or the heat bath rule. If the proposed configuration was accepted, the new Green's function should be updated accordingly to Eq. 57. This implies the following

$$G'_{pn} = G_{pn} + (G_{pm} - \delta_{pm})(e^{\sigma\alpha(s'_m - s_m)} - 1) G'_{mn} \tag{61}$$

$$G'_{mn} = G_{mn} + (G_{mm} - 1)(e^{\sigma\alpha(s'_m - s_m)} - 1) G'_{mn} \tag{62}$$

$$G'_{mn} = \frac{G_{mn}}{1 - (G_{mm} - 1)(e^{\sigma\alpha(s'_m - s_m)} - 1)} \ , \tag{63}$$

which results in

$$G'_{pn} = G_{pn} + \frac{(G_{pm} - \delta_{pm})(e^{\sigma\alpha(s'_m - s_m)} - 1)}{1 - (G_{mm} - 1)(e^{\sigma\alpha(s'_m - s_m)} - 1)} G_{mn} \ . \tag{64}$$

$$R = \frac{W(s')}{W(s)} = \prod_{\sigma} \det G'^{-1}_{\sigma}(s') \det G_{\sigma}(s) = \prod_{\sigma} [1 - (G_{\sigma mm} - 1)(e^{\sigma\alpha(s'_m - s_m)} - 1)]. \tag{65}$$

Eq. 65 results directly from Eq. 58. Note that only the Green's function defined between

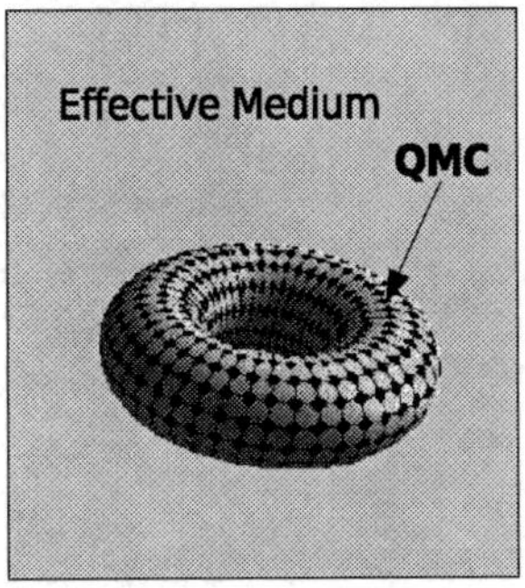

FIGURE 18. Whereas the HF Hamiltonian has degrees of freedom on the cluster and in the effective medium, the algorithm may be restricted to interacting (cluster) sites only.

clusters sites is required for the QMC procedure (Fig 18). This is also enough for measuring the cluster observables and for determining the irreducible quantities as self-energy, Σ, and two-particle vertices, Γ's.

In order to initialize the QMC process, first the HHS fields are set to zero and G_σ is set to $\mathcal{G}_\sigma$. Here, $\mathcal{G}_\sigma$ is the noninteracting Green's function (i.e., the one obtained when $U = 0$) if we mean to simulate a finite size cluster, or it is the cluster excluded Green's function (i.e. the one obtained when we subtract the self energy from the cluster) if we mean to simulate an embedded cluster. Afterwards, by turn, the value of the HHS fields are changed to 1 or -1 at all space-time points and the Green's function is updated in accordance to Eq. 64. The obtained Green's function is used to initialize the QMC procedure. A "sweep" of the QMC procedure consists of proposing a flip of the HHS field for every time slice, calculating (Eq. 65) the ratio R between the weights of the proposed and the present configuration and accepting the flip according to the Metropolis or the heat-bath algorithm. After a certain number of warm-up sweeps through the space-time points (usually between twenty and one hundred), the system reaches equilibrium and the measurements can start. It is a good idea to consider a few update sweeps between the measurements, in order to eliminate the correlation between them.

Numerical round-off errors accumulate in the Green's function during the updating process, due to the repeated use of Eq. 64. In order to eliminate them, after a certain number of iterations the Green's function should be refreshed. This is done first by setting G_σ to $\mathcal{G}_\sigma$ and afterwards updating it using Eq. 64 repeatedly until it corresponds to the present HHS field configuration.

3.2.1. *Combining HFQMC with Quantum Cluster Methods*

The Hirsch-Fye algorithm naturally produces Green's functions in Matsubara (imaginary) time. However, the self-consistency cycle of quantum cluster methods like the DMFA and DCA require imaginary frequency Green's functions. Thus the algorithm requires Fourier transforms from both the imaginary frequency to the imaginary time and from the imaginary time to the imaginary frequency. The transform from frequency to time is given by

$$G(\tau) = \frac{1}{\beta} \sum_{n=-\infty}^{\infty} G(i\omega_n)\, e^{-i\omega_n \tau} \tag{66}$$

and the inverse one, from time to frequency, by

$$G(i\omega_n) = \int_0^\beta d\tau\, G(\tau)\, e^{i\omega_n \tau} \ . \tag{67}$$

The Matsubara frequencies for the fermionic Green's function are defined as

$$\omega_n = \frac{(2n+1)\pi}{\beta} \ . \tag{68}$$

Due to the finite $\Delta\tau$, the numerical implementation of these transformations requires special care. We will discuss both cases next.

Transform from frequency to time. In Eq. 66 the frequency summation is taken from minus to plus infinity. The contribution at large (both positive and negative) frequency is important, thus a truncation of the sum is not possible. The trick is to subtract and add back a function which has at large frequency the same asymptotic behavior as the Green's function. For example, if $f(i\omega_n)$ fulfills this condition, Eq. 66 can be approximated by

$$G(\tau) \approx \frac{1}{\beta} \sum_{n=-n_c}^{n_c} (G(i\omega_n) - f(i\omega_n))e^{-i\omega_n \tau} + \frac{1}{\beta} \sum_{n=-\infty}^{\infty} f(i\omega_n)\, e^{-i\omega_n \tau} \ , \tag{69}$$

where n_c is a cutoff number chosen large enough to have negligible numerical errors.

In order to determine the Green's function behavior at large frequency we integrate Eq. 67 by parts as suggested in [26]

$$\begin{aligned} G(i\omega_n) &= \tfrac{1}{i\omega_n} G(\tau) e^{i\omega_n \tau} \,|_0^\beta - \tfrac{1}{i\omega_n} \textstyle\int_0^\beta G'(\tau) e^{i\omega_n \tau} d\tau = \\ &= \tfrac{1}{i\omega_n} G(\tau) e^{i\omega_n \tau} \,|_0^\beta - \tfrac{1}{(i\omega_n)^2} G'(\tau) e^{i\omega_n \tau} \,|_0^\beta + \mathcal{O}((i\omega_n)^{-3}) \ . \end{aligned} \tag{70}$$

We find that, for cutoff values of about ≈ 500, it is necessary to consider the asymptotic behavior of G up to second order in ω_n^{-1}. If only the first order term in ω_n^{-1} is considered, in order to get negligible numerical errors the cutoff number n_c should be of order 10^6. Thus, the second order term in Eq. 70 has a major practical importance and therefore $f(i\omega_n)$ should be taken as

$$f(i\omega_n) = \frac{a_1}{i\omega_n} + \frac{a_2}{(i\omega_n)^2} \tag{71}$$

with

$$a_1 = -G(\beta^-) - G(0^+) = G(0^-) - G(0^+) \tag{72}$$

and

$$a_2 = G'(\beta^-) + G'(0^+) = -G'(0^-) + G'(0^+)\ . \tag{73}$$

The last sum in Eq. 69 can be easily calculated by using the following relations

$$\frac{1}{\beta}\sum_{n=-\infty}^{\infty}\frac{e^{-i\omega_n\tau}}{i\omega_n} = \begin{cases} \frac{1}{2} & \tau<0 \\ -\frac{1}{2} & \tau>0 \end{cases} \tag{74}$$

and

$$\frac{1}{\beta}\sum_{n=-\infty}^{\infty}\frac{e^{-i\omega_n\tau}}{(i\omega_n)^2} = \begin{cases} -\frac{1}{4}(\beta+2\tau) & \tau<0 \\ -\frac{1}{4}(\beta-2\tau) & \tau>0 \end{cases}\ . \tag{75}$$

The first of these relations (Eq. 74) is the well known Fourier transform of a noninteracting Green's function [22]

$$\frac{1}{\beta}\sum_{n=-\infty}^{\infty}\frac{e^{-i\omega_n\tau}}{i\omega_n - E} = \begin{cases} \frac{1}{e^{\beta E}+1}e^{-E\tau} & \tau<0 \\ -(1-\frac{1}{e^{\beta E}+1})e^{-E\tau} & \tau>0 \end{cases}\ . \tag{76}$$

with the energy pole $E=0$. Eq. 75 results from differentiating Eq. 76 with respect to E and setting $E=0$ afterwards.

The coefficients a_1 and a_2 in Eq. 71 can also be easily determined. Starting with the Green's function definition

$$G_{ij}(\tau) = -T_\tau\langle c_j(\tau)c_i^\dagger\rangle\ , \tag{77}$$

where i and j are space or momentum indices and using Eq. 72, a_1 becomes

$$a_1 = \langle c_i^\dagger c_j\rangle + \langle c_j c_i^\dagger\rangle = \delta_{ij}\ . \tag{78}$$

Considering

$$\frac{dc_j}{d\tau} = [H, c_j] \tag{79}$$

the coefficient a_2 is

$$a_2 = -\langle c_i^\dagger[H,c_j]\rangle - \langle [H,c_j]c_i^\dagger\rangle\ , \tag{80}$$

and can be evaluated to

$$a_2 = K_{ij} + K_{ij}^{HF}\ . \tag{81}$$

K_{ij} was defined in Eq. 46 and represent the bilinear (noninteracting) part of the Hamiltonian. K_{ij}^{HF} results from the interacting part of the Hamiltonian when the corresponding Hartree-Fock factorization is done.

Transform from time to frequency. The difficulties associated with this transform are related with the fact that in the QMC process the Green's function is calculated and stored only in a discrete set of points τ_i evenly spaced by $\Delta\tau$. This implies a periodic Fourier transform with the period equal to the Nyquist critical frequency

$$\omega_c = \frac{\pi}{\Delta\tau}, \tag{82}$$

which is evidently unphysical, since the Green's function goes as ω_n^{-1} at large frequency (see Eq. 70). Besides that, due to the fact that part of the spectral density at high frequency ($|\omega| > \omega_c$) is translated into the low frequency domain ($|\omega| < \omega_c$), a phenomenon called *aliasing* [27], the Fourier transform is inaccurate even at frequencies smaller than the Nyquist frequency.

In order to cure [28] the pathology discussed above we again take advantage of our knowledge of the Green's function behavior at high frequency. A function which has the right large frequency asymptotic behavior is the Green's function G_p obtained from second order perturbation theory (or any other method which becomes exact at high frequency). The Fourier transform can be written as

$$G(i\omega_n) = G_p(i\omega_n) + \int_0^\beta d\tau (G(\tau) - G_p(\tau)) e^{i\omega_n \tau} . \tag{83}$$

It is important that the last term in Eq. 83 does not produce spurious high frequency contributions. Therefore before integration the Akima spline [29] is used to interpolate $G(\tau) - G_p(\tau)$. The Akima spline produces smooth curves and therefore acts as a low-pass filter and eliminates the high frequency noise.

3.3. Continuous Time Quantum Monte Carlo

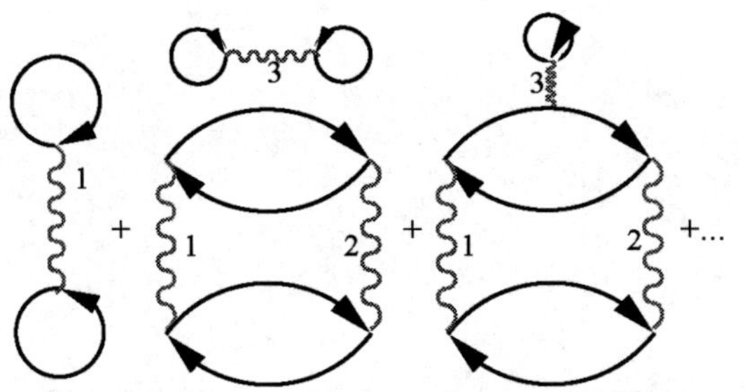

FIGURE 19. In contrast to HFQMC which employs a path-integral formalism, CTQMC uses QMC to stochastically sample the graphs in the partition function.

In this subsection, we will derive a Continuous time QMC (CTQMC) algorithm for the Hubbard model, Eq. 1 following closely the derivation by Rubtsov [30]. In contrast to the path integral formalism of HFQMC, CTQMC employs the same tricks used to derive Feynman-Dyson perturbation theory (the interaction representation, the time-ordered

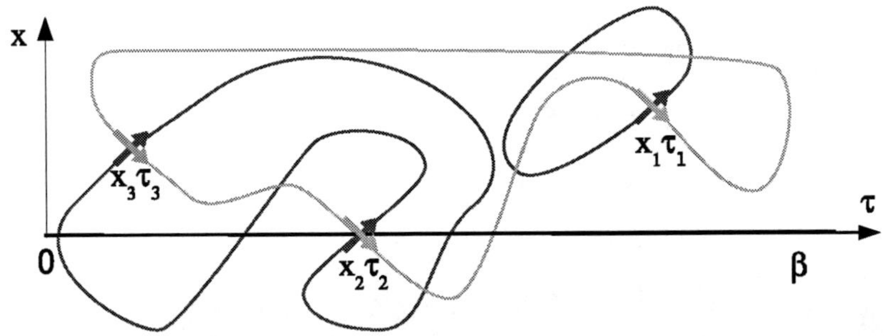

FIGURE 20. An example of a third-order ($k=3$) CTQMC graph. Many graphs may be drawn to third order in U, one example is shown.

S-matrix expansion, Wick's theorem, etc.) to stochastically generate diagrams for the partition function (Fig. 19).

Starting with

$$\exp(-\beta(H_0+H_1)) = \exp(-\beta H_0)S(\beta), \tag{84}$$

where

$$S(\beta) = T_\tau \exp\left(-\int_0^\beta H_1(\tau)d\tau\right), \tag{85}$$

where T_τ is the time-ordering operator, and

$$H_1(\tau) = e^{\tau H_0}H_1 e^{-\tau H_0} \tag{86}$$

is the quartic part of H in the interaction picture.

Specializing now on the Hubbard model, Eq. 1, the partition function becomes

$$Z = Z_0 \sum_k \frac{(-U)^k}{k!} \int \cdots \int d1 \cdots dk \quad \left\langle T_\tau n_\uparrow(1)\cdots n_\uparrow(k)\right\rangle_0 \\ \left\langle T_\tau n_\downarrow(1)\cdots n_\downarrow(k)\right\rangle_0 \tag{87}$$

where $1 \to (x_1, \tau_1)$, etc. and $\langle\rangle_0$ indicates a thermodynamic average with respect to $\exp(-\beta H_0)$. Since H_0 is noninteracting, we can apply the Wick's theorem to evaluate the expectation values in Eq. 87 (see, for instance, [22]), by evaluating and summing over all closed Feynman graphs, see for example Fig. 20. For each order in k, there are $k!$ graphs that can be drawn for each of the up and down electrons. Note that the graphs can include both connected and disconnected parts, as shown in the figure. According to Wick's theorem, each graph corresponds to a product of non-interacting Green's functions G^0 (i.e. Green's functions from the solution of H_0 only) with arguments determined by the vertex labels, etc., and sign determined by the number of line crossings. For example, the graph shown is

$$G^0(2,3)G^0(3,2)G^0(1,1)G^0(3,2)G^0(2,1)G^0(1,3) \tag{88}$$

where G^0 with the same argument are to be interpreted as having the right time argument slightly greater. I.e., $G^0(1,1) = G^0(x_1\tau_1; x_1\tau_1+)$. There will be 36 such graphs for each $k = 3$ vertex configuration, and the number of graphs will increase with order like $k!^2$. Fortunately, we may conveniently represent them as the product of up and down determinants $\det D_\uparrow(k) \det D_\downarrow(k)$, where

$$D_\sigma(k) = \begin{pmatrix} G^0_\sigma(1,1) & G^0_\sigma(1,2) & \cdots & G^0_\sigma(1,k) \\ G^0_\sigma(2,1) & G^0_\sigma(2,2) & \cdots & G^0_\sigma(2,k) \\ G^0_\sigma(3,1) & G^0_\sigma(3,2) & \cdots & G^0_\sigma(3,k) \\ \cdot & \cdot & \cdot & \\ \cdot & \cdot & \cdot & \cdot \\ \cdot & \cdot & \cdot & \cdot \\ G^0_\sigma(k,1) & G^0_\sigma(k,2) & \cdots & G^0_\sigma(k,k) \end{pmatrix} \tag{89}$$

where, in lieu of an external field the bare Green's function does not need a spin label, $G^0_\sigma(k,2) = G^0(k,2)$. Note that $D_\sigma(k)$ is a $k \times k$ matrix.

In CTQMC, the sum of integrals in the Eq. 87 is evaluated using an importance sampling Monte Carlo algorithm. Suppose the cluster has N sites and $L = \beta/\Delta\tau$ time locations (in the end we may let $\Delta\tau \to 0$). We will consider Monte Carlo moves of adding vertices at a random location in space-time or subtracting existing vertices. By combining such moves, we can clearly access all possible vertex configurations. Consider two configurations, one with k and the other with $k+1$ vertices, then the weights W of these two configurations are given by the integrand of Eq. 87, or, in terms of the $D_\sigma(k)$,

$$\begin{aligned} W_k &= (-\Delta\tau U)^k \det D_\uparrow(k) \det D_\downarrow(k) \\ W_{k+1} &= (-\Delta\tau U)^{k+1} \det D_\uparrow(k+1) \det D_\downarrow(k+1)\,. \end{aligned} \tag{90}$$

If we let the probability to add and remove a vertex be P_a and P_r respectively, where $P_a + P_r = 1$, then the detailed balance condition requires that

$$P_a \frac{1}{L}\frac{1}{N} W_k P_{k\to k+1} = P_r \frac{1}{k+1} W_{k+1} P_{k+1\to k}\,. \tag{91}$$

Here $\frac{1}{L}\frac{1}{N}$ is the probability to choose a position in time and space for the vertex you intend to add while $\frac{1}{k+1}$ is the probability to choose one vertex you intend to remove of from the existing $k+1$ ones. If we accept or reject this change using a Metropolis algorithm [25], then we need the Metropolis ratio R.

Suppose we propose to add vertex, then

$$R = P_{k\to k+1}/P_{k+1\to k} = -\frac{U\beta N}{k+1} \prod_\sigma \det D_\sigma(k+1) / \prod_\sigma \det D_\sigma(k)\,. \tag{92}$$

We need an efficient way to calculate the ratio of determinants

$$\begin{aligned} \det D_\sigma(k+1)/\det D_\sigma(k) &= \det D_\sigma(k+1) M_\sigma(k) \\ &= \det\left(I + (D_\sigma(k+1) - D_\sigma(k)) M_\sigma(k)\right) \end{aligned} \tag{93}$$

where $M_\sigma(k) = D_\sigma(k)^{-1}$, and we must pad the matrices $M_\sigma(k)$ and $D_\sigma(k)$ with an extra row and column in the $k+1$ location with all zeroes, except for a 1 in the $k+1$, $k+1$ location.

$$\det(I + (D_\sigma(k+1) - D_\sigma(k))M_\sigma(k)) = \det \tag{94}$$

$$\begin{pmatrix} 1 & 0 & \cdots & G^0_\sigma(1,k+1) \\ 0 & 1 & \cdots & G^0_\sigma(2,k+1) \\ 0 & 0 & \cdots & G^0_\sigma(3,k+1) \\ . & . & . & \\ . & . & . & . \\ . & . & . & . \\ G^0_\sigma(k+1,i)M(k)_{i,1} & G^0_\sigma(k+1,i)M(k)_{i,2} & \cdots & G^0_\sigma(k+1,k+1) \end{pmatrix}$$

A cofactor expansion (first on the last column and then on the last row) yields

$$\det D_\sigma(k+1) / \det D_\sigma(k) = G^0_\sigma(k+1,k+1) - G^0_\sigma(k+1,i)M(k)_{ij}G^0_\sigma(j,k+1) \tag{95}$$

with an implied sum over repeated indices. Clearly, the matrices M are important for the update, and they, not D are stored. The change is accepted if R is greater than a random number between zero and one. Then, we need an equation used to update M which may be derived using the Inversion by Partitioning described in Numerical Recipes [27].

$$M_\sigma(k+1) = \begin{pmatrix} . & . & . & -\lambda^{-1}L_{1k+1} \\ . & . & . & -\lambda^{-1}L_{2k+1} \\ . & M_\sigma{}' & . & -\lambda^{-1}L_{3k+1} \\ . & . & . & \\ . & . & . & . \\ -\lambda^{-1}R_{k+1,1} & -\lambda^{-1}R_{k+1,2} & \cdots & -\lambda^{-1} \end{pmatrix} \tag{96}$$

where the first k by k section is filled by the matrix

$$M_\sigma{}'_{ij} = M(k)_{ij} + L_{ik+1}\lambda^{-1}R_{k+1,j} \tag{97}$$

and

$$R_{ij} = G^0_\sigma(i,n)M(k)_{nj} \; L_{ij} = M(k)_{in}G^0_\sigma(n,j) \tag{98}$$

and λ is given by Eq. 95.

If we propose to remove vertex at location n, then to determine whether to accept the move, we need

$$\det D_\sigma(k-1) / \det D_\sigma(k) = \det((M_\sigma(k) - M_\sigma(k-1))D_\sigma(k-1) + I) \,. \tag{99}$$

Again, using a cofactor expansion, it is easy to show that this is just $M_{\sigma nn}$. Then if

$$R = \frac{-k}{U\beta N} \prod_\sigma \det D_\sigma(k-1) / \prod_\sigma \det D_\sigma(k) \tag{100}$$

is greater than a random number between zero and one, we accept the change and must update M_σ using

$$M_{\sigma ij}(k-1) = M_{\sigma ij}(k) - M_{\sigma in}(k) M_{\sigma nj}(k) / M_{\sigma nn}(k)\,. \tag{101}$$

One natural question is to ask whether this QMC is convergent, which would require that the average value of k be finite.

$$\begin{aligned} \langle k \rangle &= \frac{Z_0}{Z}\sum_0^\infty \frac{(-1)^k}{k!} k \int_0^\beta d\tau_1 \cdots d\tau_k \langle T_\tau H_1(\tau_1)\cdots H_1(\tau_k)\rangle_0 \\ &= -\frac{Z_0}{Z}\sum_0^\infty \frac{(-1)^k}{k!} k \int_0^\beta d\tau_1 \cdots d\tau_k d\tau \langle T_\tau H_1(\tau_1)\cdots H_1(\tau_k) H_1(\tau)\rangle_0 \\ &= -\int_0^\beta d\tau \langle H_1(\tau)\rangle \\ &= -\beta \langle H_1 \rangle \end{aligned} \tag{102}$$

where the brackets $\langle\rangle$ denote the average with respect to full interacting Hamiltonian (Eq. 33), which is also the average value of Monte Carlo measurements. The last line in Eq. 103 follows if H_1 is independent of τ. Since $\langle H_1\rangle$ is extensive, this means that $\langle k\rangle = -\beta N \langle h_1\rangle$ where h_1 is the average potential energy per site.

As will be described below in Sec. 3.4.1, the minus sign problem emerges when the Metropolis ratio R is not positive definite. It is easy to see from Eqs. 92 and 100 that when $U < 0$ there is no minus sign problem since the up and down determinants the same and real, so their product is positive. However, the minus sign problem can be severe when $U > 0$ since the sign of the sampling weight will fluctuate as k changes by ± 1. Some tricks can be used to control the sign problem. For example, at half filling where there is particle-hole symmetry, we rewrite the interaction as

$$H_1 = U\sum_i \left(n_{\uparrow_i} - \frac{1}{2}\right)\left(n_{\downarrow_i} - \frac{1}{2}\right) \tag{103}$$

then the particle-hole transformation $c_{j\downarrow} \to c^\dagger_{j\downarrow}$ changes the sign of U eliminating the sign problem. Away from half filling, this trick is useless. However, here we introduce an additional Ising-like auxiliary field $s_i = \pm 1$ on each site and rewrite the interaction as [31]

$$H_1 = \frac{U}{2}\sum_{i,s_i} \left(n_{\uparrow_i} - \frac{1}{2} - s_i\alpha\right)\left(n_{\downarrow_i} - \frac{1}{2} + s_i\alpha\right). \tag{104}$$

At least in 0 or 1 dimension, the choice $\alpha > 1/2$ eliminates the sign problem. The cost is that the Ising field is now associated with each point in space-time and must be sampled using the QMC algorithm (Fig. 21).

Since Wick's theorem applies to the configurations of CTQMC, most measurements are simply accomplished by forming the appropriate Wicks contractions of the Green's function

$$G_\uparrow(i,j) = -\left\langle T_\tau c_\uparrow(i) c^\dagger_\uparrow(j)\right\rangle. \tag{105}$$

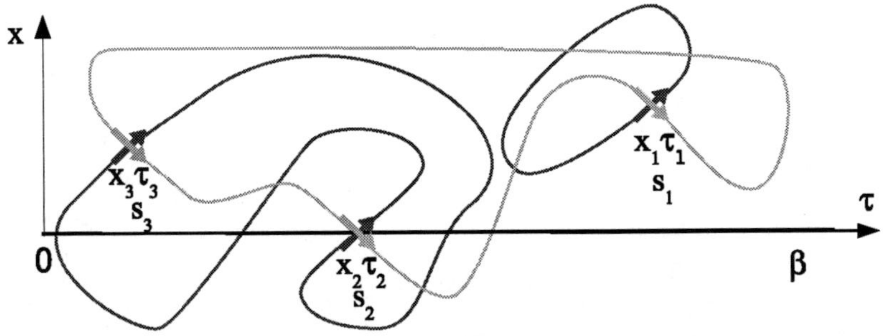

FIGURE 21. An example of a third-order ($k = 3$) CTQMC graph with the Rubtsov auxiliary field used to suppress the minus sign configuration. Now each vertex also carries a field $s_i = \pm 1$ which must be updated along with the vertex locations in the QMC.

corresponding to a configuration of vertices (and auxiliary fields when appropriate).

$$G_\uparrow(i,j) = -\frac{Z_0}{Z}\sum_k \frac{(-U)^k}{k!}\int\cdots\int d1\cdots dk \quad \left\langle T_\tau c_\uparrow(i)c_\uparrow^\dagger(j)n_\uparrow(1)\cdots n_\uparrow(k)\right\rangle_0 \left\langle T_\tau n_\downarrow(1)\cdots n_\downarrow(k)\right\rangle_0 . \tag{106}$$

However, this formula is very similar to the one we encountered when we added a vertex. In fact,

$$G_\uparrow(i,j) = \det D_\uparrow(k+1)/\det D_\uparrow(k)\,, \tag{107}$$

where the $k\times k$ matrix $D_\uparrow(k)$ is given by Eq. 89 and

$$D_\uparrow(k+1) = \begin{pmatrix} G_\uparrow{}^0(1,1) & G_\uparrow{}^0(1,2) & \cdots & G_\uparrow{}^0(1,k) & G_\uparrow{}^0(1,j) \\ G_\uparrow{}^0(2,1) & G_\uparrow{}^0(2,2) & \cdots & G_\uparrow{}^0(2,k) & G_\uparrow{}^0(2,j) \\ G_\uparrow{}^0(3,1) & G_\uparrow{}^0(3,2) & \cdots & G_\uparrow{}^0(3,k) & G_\uparrow{}^0(3,j) \\ \cdot & \cdot & \cdot & \cdot & \\ \cdot & \cdot & \cdot & \cdot & \cdot \\ \cdot & \cdot & \cdot & \cdot & \cdot \\ G_\uparrow{}^0(k,1) & G_\uparrow{}^0(k,2) & \cdots & G_\uparrow{}^0(k,k) & G_\uparrow{}^0(k,j) \\ G_\uparrow{}^0(i,1) & G_\uparrow{}^0(i,2) & \cdots & G_\uparrow{}^0(i,k) & G_\uparrow{}^0(i,j) \end{pmatrix} . \tag{108}$$

We obtain the familiar form for

$$G_\uparrow(i,j) = G_\uparrow{}^0(i,j) - G_\uparrow{}^0(i,p)M_{\uparrow pq}G_\uparrow{}^0(q,j)\,. \tag{109}$$

Of course, a similar equation holds for $G_\downarrow$.

All other measurements are then formed from Wick's contractions of these Green's functions, as described in standard many-body texts [22]. This is worth illustrating for one example, such as the transverse component of the spin susceptibility

$$\chi^\pm(i,j) = -\left\langle T_\tau c_\uparrow^\dagger(i)c_\downarrow(i)c_\downarrow^\dagger(j)c_\uparrow(j)\right\rangle . \tag{110}$$

The form for this is strikingly similar to what we did above for $G_\uparrow(i,j)$, the main difference being that we now have both spin components

$$\chi^{\pm}(i,j) = \frac{Z_0}{Z}\sum_k \frac{(-U)^k}{k!}\int d1\cdots dk \quad \left\langle T_\tau c_\uparrow^\dagger(i) c_\uparrow(j) n_\uparrow(1)\cdots n_\uparrow(k)\right\rangle_0 \left\langle T_\tau c_\downarrow^\dagger(j) c_\downarrow(i) n_\downarrow(1)\cdots n_\downarrow(k)\right\rangle_0 . \quad (111)$$

So the estimator is

$$\chi^{\pm}(i,j) = \left\langle \left(G_\uparrow{}^0(j,i) - G_\uparrow{}^0(j,p) M_{\uparrow pq} G_\uparrow{}^0(q,i)\right) \left(G_\downarrow{}^0(i,j) - G_\downarrow{}^0(i,p) M_{\downarrow pq} G_\downarrow{}^0(q,j)\right)\right\rangle_{QMC} . \quad (112)$$

Similar estimators may be formed from $G_\uparrow(i,j)$ and $G_\downarrow(i,j)$ for the longitudinal spin and pair susceptibilities. These will be discussed below.

In order to initialize the QMC process, the number of vertices can be simply set to zero. Subsequent Monte Carlo steps will tend to add vertices until the configuration is thermalized. Alternatively, one can take any other k vertex configuration and use $\mathscr{G}_\sigma$ to construct the D_σ matrices (which are then inverted to obtain M_σ matrices). Here, $\mathscr{G}_\sigma$ is the noninteracting Green's function (the one obtained when $U = 0$) if we mean to simulate a finite size cluster, or its the cluster excluded Green's function (the one obtained when we subtract the self energy from the cluster) if we mean to simulate an embedded cluster.

Numerical round-off errors accumulate in the M_σ matrices during the updating process, Eqs. 97 and 101. In order to eliminate them, after a certain number of iterations they should be refreshed. This may be done by recalculating the D_σ matrices and inverting for the M_σ.

3.3.1. Combining CTQMC with Quantum Cluster Methods

Quantum cluster methods generally require the Green's function in terms of Matsubara frequency and wavevector, $G(k,i\omega_n)$. However, unlike HFQMC, there is no difficulty in performing the Fourier transform in CTQMC. In fact, this measurement may be made efficiently directly in terms of Matsubara frequency and wavevector, simply by performing a double Fourier transform of the Green's function estimator in Eq. 109

$$G_\uparrow(K_i) = G_\uparrow{}^0(K_i) - G_\uparrow{}^0(K_i) M_\uparrow(K_i) G_\uparrow{}^0(K_i) , \quad (113)$$

where $K = (\mathbf{K}, i\omega_n)$ is now a frequency-wavevector label, and $M_\uparrow$ is

$$M_\uparrow(K_i) = M_\uparrow(\mathbf{K}, i\omega_n) = \sum_{i,j} \exp\left(i\mathbf{K}\cdot(\mathbf{X}_i - \mathbf{X}_j) - i\omega_n(\tau_i - \tau_j)\right) M_{\uparrow ij} \quad (114)$$

where i and j label the space-time locations of the vertices.

3.4. Making and Conditioning Measurements

The natural byproduct of the QMC algorithm are the Green's functions. These may be used to make measurements of most one and two-particle properties using standard diagrammatic techniques. In doing so, several points must be remembered:

- In the HF QMC algorithm, the Hubbard-Stratonovich transformation reduces the problem to one of free electrons moving in a time-dependent Hubbard-Stratonovich field. In the CTQMC, an interaction representation together with an S-matrix expansion is used. Thus, for each field configuration in the HFQMC or any vertex configuration in the CTQMC, measurements may be formed by summing all allowed Wick's contractions. The full interacting quantity is recovered by QMC averaging this over all configurations.
- It is important to use *all* allowed Wick's contractions, both connected and disconnected, in this series.
- If your Hamiltonian is invariant under translations in space and time, and you are performing a simulation in an unbroken symmetry phase, it is important to average your measurement over all time and space differences in order to achieve the lowest variance estimator.
- It is also important to average over other symmetries of the Hamiltonian, which may not be preserved by the CTQMC or HFQMC algorithms (e.g., spin symmetry).

For example, consider the local impurity magnetic correlation function

$$\begin{aligned}\chi_{ii}(\tau) &\approx \langle S_i^+(\tau)S_i^-(0)\rangle \\ &\approx \left\langle C_{i\uparrow}^\dagger(\tau)C_{i\downarrow}(\tau)C_{i\downarrow}^\dagger(0)C_{i\uparrow}(0)\right\rangle \\ &\approx \frac{T}{2N}\sum_{i\sigma}\int_0^\beta d\tau' \langle G_\sigma(x_i,\tau+\tau';x_i\tau')G_{-\sigma}(x_i\tau';x_i\tau+\tau')\rangle_{QMC}\end{aligned} \tag{115}$$

where the *QMC* subscript means that the Monte Carlo average over the Hubbard-Stratonovich fields or vertex configurations is still to be performed. Note that in the last step in Eq. 116 we form all allowed Wick's contractions and average over all equivalent time differences, spins, and sites to reduce the variance of this estimator.

At this point the measurements for CTQMC and HFQMC differ. For CTQMC we essentially have a continuum of Matsubara time, so the time integral in Eq. 116 may be completed very accurately. However, in HFQMC care must be used to reduce the time-step error. Due to time translational invariance, the integral over τ' is not terribly sensitive to $\Delta\tau$ error, so we approximate as a sum using a rectangular approximation. For $\tau > 0$

$$\chi(\tau_l) \approx \frac{1}{2L}\sum_{\sigma,l'}\langle G_\sigma(i,ind(l+l');i,l')G_{-\sigma}(i,l';i,ind(l+l'))\rangle_{QMC}, \tag{116}$$

where $ind(l)$ is the smaller nonnegative value of either l or $l-L$. For $\tau = 0$ the fact that in HFQMC we always store $G_\sigma(l',l') = G_\sigma(\tau_{l'}+0^+,\tau_{l'})$ requires us to modify the measurement

$$\chi(\tau=0) \approx \frac{1}{2L} \sum_{\sigma,l'} \langle G_\sigma(l',l') \left(G_{-\sigma}(l',l') - 1\right) \rangle_{QMC} . \tag{117}$$

Finally the susceptibility may be calculated by

$$\chi(T) = \int_0^\beta d\tau \chi(\tau) \approx \sum_l sf(l) \Delta\tau \chi(\tau_l) , \tag{118}$$

where the Simpson factor $sf(l) = 2\Delta\tau/3$ $(4\Delta\tau/3)$ for odd (even) l is used to reduce the systematic error of the integral.

As a final example, consider the cluster particle-particle Green's function matrix $\chi_c(q,K,K')$ $(K = (\mathbf{K}, i\omega_n))$ which in the cluster space-time takes the form

$$\chi_c(X_1,X_2,X_3,X_4) = \left\langle T_\tau c_\uparrow(X_1) c_\downarrow(X_2) c_\downarrow^\dagger(X_3) c_\uparrow^\dagger(X_4) \right\rangle . \tag{119}$$

Here X_i is in the space-(imaginary)time notation $X_i = (\mathbf{X}_i, \tau_i)$, where the points $\mathbf{X}_i$ are on the corresponding reciprocal cluster of $\mathbf{K}$ in real space.

Since the storage associated with this quantity is quite large, it cannot be measured for many times. Thus the measurement in CTQMC and HFQMC are quite similar. First, using Wick's theorem, its value is tabulated for each field configuration and then transformed into the cluster Fourier space. Second, we Monte Carlo average over these configurations. After the first step, the expression for the above two-particle Green's function in the cluster momentum-frequency space becomes

$$\begin{aligned} \chi_c(\mathbf{Q}, i\nu_n; \mathbf{K}, i\omega_n; \mathbf{K}', i\omega_{n'}) &= \Bigg\langle \sum_{X_1,X_4} e^{iK'X_1} G_{c\uparrow}(X_1,X_4) e^{-iKX_4} \\ &\quad \sum_{X_2,X_3} e^{i(Q-K')X_2} G_{c\downarrow}(X_2,X_3) e^{-i(Q-K)X_3} \Bigg\rangle_{QMC} . \end{aligned} \tag{120}$$

where K is the momentum-frequency point $K = (\mathbf{K}, i\omega_n)$.

The sums (integrals) over τ in Eq. 120 require special consideration. Since the Green's functions change discontinuously when the two time arguments intersect, the best applicable integral approximation is the trapezoidal approximation. Using this, we will run into Green's functions $G_c(\mathbf{X},\tau;\mathbf{X},\tau)$ with both time and space arguments the same. In the HFQMC algorithm, this is stored as $G_c(\mathbf{X},\tau^+;\mathbf{X},\tau)$ (i.e. it is assumed that the first time argument is slightly greater than the second) and in CTQMC, the other time ordering is assumed; however, if we replaced the equal time Green's function to be the average $\{G_c(\mathbf{X},\tau^+;\mathbf{X},\tau) + G_c(\mathbf{X},\tau;\mathbf{X},\tau^+)\}/2 = G_c(\mathbf{X},\tau^+;\mathbf{X},\tau) - 1/2 = G_c(\mathbf{X},\tau;\mathbf{X},\tau+) + 1/2$ then a trapezoidal approximation of the integrals results. If we call the matrix $\mathbf{G}_c$, with $1/2$ subtracted (added) from its diagonal elements as appropriate for HFQMC (CTQMC), as $\hat{\mathbf{G}}_c$ (note that we can treat one of the three independent momenta involved in χ_c as a variable Q outside the matrix structure), then we can write the two-particle Green's function in a matrix form

$$\chi_{c_{ij}}(\mathbf{Q}) = \left\langle \left(\mathbf{F}^{\dagger}_{\mathbf{Q}=0}\hat{\mathbf{G}}_{c\uparrow}\mathbf{F}_{\mathbf{Q}=0}\right)_{ij}\left(\mathbf{F}^{\dagger}_{\mathbf{Q}}\hat{\mathbf{G}}_{c\downarrow}\mathbf{F}_{\mathbf{Q}}\right)^{*}_{ij}\right\rangle_{QMC}, \tag{121}$$

where $(\mathbf{F_Q})_{ij} = \Delta\tau e^{-i(\mathbf{K}_j-\mathbf{Q}).\mathbf{X}_i - i\omega_j\tau_i}$ where we have chosen ι and j to index the cluster momentum-frequency space.

This measurement may be performed efficiently if the product of three matrices in each set of parenthesis is tabulated as two sequential matrix-matrix products and stored before the direct product between the terms in parenthesis is calculated. When done this way, the calculation time required for this process scales like $(NL)^3$ rather than $(NL)^4$ as would result from a straight-forward evaluation of the sums implicit in Eq. 121. Greater efficiency can be obtained if we perform the Fourier transforms as a two-step process; i.e., first doing the transform in time and then in space. Then the measurement scales like $(N+L)(NL)^2$, of course, these reductions in FLOPs require an increase in memory needed to store the intermediate results.

CTQMC presents the possibility of measuring these two-particle susceptibilities directly in the cluster momentum-frequency space without the need to perform the discrete Fourier transform in Matsubara time presented in Eq. 120. Again, this measurement may be made efficiently again performing a double Fourier transform of the Green's function estimator Eq. 109

$$G_{\uparrow}(K_i,K_j) = G_{\uparrow}{}^{0}(K_i)\delta_{ij} - G_{\uparrow}{}^{0}(K_i)M_{\uparrow}(K_i,K_j)G_{\uparrow}{}^{0}(K_j), \tag{122}$$

and $M_{\uparrow}$ is

$$\begin{aligned} M_{\uparrow}(K_n,K_m) &= M_{\uparrow}(\mathbf{K}_n, i\omega_n; \mathbf{K}_m, i\omega_m;) \\ &= \sum_{i,j} \exp\left(i\mathbf{K}_n\cdot\mathbf{X}_i - i\omega_n\tau_i)\right) M_{\uparrow_{ij}} \exp\left(-i\mathbf{K}_m\cdot\mathbf{X}_j + i\omega_m\tau_j)\right), \end{aligned} \tag{123}$$

where i and j label the space-time locations of the vertices. The last step is the numerical bottleneck. However,using the methods described above of performing the Fourier transforming steps and storing the intermediate result, this step may be performed in $\mathcal{O}2(L+N)k^2)$ FLOPs. The estimator of the cluster particle-particle then becomes

$$\chi_c(Q,K,K') = \langle G_{\uparrow}(K'+Q,K+Q)G_{\downarrow}(-K',-K)\rangle_{QMC}. \tag{124}$$

3.4.1. The minus sign problem

Except in a few and fortunate situations, the calculation of physical quantities with QMC reduces to the calculation of integrals/sums with an integrand which can take both positive and negative values. This is the origin of the *sign problem* in the Monte Carlo simulations.

For example, the average of the observable A is formally given by

$$\langle A \rangle = \frac{\int W(x)A(x)dx}{\int W(x)dx} \tag{125}$$

where $\int dx$ is a integral over all the possible configurations x of the system of weight $W(x)$. For a classical system $W(x)$ is positive definite and a Monte Carlo process consists of stochastically generating configurations x_i according to their weight $W(x_i)$ and taking

$$\langle A \rangle = \frac{1}{N_d} \sum_{i=1}^{N_d} A(x_i) \,. \tag{126}$$

Where N_d is the number of measurements. For quantum systems $W(x)$ is not necessarily positive definite and therefore cannot be considered a probability distribution. The solution is to rewrite Eq. 125 as

$$\langle A \rangle = \frac{\int \mid W(x) \mid A(x)s(x)dx}{\int \mid W(x) \mid dx} \frac{\int \mid W(x) \mid dx}{\int \mid W(x) \mid s(x)dx} \tag{127}$$

where $s(x)$ is the sign of $W(x)$, i.e.

$$W(x) = \mid W(x) \mid s(x) \,. \tag{128}$$

Thus, we can sample the configurations space with the probability $\mid W(x) \mid$ and

$$\langle A \rangle = \frac{\sum_{i=1}^{N_d} A(x_i)s(x_i)}{\sum_{i=1}^{N_d} s(x_i)} = \frac{\langle As \rangle_{|W|}}{\langle s \rangle_{|W|}} \,. \tag{129}$$

Thus, we may use the absolute value of $|W|$ as the sampling weight as long as we measure the product of the sign and measurement and divide by the average sign. If the average sign $\langle s \rangle_{|W|}$ is finite and not very small, then this approach is successful. However, a problem occurs when $\langle s \rangle_{|W|}$ is extremely small, because a small sign strongly amplifies the error bar associated with $\langle A \rangle_{|W|}$ ($\propto \delta s / \langle s \rangle^2$). In fact, for many quantum systems $\langle s \rangle$ is found to fall exponentially to zero with decreasing temperature or increasing cluster size [55]. This makes Monte Carlo calculations impractical at low temperatures.

The average sign obtained from a simulation of the 2D Hubbard model is shown in Fig. 22. The average sign from HFQMC and CTQMC are similar but the CTQMC result is slightly worse. The figure on the right shows that increasing the value of α from Eq. 104 does not help to increase the average sign significantly and it also increases the error bar on the sign measurement due to an increase in the average order $\langle k \rangle$. The error bar increases, since the average order $\langle k \rangle$ increases like α^2 for large α as may be seen from Eq. 103, or here (for some site i).

$$\begin{aligned} \langle k \rangle &= -\beta \langle H_1 \rangle \\ &= -N\beta \left\langle \sum_{s_i} \left(n_{\uparrow i} - \frac{1}{2} - s_i\alpha \right) \left(n_{\downarrow i} - \frac{1}{2} + s_i\alpha \right) \right\rangle \end{aligned} \tag{130}$$

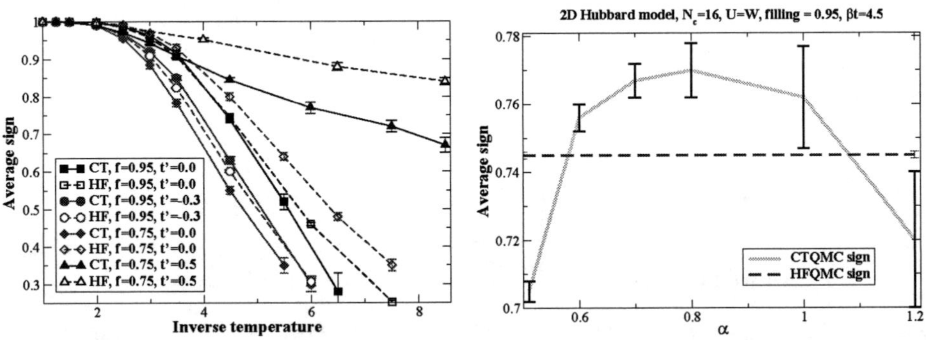

FIGURE 22. Comparison of the average sign from a CTQMC and HFQMC DCA simulation of the two-dimensional Hubbard model. (Left) Average sign versus inverse temperature for 4-site cluster with $U = W = 8t$ and different fillings. (Right) the average sign for 16-site cluster with $U = W = 8t$ versus α from Eq. 104 at fixed filling and temperature.

On the other hand, the average sign obtained from a simulation of the 1D Hubbard model is shown in Fig. 23. Whereas the average sign from HFQMC depends weakly with cluster size, that from CTQMC actually *increases* strongly with increasing cluster size. Also studied is an additional coupling between chains t_{perp}/t which is a hopping between equivalent sites on adjacent chains. A finite t_{perp}/t causes the average sign to increase strongly.

We find that the average sign depends on many things: fillings, dimensionality model parameters, etc. It is difficult to predict *a priori* whether the average sign will be problematic.

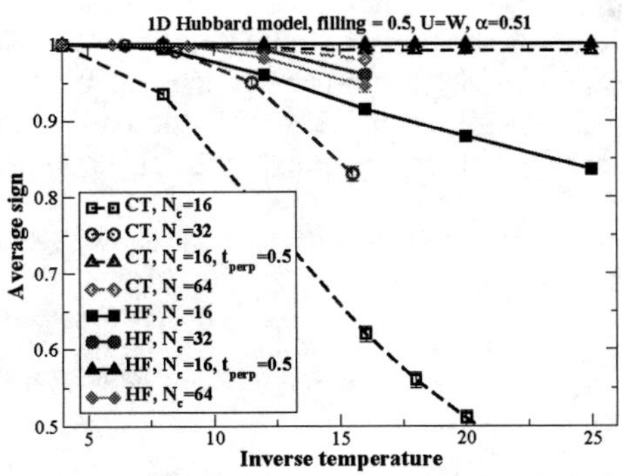

FIGURE 23. Comparison of the average sign from a CTQMC and HFQMC DCA simulation of the one-dimensional Hubbard model. The average sign of HFQMC depends weakly on the cluster size; whereas the average sign of CTQMC increases strongly with increasing cluster size.

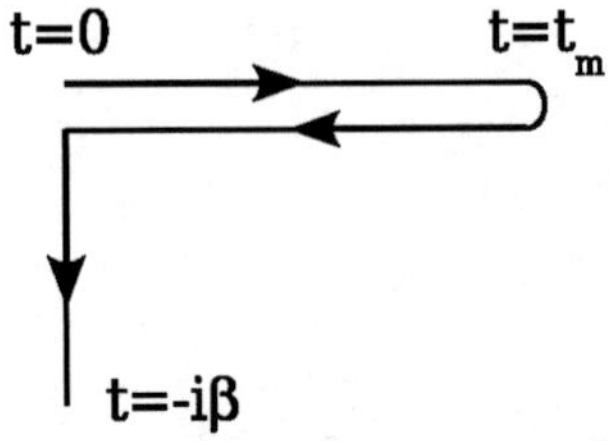

FIGURE 24. The contour for measuring the real time Green's function.

3.4.2. CTQMC and real time measurements

In this section, we briefly discuss a possibility to extend CTQMC to measure real time Green's functions or even treat the systems out of thermal equilibrium. We start with the same Hubbard model Hamiltonian [1], which has no time dependent terms, and proceed to derive expressions for partition function and real time Green's functions. In this formalism, the partition function can be expressed as an expansion of contour ordered exponent:

$$Z = Z_0 \sum_{k=0}^{\infty} \frac{(-i)^k}{k!} \int_c dt_1 \ldots \int_c dt_k \langle T_c H_1(t_1) \ldots H_1(t_k) \rangle_0 , \tag{131}$$

where

$$H_1(t) = e^{iH_0 t} H_1 e^{-iH_0 t} \tag{132}$$

is the interaction (quartic) part of Hamiltonian in the interaction representation and T_c is the contour-ordering operator [32]. The integrals are performed along the contour c, shown in Fig. 24. By writing out the contributions to the partition function coming from lowest orders in expansion (Eq. 131), one can see that all the terms containing $H_1(t)$ on real parts of contour cancel out exactly and thus the partition function in this case is the same as in Matsubara formalism. We now consider the expansion for the contour ordered Green's function:

$$\begin{aligned} G_c(t,t') &= \left\langle T_c c(t) c^\dagger(t') \right\rangle \\ &= -i \frac{Z_0}{Z} \sum_{k=0}^{\infty} \frac{(-i)^k}{k!} \int_c dt_1 \ldots \int_c dt_k \left\langle T_c c(t) c^\dagger(t') H_1(t_1) \ldots H_1(t_k) \right\rangle_0 . \end{aligned} \tag{133}$$

In this case, however, the contributions containing $H_1(t)$ on real parts of contour do not cancel out (as one can easily check by writing out $k = 1$ term in the expansion). The application of Wick's theorem yields CTQMC diagrams that contain vertices on both imaginary as well as real parts of contour (See Fig. 25). Thus, the partition function and the Green's function have dominant contributions coming from entirely different regions in the k-dimensional space spanned by time coordinates along the contour. As a consequence, it appears impossible to use the partition function as a sampling weight

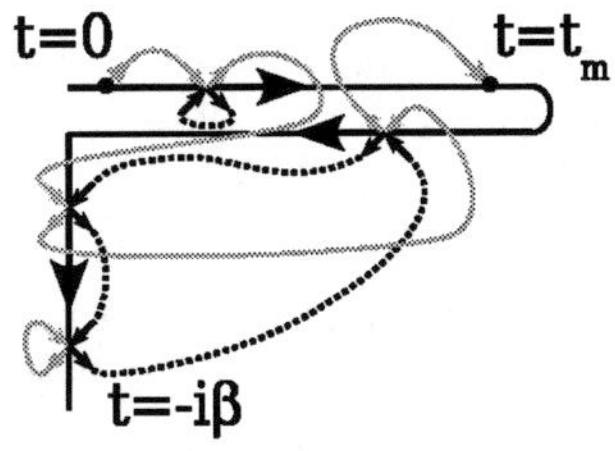

FIGURE 25. A $k=4$ diagram for the contour-ordered spin-up Green's function. The light continuous and darker dotted lines represent the propagators for spin up and spin down electrons, respectively.

in Monte Carlo algorithm to measure the Green's function. It is still possible to use the Green's function itself as a weight in the Monte Carlo importance sampling, and then measure it's value as a distribution in the space of it's arguments (t,t'). Since the Green's function in this case would be a complex number, it's modulus would be taken for the sampling weight instead. The sign problem would now translate into the phase problem, if the average value of the phase of the Green's function gets too small. By looking at Eq. 134 we see that the phase problem is likely to be very severe due to a factor of $(-i)$ attached to every vertex on the real part of the contour (unless one finds again a way to introduce an auxiliary field that remedies this problem). In any case, the oscillatory nature of real time Green's function implies that the phase problem would most likely restrict measurements of Green's function to small values of its real time arguments.

3.5. Summary

We describe two QMC algorithms for solving the embedded cluster (impurity) problem used in cluster mean-field methods. In the HFQMC algorithm the interaction term is replaced by a summation over a set of auxiliary fields, with every field configuration defining a noninteracting problem. In the CTQMC algorithm, an expansion in the interaction term in the partition function is employed. The problem is reduced to an infinite sum of integrals with different multiplicity. Measurements, the sign problem and a possible extension of the CTQMC technique to non-equilibrium processes is discussed.

4. ANALYTIC CONTINUATION OF QUANTUM MONTE CARLO DATA

4.1. Introduction

Most quantum Monte Carlo (QMC) simulations produce Green's functions $G(\tau)$ of imaginary time $\tau = it$. However, real-frequency results are crucial since most experiments probe dynamical quantities, including transport, densities of states, nuclear magnetic resonance, inelastic scattering, etc. Thus, the inability to extract real-frequency or real-time results from Matsubara (imaginary) time QMC simulations presents a significant limitation to the usefulness of the method. The relation between $G(\tau)$ and $A(\omega) = -\frac{1}{\pi}\mathrm{Im}G(\omega)$ is linear and surprisingly simple

$$G(\tau) = \int d\omega K(\tau,\omega) A(\omega). \tag{134}$$

Nevertheless, inversion is complicated by the exponential nature of the kernel. For example, for a Fermionic single-particle Green's function G, $K(\tau,\omega) = e^{-\tau\omega}/(1+e^{-\beta\omega})$ [33]. For finite τ and large ω the kernel is exponentially small, so that $G(\tau)$ is insensitive to the high frequency features of $A(\omega)$. Equivalently, if we approximate both G and A by equal-length vectors and K by a square matrix, then we find that the determinant of K is exponentially small, so that K^{-1} is ill-defined. Apparently, there are an infinite number of A with very different characters (many of them not even causal) that yield the same G.

Previous attempts to address this problem include least-squares fits, Padé approximants and regularization. In the least squares method, Schüttler and Scalapino [33] approximated the spectrum with a set of box functions. The location and weight of these functions was determined by minimizing the least-squares misfit between the spectrum and the QMC data. However, as the number of box functions is increased to better resolve features in the spectrum, the fit becomes unstable and noisy. In the Padé method [34], G (or rather its Fourier transform) is fit to a functional form, usually the ratio of two polynomials, which is then analytically continued formally by replacing $i\omega_n \to \omega + i0^+$. This technique works when the data G is very precise, as when analytic continuing Eliashberg equations, or when the fitting function is known *a priori*. However, it is generally unreliable for the continuation of less-precise QMC data to real frequencies. A more useful approach is to introduce regularization to the kernel, so that K^{-1} exists. This method was developed by G. Wahba [35], and employed by White et al.[36] and Jarrell and Biham [37]. They used similar methods to minimize $(G-KA)^2$ subject to constraint potentials which introduce correlations between adjacent points in A and impose positivity. However, these techniques tend to produce spectra A with features which are overly smeared out by the regularization.

With entropy-based methods of Bayesian data analysis, we employ a different philosophy. Using Bayesian statistics, we define the posterior probability of the spectra A given the data G, i.e., $P(A|G) \propto P(A)P(G|A)$ with the prior probability $P(A)$ defined so that A has only those correlations that are required to reproduce the data G. To define the likelihood function $P(G|A)$, we take advantage of the statistical sampling nature of the QMC process. The choice $P(A)$ is based on information theory concepts of entropy. The result is either the most probable $P(A|G)$ or the average of this function over free parameters. The latter has proven to be the most effective. Collectively, these methods are colloquially called Maximum Entropy Methods (MEM).

In this section, we will present a short pedagogical development of the MEM to analytically continue QMC data. A more detailed review has been presented previously [16], and to the extent possible, we will follow the notation used there. This section is organized as follows: In Sec. 4.2, we will present the MEM formalism including an efficient way to calculate the spectra when the sign problem is present in the QMC calculations. In

Sec. 4.3, the Bryan MEM algorithm will be sketched, which has been optimized for this type of problem. In Secs. 4.4 and 4.5 we will illustrate these techniques with the spectra of the periodic Anderson model and Hubbard model respectively. Finally in Sec. 4.6, we will conclude.

Throughout this section, we will illustrate the formalism and methods introduced with a Dynamical Field Theory (DMFT) [10, 12, 13] simulation of the periodic Anderson model (PAM) (2) and a Dynamical Cluster Approximation (DCA) [8, 9, 38] simulation of the two-dimensional Hubbard model (1). As described in the Sec. 3.2, the core of both DMFT and DCA algorithms is the Hirsch-Fye impurity algorithm [23]. Here the problem is cast into a discrete path formalism in imaginary time, τ_l, where $\tau_l = l\Delta\tau$, $\Delta\tau = \beta/L$, $\beta = 1/k_BT$, and L is the number of times slices. Matsubara-time Green's functions are measured on this discrete time domain.

4.2. Formalism

4.2.1. Green's Functions

If this system is perturbed by an external field which couples to an operator B, then the linear response to this field is described by the retarded Green's function

$$G(t) = -i\Theta(t)\left\langle \left[B(t), B^{\dagger}(0)\right]_{\pm}\right\rangle \tag{135}$$

where the negative (positive) sign is used for Boson (Fermion) operators B and $B^{\dagger}$, and makes reference to the Dirac (anti)commutator. The Fourier transform of $G(t)$, $G(z)$ is analytic in the upper half plane, and its real and imaginary parts are related by

$$G(z) = \int d\omega \frac{\frac{-1}{\pi}\mathrm{Im}G(\omega)}{z-\omega}. \tag{136}$$

The Matsubara-frequency Green's function $G(i\omega_n)$ is obtained by letting $z \to i\omega_n$ in Eq. 136. This may be Fourier transformed to yield a relation between the Matsubara-time Green's function produced by the QMC procedure, and $\frac{-1}{\pi}\mathrm{Im}G(\omega)$

$$G(\tau) = \int d\omega \frac{\frac{-1}{\pi}\mathrm{Im}G(\omega)e^{-\tau\omega}}{1 \pm e^{-\beta\omega}}. \tag{137}$$

4.2.2. Bayesian Statistics

We use our QMC algorithm to generate a set $\bar{G}_l^i$ of $i = 1,..,N_d$ estimates for the Green's function at each time slice $\tau_l = (l-1)\Delta\tau$, $l = 1, L$. Since many A correspond to the same data $\bar{G}$, we must employ a formalism to determine which $A(\omega)$ is the most probable, given the statistics of the data and any *a priori* information that we have about

A. To quantify the conditional probability of A given the data, and our prior knowledge, we use Bayesian statistics.

If we have two events a and b, then according to Bayes theorem, the joint probability of these two events is

$$P(a,b) = P(a|b)P(b) = P(b|a)P(a), \tag{138}$$

where $P(a|b)$ is the conditional probability of a given b. The probabilities are normalized so that

$$P(a) = \int dbP(a,b) \quad \text{and } 1 = \int daP(a). \tag{139}$$

In our problem, we search for the spectrum A which maximizes the conditional probability of A given the data $\bar{G}$,

$$P(A|\bar{G}) = P(\bar{G}|A)P(A)/P(\bar{G}). \tag{140}$$

Typically, we call $P(\bar{G}|A)$ the likelihood function, and $P(A)$ the prior probability of A (or the prior). Since we work with one set of QMC data at a time, $P(\bar{G})$ is a constant during this procedure, and may be ignored. The prior and the likelihood function require significantly more thought, and will be the subject of the next two subsections.

4.2.3. *Prior Probability*

We can define a prior probability for positive-definite normalizable spectra. For Bosonic Green's functions, we may define positive definite spectra if we redefine the kernel

$$K(\tau,\omega) = \frac{\omega[e^{-\tau\omega} + e^{-(\beta-\tau)\omega}]}{1-e^{-\beta\omega}} \text{ with } A(\omega) = \frac{-1}{\pi\omega}\text{Im}G(\omega) \geq 0 \text{ for Bosons}. \tag{141}$$

We modified the kernel to account for the symmetry of the Bosonic data $G(\tau) = G(\beta-\tau)$ and the spectrum $A(\omega) = A(-\omega)$. Note that the kernel is non-singular at $\omega = 0$ and the spectral density $A(\omega)$ is positive definite. For Fermionic Green's functions the spectra are already positive definite

$$K(\tau,\omega) = \frac{\exp(-\tau\omega)}{1+\exp(-\beta\omega)} \text{ with } A(\omega) = \frac{-1}{\pi}\text{Im}G(\omega) \geq 0 \text{ for Fermions}. \tag{142}$$

We may also define positive definite spectra for more exotic cases, such as for the Nambu off-diagonal Green's function G_{12}. Since the corresponding spectrum $A_{12}(\omega) = \frac{-1}{\pi}\text{Im}G_{12}(\omega)$ is not positive definite, we enforce positivity by adding a positive real constant b

$$G_{12}(\tau) + b\int d\omega K(\tau,\omega) = \int d\omega K(\tau,\omega)\left(A_{12}(\omega)+b\right). \tag{143}$$

Here, we may incorporate the symmetry of the spectrum $A_{12}(\omega) = -A_{12}(-\omega)$ and the data $G_{12}(\tau) = -G_{12}(\beta-\tau)$ by modifying the kernel

$$K(\tau,\omega) = \frac{e^{-\tau\omega} - e^{-(\beta-\tau)\omega}}{1+\exp(-\beta\omega)}. \tag{144}$$

With this kernel, the equation takes the canonical form Eq. 134, if we identify

$$A(\omega) = A_{12}(\omega) + b, \text{ and } G(\tau) = G_{12}(\tau) + b \int d\omega K(\tau, \omega) \tag{145}$$

In each of the Bosonic, Fermionic and Anomalous cases,

$$\int_{-\infty}^{\infty} d\omega A(\omega) < \infty. \tag{146}$$

These positive-definite normalized spectra A may be reinterpreted as probability densities.

Skilling [39] argues that the prior probability for such an unnormalized probability density is proportional to $\exp(\alpha S)$ where S is the entropy defined relative to some positive-definite function $m(\omega)$

$$\begin{aligned} S &= \int d\omega \left[A(\omega) - m(\omega) - A(\omega) \ln\left(A(\omega)/m(\omega)\right)\right] \\ &\approx \sum_{i=1}^{N_f} A_i - m_i - A_i \ln\left(A_i/m_i\right), \end{aligned} \tag{147}$$

where $A_i = A(\omega_i)d\omega_i$, $i = 1, N_f$. Thus, the prior is conditional on two as yet unknown quantities $m(\omega)$ and α

$$P(A|m, \alpha) = \exp(\alpha S). \tag{148}$$

$m(\omega)$ is called the default model since in the absence of data $\bar{G}$, $P(A|\bar{G}, m, \alpha) \propto P(A|m, \alpha)$, so the optimal $A = m$. The choice of α will be discussed in Sec. 4.2.6.

Rather than try to repeat Skilling's arguments here for the entropic form of $P(A|m, \alpha)$, we argue that this form yields the desired effects:

1. it enforces positivity of A,
2. it requires that A only have correlations which are required to reproduce the data $\bar{G}$, and
3. it allows us to introduce prior knowledge about the the spectra (i.e. exact results at high frequencies) in the default model.

The first effect follows from the form of $P(A|m, \alpha)$, assuming that m is positive definite. The third effect will be discussed in Sec. 4.4.5.

To illustrate the second effect, Gull and Skilling use their kangaroo argument [40]. Imagine we have a population of kangaroos. We know that one third of them are left handed and one third have blue eyes. The joint probabilities of left-handedness and eye color may be represented in a contingency table.

		Left T	Handed F
Blue	T	p_1	p_2
Eyes	F	p_3	p_4

We are given that $p_1 + p_2 = p_1 + p_3 = 1/3$, what is the fraction that are both blue eyed and left handed, p_1? Clearly, there is not enough information to answer this question. We must make some additional assumptions. If we assume that there is a maximum positive correlation between left handedness and blue eyes, then

		Left Handed T	Left Handed F
Blue	T	1/3	0
Eyes	F	0	2/3

If these events have a maximum negative correlation, then

		Left Handed T	Left Handed F
Blue	T	0	1/3
Eyes	F	1/3	1/3

However, if we are forced to answer this question without the use of further information, a more natural assumption to make is that the events of handedness and eye color are uncorrelated, so that $1/9$ of the kangaroos are both blue eyed and left handed.

		Left Handed T	Left Handed F
Blue	T	1/9	2/9
Eyes	F	2/9	4/9

This final answer is the one obtained by maximizing the entropy $S = -\sum_{i=1}^{4} p_i \ln p_i$ subject to the Lagrange constraints $\sum_{i=1}^{4} p_i = 1$, $p_1 + p_2 = 1/3$ and $p_1 + p_3 = 1/3$. All other regularization functions yield either positive or negative correlations between handedness and eye color.

To relate this to the analytic continuation problem, imagine that each A_i is an independent event. If we maximize the entropy of A, subject to the constraint of reproducing the data $G = KA$, then the resulting spectrum is the one with the least correlations that is consistent with $\bar{G}$. If we identify a feature in the spectrum as a region of correlated A_i (such as a peak) in deviation from the default model m_i, and such a feature emerges in the spectrum $A(\omega)$ and persists as the data $\bar{G}$ becomes more precise, then we have reason to believe that this feature is real. The choice of any other regularization function would produce artificial features in the data.

4.2.4. *Likelihood function*

The form of the likelihood function is dictated by the central limit theorem, which for the purposes of this section may be illustrated with the following example. Suppose we use our QMC algorithm to generate N_d measurements of the Green's function G_l^i (where l is an integer between 1 and L, and i an integer between 1 and N_d). According to the

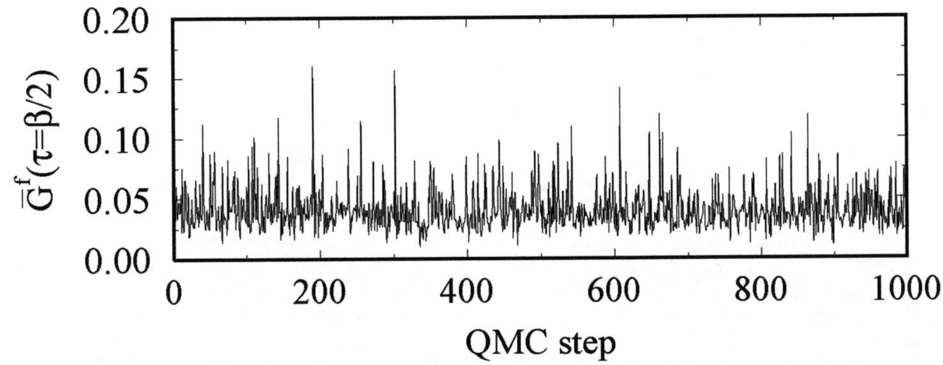

FIGURE 26. *Symmetric PAM f-electron local Green's function $\bar{G}^f(\tau=\beta/2)$ plotted as a function of the QMC step for $U=2$, $V=0.6$, and $\beta=20$.*

central limit theorem, if each of these measurements is completely independent of the others, then in the limit of large N_d, the distribution of $\bar{G}_l$ will approach a Gaussian, and the probability of a particular value G_l is given by

$$P(G_l) = \frac{1}{\sqrt{2\pi}\sigma} e^{-\chi^2/2}, \tag{149}$$

where $\chi^2 = \frac{1}{\sigma^2}\left(\frac{1}{N_d}\sum_{i=1}^{N_d}\bar{G}_l^i - G_l\right)^2 = \frac{1}{\sigma^2}\left(\langle\bar{G}_l\rangle - G_l\right)^2$, $\sigma^2 = \frac{1}{N_d(N_d-1)}\sum_i\left(\langle\bar{G}_l\rangle - \bar{G}_l^i\right)^2$ and the angular brackets indicate an average over the bins of data.

Of course, in the QMC process each of the measurements is not independent of the others. Correlations exist between adjacent measurements ($\bar{G}_l^i$ and $\bar{G}_l^{i+1}$) in the QMC process, and between the errors of the Green's function at adjacent time slices ($\bar{G}_l^i$ and $\bar{G}_{l+1}^i$) at the same QMC step. The removal of these correlations is the most critical step in the MEM analytic continuation procedure.

Correlations between adjacent measurements are illustrated in Fig. 26 where measurements of $\bar{G}^f(\tau=\beta/2)$ are plotted versus the QMC step. The data from adjacent QMC steps are correlated and the data are skewed since the Green's function is bounded from below ($\bar{G}_l^i > 0$). As a result the data are not Gaussianly distributed, as shown in Fig. 27(a). Here, a histogram of the data is compared to a Gaussian fit. The deviations from a Gaussian are quantified by the moments of the distribution. The most relevant ones in the present case are the skewness (third moment) and kurtosis (fourth moment) which measure the degree of asymmetry around the mean and the pointedness (or flatness) of the distribution relative to the Gaussian [41]. The data are clearly not Gaussianly distributed, and display significant skew and kurtosis. To deal with this difficulty, we rebin the data. For example, we set $\bar{G}_l^1$ equal to the average of the first 30 measurements, $\bar{G}_l^2$ equal to the average of the next 30 measurements, etc. The distribution of this rebinned data is shown in Fig. 27b. It is well approximated by a Gaussian fit (the solid line).

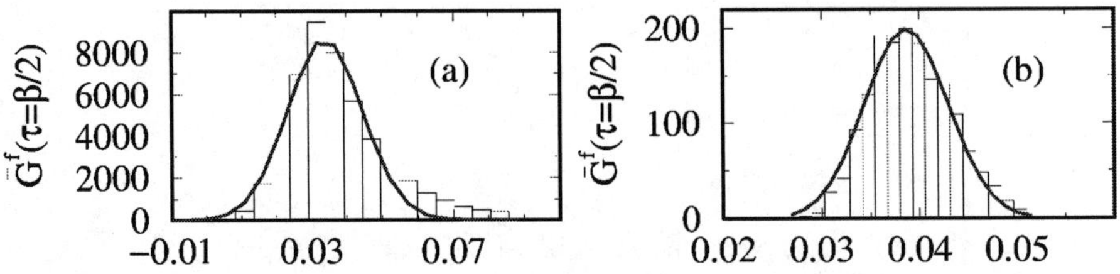

FIGURE 27. *Distribution of the data shown in Fig. 26 (a) and after rebinning (b). The solid line is a Gaussian fit. In (b) the data was processed by packing it sequentially into bins of 30 measurements each.*

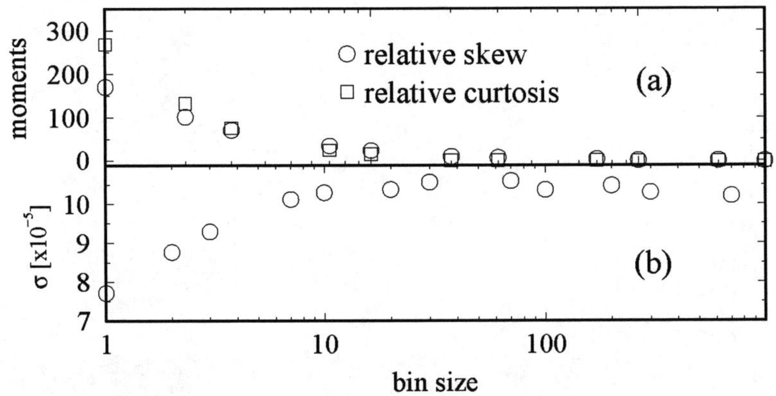

FIGURE 28. *Relative kurtosis and skew (a) and error bar (b) of the data shown in Fig. 26 as a function of bin size. Here the total amount of data is fixed, so increasing the bin size decreases N_{bins} proportionately. As the bin size increases to about 30, the relative kurtosis and skew decrease to roughly zero and the error bar saturates, indicating that the bins are uncorrelated samples and that the data has become Gaussianly distributed.*

The chosen bin size (here, 30 measurements) must be large enough so that the bin averages are uncorrelated, but small enough so that sufficient bins remain to calculate the likelihood function. To determine the smallest bin size that yields uncorrelated data we quantify the deviation of the distribution from a Gaussian by measuring moments of the distribution. Of course, because the data are a finite set, each of these measured moments has some standard deviation (proportional to $1/\sqrt{N_{bins}}$). Thus, one way to determine if the skewness and kurtosis of a distribution are acceptably small is to measure these values relative to what is expected from a Gaussian distribution. We will use such relative values.

As the bin size increases, the relative kurtosis and skewness decrease monotonically, indicating the convergence of the distribution to a Gaussian. This behavior is shown in Fig. 28a for the $G(\tau = \beta/2)$ data.

In addition, Fig. 28b shows that the error estimate also converges as the bin size increases. Here, the error estimate is given by

$$\sigma = \sqrt{\left(\langle \bar{G}^2 \rangle - \langle \bar{G} \rangle^2\right) / (N_{bins} - 1)} \tag{150}$$

where angular brackets indicate an average over the bins of data. Because correlations between successive Monte Carlo measurements always make this error estimate smaller than the actual value, this error estimate should initially increase monotonically with bin size, as shown. This behavior is easily understood by considering a perfectly correlated sample where the data in each bin is identical. Clearly, for this perfectly correlated sample, the error estimate would be zero. As the bins become uncorrelated, the error estimate increases. With independent data and a large number of equally sized bins, eventually $\sigma^2 \sim 1/N_{bins}$. However, with a fixed amount of data, as is typical with a QMC simulation, increasing the bin size decreases N_{bins} proportionally, and the error estimate can saturate as illustrated in Fig. 28b. Thus, the saturation of the error estimate indicates that the correlations between Monte Carlo measurements, i.e., between bin averages, have been removed. The point at which saturation occurs in a plot like Fig. 28b provides a useful first estimate of the minimum bin size required to remove correlations between the bins. In general, one should perform this test for the Green's function at all times τ_l; however, we have found it is often sufficient to perform this test at only a few times. For the remainder of this section, we will assume that the bin size is sufficiently large so that both the error estimate and the moments of the distribution have converged to values which indicate that the data are both statistically independent and Gaussian-distributed.

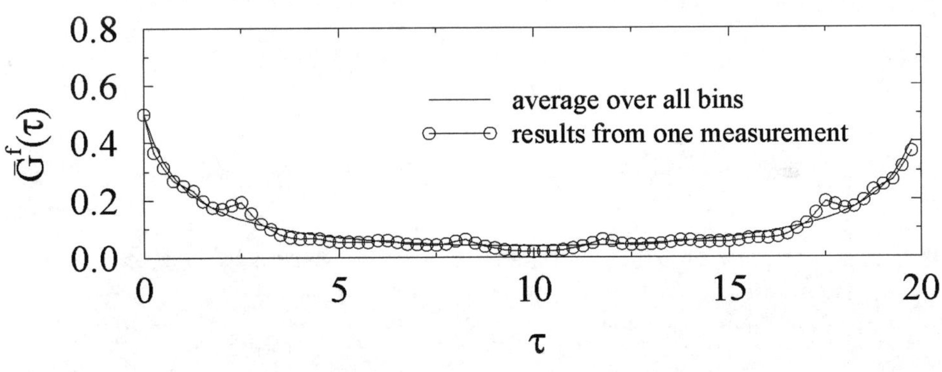

FIGURE 29. *$\bar{G}^f(\tau)$ from one measurement compared to $\bar{G}^f(\tau)$ obtained from the average over 800 bins of data, each containing 1520 measurements. If the result from a single measurement at a certain point differs from the essentially exact result obtained by averaging over many bins, then the results at adjacent points also deviate from the exact results.*

Now, only the errors in the Green's function $\bar{G}$ at adjacent time slices remain correlated. This correlation may be seen by comparing the results from a single measurement with those essentially exact values obtained from averaging over many measurements. Such a comparison is shown in Fig. 29 where if the result from a single measurement differs from the essentially exact result at a certain value of τ, then the results at adjacent values of τ also tend to deviate from the exact results in a similar way. These correlations of the error in Matsubara time are characterized by the covariance

$$C_{lk} = \frac{1}{N_{bins}(N_{bins}-1)} \sum_{j=1}^{N_{bins}} (\langle \bar{G}_l \rangle - \bar{G}_l^j)(\langle \bar{G}_k \rangle - \bar{G}_k^j) . \tag{151}$$

If C is diagonal, then according to the central limit theorem, the likelihood function is $P(\bar{G}|A) = \exp[-\chi^2/2]$ where

$$\chi^2 = \sum_{l=1}^{L} \left(\frac{\bar{G}_l - \sum_j K_{l,j} A_j}{\sigma_l} \right)^2 , \tag{152}$$

and σ_l^2 are the diagonal elements of C. However, in general, the covariance matrix C_{ij} is not diagonal because errors at different values of τ are correlated. To define a meaningful measure of how well A_i reproduces the data, we must find the transformation $\mathbf{U}$ which diagonalizes the covariance matrix

$$\mathbf{U}^{-1}\mathbf{C}\mathbf{U} = \sigma_i'^2 \delta_{ij} . \tag{153}$$

Both the data and kernel are now rotated into this diagonal representation

$$\mathbf{K}' = \mathbf{U}^{-1}\mathbf{K} \quad \bar{\mathbf{G}}' = \mathbf{U}^{-1}\bar{\mathbf{G}}, \tag{154}$$

and each measurement $\bar{G}_i'$ is statistically independent. Therefore, we can use

$$\chi^2 = \sum_{l} \left(\frac{\bar{G}_l' - \sum_j K_{l,j}' A_j}{\sigma_l'} \right)^2 \tag{155}$$

to measure the misfit between the spectrum and the data and to define the likelihood function.

Cautionary Notes. We find that proper preparation of the data, removing correlations, is the most critical step in the MEM procedure. If the data are uncorrelated and the covariance is calculated and diagonalized correctly, then the resulting spectra will be reliable (however, for weak data, it will show a significant bias towards the default model). If not, then the Gaussian form of the likelihood function is unjustified and the resulting spectra will generally have spurious features.

However, care must be taken when calculating and diagonalizing the covariance. First, Since the set of data is finite, it is necessary to balance the need of removing the correlations in imaginary-time with the need of removing the correlations between Monte Carlo steps. To remove the correlations in Monte Carlo steps the bin size must be large; however, to calculate the covariance accurately, many bins of data are required. If there are not enough bins of data, then the covariance and (as shown in Fig. 30) its eigenvalue spectrum can become pathological. The reason for this pathology is that when we diagonalize the covariance matrix, we are asking for L independent eigenvectors. We must have enough bins of data to determine these directions so that

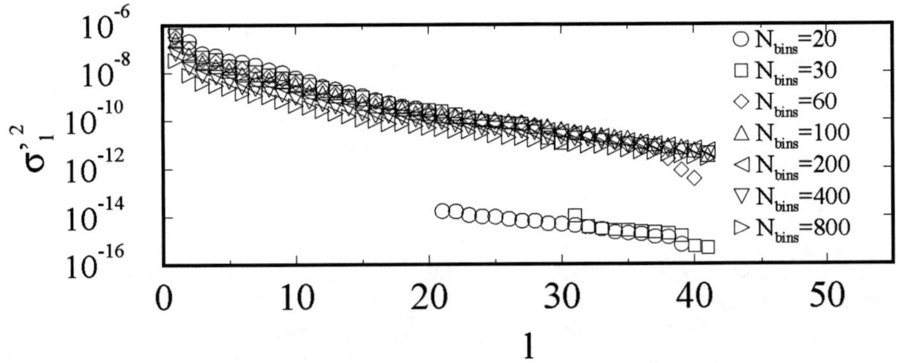

FIGURE 30. *Eigenvalue spectra of the covariance matrix of G^f for different numbers of bins of data. Each bin contains 100 measurements and $L = 41$. When $N_{bins} \sim 2L$, $\sigma_l'^2$ develops a sharp break.*

N_{bins} must be greater than or equal to L. In fact, since the information contained in a given bin of data is not completely independent from the other bins, we must have $N_{bins} > L$. Otherwise, as shown in Fig. 30, where $L = 41$, the eigenvalue spectrum displays a sharp break when $N_{bins} < L$, indicating that only a finite number of directions, less than L, are resolved. The small eigenvalues after the break are essentially numerical noise and yield artifacts in the spectra. Simply throwing away the small eigenvalues and their associated eigenvectors does not cure the difficulty since the small eigenvalues and eigenvectors contain the most precise information about the solution. Thus, the only reasonable thing to do is to increase the number of bins. Empirically, we find that we need

$$N_{bins} \geq 2L \tag{156}$$

in order to completely remove the pathology of the sharp break in the eigenvalues [42]. Second, as illustrated in Fig. 29 adjacent data in time tend to be highly correlated. These correlations grow as the time step used in the QMC calculation is reduced, making the rows and columns of the covariance more correlated. Eventually, the covariance becomes ill conditioned and cannot be diagonalized. Such severe oversampling of the Green's function data does not provide more information, but a small time step may be useful for other reasons (such as reducing Trotter errors). In this case we can fix the problem by eliminating some fraction of the data (i.e. taking the data from every other time step).

4.2.5. Data produced by QMC simulations with sign problem

Sign problem in QMC simulations. The nature of the sign problem in QMC calculations was discussed in 3.4.1. When the average sign is small the Monte Carlo calculations become impractical. Even a moderate sign problem, for which the computation of the static quantities described in Eq. 129 is still possible, introduces additional problems in the calculation of the dynamical spectra including especially strong correlations of the data resulting in non-gaussian distributions, such as those shown in Fig. 31 c) and d). However, these cannot be removed by standard techniques due to the strong correlation between the data and the averaged sign of the configurations which produce these data. This makes it essentially impossible to calculate spectra long before the minus sign problem makes the calculation of the static properties impractical. In this section we address this problem and describe a solution which greatly increases the resolution of MEM when calculating spectra from such poorly conditioned data.

Nevertheless, we should mention that we do not attempt to improve the sign problem in general. We only introduce a proper way to compute the dynamical spectra when the average sign is small, but still large enough to allow the computation of the static quantities.

Analysis of QMC data with sign problem . The difficulty in obtaining good data for G^i can easily be understood from the measurement process when the sign is not always positive. Then, to obtain the Green's function, we no longer measure G^i but rather the product of it and the sign s of the configuration, $Gs^i \equiv G^i * s$, and the sign s. At the end of the simulation, i.e. after a large number of measurements, we then obtain

$$\langle G_l \rangle = \frac{\langle Gs_l \rangle}{\langle s \rangle} = \frac{\sum_{i=1}^{N_d} Gs_l^i}{\sum_{i=1}^{N_d} s^i}. \tag{157}$$

However, besides the means (i.e. the average $\langle G_l \rangle$), in order to employ MEM we need to generate uncorrelated and independent $\bar{G}_l^{(i)}$ points normally distributed around the mean (see Eq. 151). Generating these points is problematic if the sign problem is present, as we discuss below.

First, in order to obtain good estimates of $\bar{G}_l^i$ (i.e. to have a good statistics), we need to average a large number of Gs_l^i and s^i

$$\bar{G}_l^i = \frac{\bar{Gs}_l^i}{\bar{s^i}} = \frac{\sum_{i=1}^{Bs} Gs_l^i}{\sum_{i=1}^{Bs} s^i}. \tag{158}$$

Here $Bs = \frac{N_d}{N_{bins}}$ is the bin size or the number of points in the bin i, thus $\bar{Gs}_l^i$ and $\bar{s^i}$ are averages over the measurements that form the bin i. This large number of measurements for every $\bar{G}_l^i$ requires large computational effort.

Second, even if the bin size is chosen large enough to produce good estimates of the $\bar{G}_l^i$, the distribution of these will not be Gaussian. This can be easily understood if one notices that the $\bar{G}_l^i$ distribution is a ratio of two normal distributions, $\bar{Gs}_l^i$ and s^i respectively. However, the ratio distribution of two normal distributions is not a normal distribution. Therefore in order to obtain a normal distribution for $\bar{G}_l^i$ data one has, in the virtue of the central limit theorem, to further rebin the points $\bar{G}_l^i$. Of course, this rebinning will result in an additional computational effort.

Third, as within the same bin i there is a strong correlation between different data points $\bar{Gs}_l^i$ corresponding to adjacent time slices, there is also a strong correlation

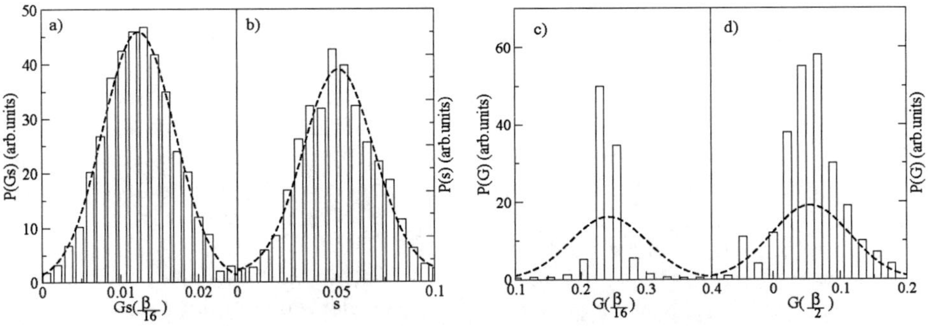

FIGURE 31. a) and b) histograms representing the distribution of $\bar{Gs}(\tau = \beta/16)$ and respectively sign $\bar{s}$ for 3000 measurements per bin. c) histogram representing the distribution of $\bar{G}(\tau = \beta/16)$ when a $\bar{G}$ point is obtained as a ratio of 3000 averaged $\bar{Gs}$ and 3000 averaged $\bar{s}$ points. d) histogram representing the distribution of $\bar{G}(\beta/2)$ when every $\bar{G}$ point is obtained by rebinning 30 initial G points obtained as a ratio of 600 averaged $\bar{Gs}$ and $\bar{s}$ points. The dashed lines represent the best Gaussian fit to the data.

between data $\bar{Gs}_l^i$ and $\bar{s}^i$. The $\bar{G}_l^i$ are a result of a nonlinear operation between two correlated quantities. Thus the average of $\bar{G}_l^i$

$$\overline{G_l} = \frac{1}{N_{bins}} \sum_{i=1}^{N_{bins}} \bar{G}_l^i \tag{159}$$

will be different from $\langle G_l \rangle$ obtained in Eq. 157. Therefore, one ends with a set of G_l data normally distributed around $\overline{G_l}$ which is not a good estimation for G_l if the correlations are strong.

In order to exemplify the problems discussed above, we employed a QMC based DCA algorithm to produce a very large amount of data for the single-particle Green's function of the two-dimensional Hubbard model (Eq. 1) on a square lattice. We choose $t = 0.25$ so that the bandwidth $W = 2$ and set $U = W$. To make the sign-problem worse, we add a next-nearest neighbor hopping $t' = -0.3t$ to frustrate the lattice. We perform calculations on a 16-site 4×4 cluster at 15% doping, down to temperatures $T = 0.125t$ where we experience a severe sign-problem, $\bar{s} = 0.051$. We calculated 8000 data points (bins) $(\bar{Gs}_l^i, \bar{s}^i)$, and for every data point we averaged 600 QMC measurements.

In Fig. 31 a) and b) we show histograms of the $\bar{Gs}(\tau = \beta/16)$ and respectively $\bar{s}$ when the bin size is increased five times, which corresponds to an average of 3000 measurements per bin. It can be seen that the data are normally distributed to a very good approximation. This is unlike the situation shown in Fig. 31 c) where the $\bar{G}(\tau = \beta/16)$ distribution, obtained as a ratio distribution of the previous $\bar{Gs}(\tau = \beta/16)$ and $\bar{s}$, is shown. The $\bar{G}(\tau = \beta/16)$ data are strongly peaked, being characterized by a large positive kurtosis. Similar distributions of data are observed (not shown) for the other values of the imaginary time. The worst situation is observed at imaginary times close to $\beta/2$ which affects mainly the low-energy properties of the spectra. In Fig. 31 d) we show the distribution of $\bar{G}(\tau = \beta/2)$ when every point is obtained after rebinning 30 initial $\bar{G}(\tau = \beta/2)$ points, with each initial point being obtained as a ratio of 600 averaged $\overline{Gs}$ and $\bar{s}$ points. Even if now a large number of measurements (18,000) is used for producing one data point, the distribution is still not Gaussian. However, this is mainly because we used a small amount of measurements (only 600) for obtaining the initial $\bar{G}$ points (i.e. the ones before rebinning). We found that much better data can be obtained if initially one uses a large number of measurements to calculate the $\bar{G}$ points and afterwards a rebinning of a few successive $\bar{G}$ points is employed. Nevertheless the number of measurements necessary to obtain normal distributed $\bar{G}$ points is much larger than the one required to obtain normal distributed $\overline{Gs}$ and $\bar{s}$ points. In our case this number is about five times larger but this value is dependent on the problem considered, being determined by both the magnitude of the correlations and the value and the distribution of the sign.

We propose the following modified approach to MEM for QMC data with sign problem.

i) Rebin $(\overline{Gs}_l^i, \bar{s})$ until they become normal distributed. As it was shown in Fig. 31 a) and b) this can be achieved with a relatively small amount of data. *ii)* Employ a rotation in the space (Gs_l, s) in order to obtain the statistically independent vectors. *iii)* Calculate χ^2 in the space spanned by (Gs_l, s). The calculation of χ^2 in a space which now includes the extra sign dimension will be discussed below in Sec. 4.2.5.

Modified likelihood function. Denoting $h \equiv (Gs_l, s)$, the *modified* likelihood function is defined as $P[h|A]$, since the measured quantities in the QMC process are the h points (and not G). As we showed in the previous section, for acceptable values of the bin size, the data h are to a good approximation Gaussianly distributed. Therefore, the *modified* likelihood function also will have the usual form as $e^{-\chi^2/2}$, with

$$\chi^2 = \sum_{l,k=1}^{L+1} (\bar{h}_l - h_l(A))[C_h^{-1}]_{lk}(\bar{h}_k - h_k(A)) \tag{160}$$

The new covariance matrix has the dimension $(L+1) \times (L+1)$,

$$C_{hlk} = \frac{1}{N_{bins}(N_{bins}-1)} \sum_{i=1}^{N_{bins}} (\langle h \rangle_l - \bar{h}_l^i)(\langle h \rangle_k - \bar{h}_k^i) . \tag{161}$$

The only problem which remains to be solved is finding an equation for $h(A)$, since Eq. 134 only provides a relation for $G(A)$. In order to achieve this we do the following: First we absorb the sign into the spectrum, i.e. we define $\mathscr{A}$ as

$$\mathscr{A}(\omega) = sA(\omega) . \tag{162}$$

Instead of searching for a spectrum A which satisfies Eq. 134 we search for $\mathscr{A}$ which satisfies

$$Gs(\tau) = \int K(\tau, \omega)\mathscr{A}(\omega)d\omega . \tag{163}$$

Second, we consider the spectrum normalization sum-rule

$$B = \int A(\omega)d\omega, \tag{164}$$

which implies

$$s = \int \frac{1}{B} \mathscr{A}(\omega) d\omega \,. \tag{165}$$

Here B is a constant, equal to one for the the one-particle spectra and equal to the static susceptibility $\chi(T)$ for the two-particle case. We relate the sign fluctuations to the norm of the new spectrum because of the sign s was absorbed into the definition of $\mathscr{A}$.
Both Eq. 163 and Eq. 165 can be written as

$$\langle h_l \rangle = \sum_{\nu=1}^{N_f} K_{hl\nu} \mathscr{A}_\nu, \quad K_{hl\nu} = \begin{cases} K_{i\nu} & l \le L \\ \frac{1}{B} & l = L+1 \end{cases} \,. \tag{166}$$

This is the basic equation which relates h to $\mathscr{A}$ and determines the likelihood function $P[h|A] \equiv P[h|\mathscr{A}]$. MEM will produce the most probable spectrum $\mathscr{A}$ normalized to $\langle s \rangle$ which minimizes the χ^2 function in Eq. (160) subject to the entropy constraint.

We note that for the one-particle case, where $B = 1$, Eq. 165 is equivalent to

$$Gs(0) + Gs(\beta) = s \,. \tag{167}$$

By using Eq. 165 in the calculation of the likelihood function we impose

$$G(0) + G(\beta) = \frac{Gs(0)}{s} + \frac{Gs(\beta)}{s} = 1 \tag{168}$$

at every measurement. Since Eq. 168 results solely from the anti-commutation relation of the one-particle operators it should be satisfied in every possible configuration and implicitly in every measurement. Therefore, this way of implementing the normalization sum-rule is more natural than the usual way based on Lagrange multipliers where the constraint is globally imposed, i.e. not at every measurement but only for the final Green's function obtained at the end of the QMC process.

For the two-particle case, where $B = \chi(T)$, the sum-rule Eq. 164 is not an independent equation as in the one-particle case, but merely an integration over τ of Eq. 137. Therefore it is essential to treat B as a constant (equal to the final, averaged over all QMC configurations, $\langle \chi(T) \rangle$) and to disregard measurement dependent fluctuations in $\chi(T)$. This way we relate the norm of $\mathscr{A}$ only to the fluctuation of the sign s.

4.2.6. Details of the MEM Formalism

We will now construct the formalism to locate the most likely spectrum $\hat{A}$ and set the value of α. The first step is to normalize the likelihood function $P(\bar{G}|A)$ and the prior $P(A|\alpha,m)$. Here it will be necessary to integrate over the space of all spectra A_i. This is done with Gaussian approximations to the integrals. Following Skilling and Bryan [43], we employ a measure $d^N A / \prod_i \sqrt{A_i}$ which amounts to a change of variables to a space where S has no curvature [16].

For example, the normalized prior probability is

$$P(A|\alpha,m) = \frac{1}{Z_S} \exp\left\{\alpha\left(-\sum A_i \ln A_i/m_i - A_i + m_i\right)\right\} \,, \tag{169}$$

where

$$Z_S = \int \frac{d^N A}{\Pi_i \sqrt{A_i}} \exp\left\{\alpha\left(-\sum A_i \ln A_i/m_i - A_i + m_i\right)\right\}. \tag{170}$$

The integrand is maximized when $S = 0$, i.e. when $A = m$. We approximate the integral by expanding the argument of the exponent to second order around this maximum, $S \approx \frac{1}{2}\delta A^T \, \nabla\nabla S|_{A=m} \, \delta A = -\frac{1}{2}\delta A^T \,\{1/m\}\, \delta A$, where $\{1/m\}$ is the diagonal matrix with finite elements composed of $1/m_i$, and δA is the vector $A - m$.

$$Z_S \approx \int \frac{d^N A}{\Pi_i \sqrt{A_i}} \exp\left\{\alpha\left(-\frac{1}{2}\delta A^T \,\{1/m\}\, \delta A\right)\right\}. \tag{171}$$

We define a change of variables, so that $dy_i = dA_i/\sqrt{A_i}$ and find

$$Z_S \approx \int d^N y \exp\left\{\alpha\left(-\frac{1}{2}\delta y^T \{m\}^{1/2} \{1/m\} \{m\}^{1/2} \delta y\right)\right\} = (2\pi/\alpha)^{N/2} \tag{172}$$

The likelihood function must also be normalized

$$P(\bar{G}|A) = e^{-\chi^2/2}/Z_L\,, \tag{173}$$

where

$$\chi^2 = \sum_l \frac{\left(\bar{G}'_l - \Sigma_i K'_{li} A_i\right)^2}{\sigma'^2_l}\,. \tag{174}$$

where K' and $\bar{G}'$ are the kernel and data rotated into the data space where the covariance is diagonal, and σ'^2_l are the eigenvalues of the covariance. If we let $G_l = \Sigma_i K'_{li} A_i$, then

$$Z_L = \int d^L G \exp\left\{\frac{1}{2}\sum_{l=1}^{L} \frac{\left(\bar{G}'_l - G_l\right)^2}{\sigma'^2_l}\right\} = (2\pi)^{L/2} \prod_l \sigma'_l\,. \tag{175}$$

Using Bayes theorem, we find

$$\begin{aligned} P(A,G|m,\alpha) &= P(G|A,m,\alpha)P(A|m,\alpha) \\ &= P(A|G,m,\alpha)P(G|m,\alpha) \end{aligned} \tag{176}$$

or

$$P(A|G,m,\alpha) \propto P(G|A,m,\alpha)P(A|m,\alpha) = \frac{\exp(\alpha S - \chi^2/2)}{Z_S Z_L}\,. \tag{177}$$

Since the normalization factors Z_S and Z_L are independent of the spectrum, for fixed α and data, the most probable spectrum $\hat{A}(\alpha)$ is the one which maximizes $Q = \alpha S - \chi^2/2$. An algorithm to find this spectrum is discussed in Sec. 4.3. However, the question of how to select α and the default model remains.

Selection of α. The selection of α strongly effects the choice of the optimal spectrum [44] since α controls the competition between S and χ^2. If α is large, then the entropy term is emphasized and the data cannot move the spectrum far from the model. If α is small, then the least square misfit between the spectrum and the data is minimized so that $\chi^2 \ll L$. The numerical error in the QMC data then begins to dominate the solution and the spectra displays random oscillations and noise. Thus, it is important to find a sensible way of selecting α. Typically, α is selected in one of three ways described below.

Historic MEM. In the historic method [45, 43], α is adjusted so that $\chi^2 = L$. The justification for this is that if the spectrum is known and the data was repeatedly measured, then the misfit between the data and the spectrum $\chi^2 = L$ on average. However, the data are only measured once and the spectrum is not known *a priori*. Also, setting $\chi^2 = L$ tends to under fit the data since good data can cause structure in the spectrum which reduces χ^2 from L. Thus, there is little reason to believe that α can be chosen without input from the data itself.

Classic MEM. A more appropriate method [44] of setting α is to choose the most probable value, defined by maximizing

$$P(\alpha|\bar{G},m) = \int \frac{d^N A}{\prod_i \sqrt{A_i}} P(A,\alpha|\bar{G},m). \tag{178}$$

The integrand

$$P(A,\alpha|\bar{G},m) = P(A|\bar{G},m,\alpha)P(\alpha) \propto \frac{\exp(\alpha S - \chi^2/2)}{Z_S Z_L} P(\alpha) \tag{179}$$

involves the prior probability of α. Jeffreys [46] argues that since χ^2 and S have different units, α is a scale factor. He asserts that in lieu of prior knowledge, it should have the simplest scale invariant form $P(\alpha) = 1/\alpha$. Thus,

$$P(\alpha|\bar{G},m) = \int \frac{d^N A}{\prod_i \sqrt{A_i}} \frac{\exp(\alpha S - \chi^2/2)}{Z_S Z_L \alpha} = \frac{Z_Q}{Z_S Z_L \alpha}. \tag{180}$$

Z_Q is calculated in a similar fashion to Z_S. We expand about the maximum of Q at $A = \hat{A}$ so that $\exp\{\alpha S - \chi^2/2\} \approx \exp\{Q(\hat{A}) + \frac{1}{2}\delta A^T \, \nabla\nabla Q|_{\hat{A}} \, \delta A\} = \exp\{Q(\hat{A}) + \frac{1}{2}\delta A^T \{\frac{1}{2}\nabla\nabla\chi^2|_{\hat{A}} - \{\alpha/\hat{A}\}\}\delta A\}$. We again make a Gaussian approximation to the integral, and if λ_i are the eigenvalues of $\frac{1}{2}\{A^{1/2}\}\,\nabla\nabla\chi^2|_{\hat{A}}\,\{A^{1/2}\}$, then

$$P(\alpha|\bar{G},m) = \frac{1}{W_\alpha} \prod_i \left(\frac{\alpha}{\alpha+\lambda_i}\right)^{1/2} \frac{e^{Q(\hat{A})}}{\alpha}, \tag{181}$$

where

$$W_\alpha = \int \frac{d\alpha}{\alpha} \prod_i \left(\frac{\alpha}{\alpha+\lambda}\right)^{1/2} e^{Q(\hat{A})}. \tag{182}$$

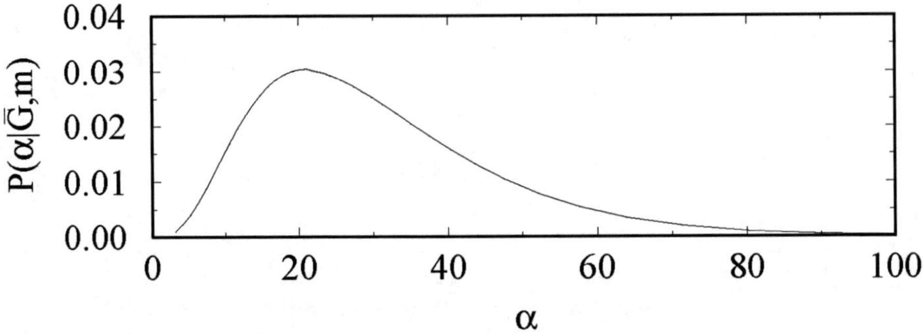

FIGURE 32. The posterior probability $P(\alpha|\bar{G},m)$ as a function of α for the periodic Anderson model data presented in Fig. 1. Since $P(G|I)$ is unknown, the magnitude of the ordinate is also unknown. The distribution is wide, so many reasonable values of α exist. The distribution is also skewed, so the value of α at the peak is not representative of the mean.

The optimal α, $\hat{\alpha}$ may be determined by the condition

$$\frac{\partial P(\alpha|\bar{G},m)}{\partial \alpha} = 0. \tag{183}$$

For strong data, $P(\alpha|\bar{G},m)$ is dominated by the product and $\exp Q(\hat{A})$ so that

$$-2\hat{\alpha}S \approx \sum_i \frac{\lambda_i}{\hat{\alpha}+\lambda_i}. \tag{184}$$

Each λ_i which is much greater than $\hat{\alpha}$ contributes one to the sum and hence one to the number of good observations in the data. If the number $N_{good} = -2\hat{\alpha}S$ is large, then $P(\alpha|\bar{G},m)$ is very sharp the spectra corresponding to $\alpha = \hat{\alpha}$ is a good approximation of the spectra which has been properly averaged over $P(\alpha|\bar{G},m)$.

Bryan's Method. However, typically we find that $N_{good} \ll L$. Then $P(\alpha|\bar{G},m)$ is a broad and highly skewed distribution. For example, $P(\alpha|\bar{G},m)$ for the data shown in Fig. 26 is plotted in Fig. 32. The distribution is wide, so many reasonable values of α exist. The distribution is also skewed, so the value of α at the peak is not representative of the mean. To deal with this, Bryan [49] calculates the optimal spectrum $\hat{A}(\alpha)$ for each α. The solution is taken to be

$$\bar{A} = \int d\alpha \hat{A}(\alpha) P(\alpha|\bar{G},m). \tag{185}$$

These three MEM methods will produce essentially identical results if the data are uncorrelated and precise. However, when the data are less precise but still uncorrelated, the method suggested by Bryan, averaging $\hat{A}(\alpha)$ weighted by $P(\alpha|G,m)$, generally produces more acceptable results and converges to a good result faster than the classic method and much faster than the historic method as the data is improved. A further advantage of the averaging is that it allows an accurate relative assessment of the posterior probability ($\int_0^\infty d\alpha P(m|G,\alpha)$) of the default model. This information is invaluable in determining which default model yields the most likely A.

Cautionary Notes. Classic MEM and Bryan's method both rely on the accuracy of Eq. 181 for $P(\alpha|\bar{G},m)$, which is calculated with a Gaussian approximation. This is only accurate if Q is a sharply peaked function in the space of all images. This approximation clearly fails when $\alpha \to 0$. Here there is no regularization and infinitely many spectra will produce the same Q. In this case, the algorithm can reach a run-away condition where it tends toward small values of α, the approximation for $P(\alpha|\bar{G},m)$ fails causing the calculation to tend towards ever smaller values of α. This condition is easily identified in the calculation, and it can be cured by choosing a better default model (a Bryan or classic MEM calculation with a *perfect* default model will always tend toward a solution with large α), using the methods described below, or the annealing method described in the example Sec. 4.4.5.

4.2.7. Model Selection

Bayesian statistics may also be employed to select the default model. If we must choose between different models, or set parameters used to define a default model function, then we choose these models or parameters based upon the posterior probability of the model

$$P(m|\bar{G}) = \int d\alpha P(\alpha|m,\bar{G})P(m) . \tag{186}$$

We see no *a priori* reason to favor one default model over an another, so we typically set the prior probability of the model $P(m)$ =constant. Then the integrand in Eq. 186 is given by Eq. 181 so that

$$P(m|\bar{G}) \propto W_\alpha . \tag{187}$$

Since the prior probability of the model is unknown, $P(m|\bar{G})$ determines only the relative probability of two models, and by inference the relative probability of their corresponding spectra.

Cautionary Notes. It can be tempting to try very informative models, such as the uncorrelated spectrum with sharp distinct features. Such default models will often have high posterior probabilities $P(m|\bar{G})$ but should nevertheless be avoided unless one can be *certain* that the sharp features are real. For example, a model with a delta function peak, has a huge amount of information, whereas the information from the QMC data is quite finite an may not be able to correct a wrong delta-function feature in the model. In this respect, again, the annealing technique described later is ideal.

4.2.8. Error Propagation

To absolutely qualify the spectrum, we need to assign error bars to it. In the quadratic approximation, the probability of the spectral density is

$$P(A|\bar{G},m,\alpha) \propto e^{-\frac{1}{2}\delta A^T \cdot \nabla\nabla Q|_{\hat{A}} \cdot \delta A}, \tag{188}$$

thus the covariance of the spectral density is

$$\langle \delta A(\omega)\delta A(\omega')\rangle = -\left(\nabla\nabla Q|_{\hat{A}}\right)^{-1}. \tag{189}$$

It is not possible to assign error bars to each point in the spectral density since this matrix is generally not diagonal. Thus errors between different points are strongly correlated. Also, A_i represents the spectral probability within some region of finite width and hence lacks meaning at a specific value of ω. However, it is possible to assign error bars to integrated functions of the spectral density such as [47],

$$H = \int d\omega A(\omega)h(\omega), \tag{190}$$

where $h(\omega)$ is an arbitrary function of ω. The error of H may be associated with the covariance of the spectral density $\langle \delta A(\omega)\delta A(\omega')\rangle$

$$\left\langle (\delta H)^2 \right\rangle = \int\int d\omega d\omega' h(\omega)h(\omega')\langle \delta A(\omega)\delta A(\omega')\rangle. \tag{191}$$

The matrix $\nabla\nabla Q|_{\hat{A}}$ is readily available because it is used as the Hessian of the Newton search algorithm typically used to find the optimal spectral density.

Cautionary Notes. Care should be taken in the interpretation of the error bars, especially if a highly informative default model is used. Suppose for example the data is weak, but a default model in essentially exact agreement with the data is used, then as discussed above, a large α solution will be found corresponding to a Q with small curvature in the space of images, and hence very small error bars. This does not necessarily mean that the resulting spectra is accurate, just that the default model is one of many which is consistent with the weak data. Unless the information in a default model is known to be accurate (such as the spectra from a higher temperature, or one which becomes exact at high frequencies), such highly informative default models should generally be avoided.

4.2.9. Integration of MEM with DMFA/DCA

Cautionary Notes. A few problems can emerge when MEM is used to analytically continue DMFA and DCA data.

For example, in multiband models, such as the periodic Anderson model Eq. 2 it may be difficult to extract the real-frequency self energy. In such models where only

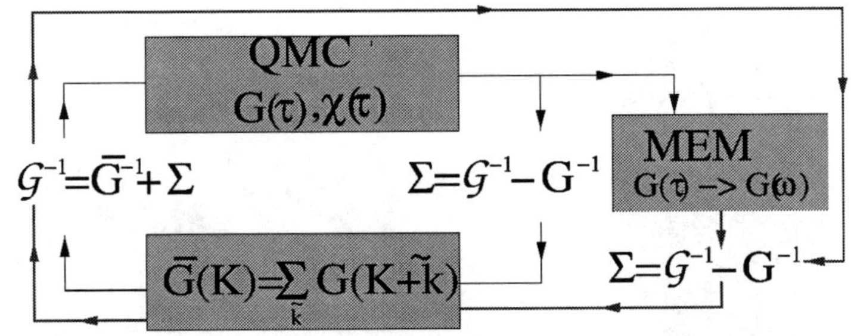

FIGURE 33. MEM may be integrated into the DMFA/DCA loop. After each QMC step, MEM is used to calculate the cluster Green's function, and the Dyson equation relating G and $\mathcal{G}$ is used to extract the self energy. This eliminate the need to invert the course-graining equation in order to solve for Σ.

a subset of bands are correlated, we typically integrate the uncorrelated band (in this example, the d-band) out of the action and only sample the Green's functions of the correlated bands (i.e. $G_f(\tau)$). In principle, the self energy may be extracted by inverting the coarse-graining equation. That is, for DMFA solution to the PAM

$$G_f(\omega) = \sum_{\mathbf{k}} G_f(\mathbf{k}, \omega) \tag{192}$$

where the DMFA cluster Green's function $G_f(\omega)$ is calculated with MEM. However, this can be very problematic since there are as many roots to this equation as there are bands, except at very high and low frequency, it can be difficult to disentangle the physical solutions from the unphysical ones. There are two ways to deal with this problem. First, we may directly sample the Green's functions for the uncorrelated bands, using the methods described in the chapter on DMFA and DCA. Second, we may build the real spectra calculation into the DMFA/DCA loop as illustrated in Fig. 33. Using MEM, we analytically continue the cluster DMFA or DCA Green's function. Of course this method requires that the real and Matsubara frequency self energies used to initialize the DMFA correspond (i.e. you may set $\Sigma = 0$ to initialize or use the self energies from a previous combined DMFA/DCA-MEM run).

Another problem, specific to self consistent methods like DCA and especially DMFA, is that the error estimate for the data described above only accounts for the error from the last DMFA/DCA iteration, but not the error in the host Green's function $\mathcal{G}$ coming from the previous iteration (Fig. 34). The simplest way to deal with this, is to make sure that the penultimate iteration is rather precise. Another way is to sample the DCA procedure. I.e. performing thirty or so statistically independent DMFA/DCA simulations with different random number seeds and then use jackknife or bootstrap [48] statistical methods to calculate the covariance.

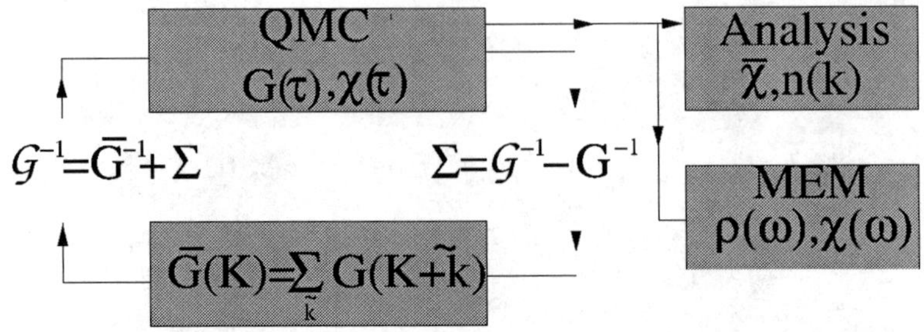

FIGURE 34. The DCA algorithm. QMC is used as a cluster solver. Once convergence is reached, $G = \bar{G}$, and the irreducible quantities are used in the Maximum Entropy Method (MEM) codes to calculate the spectra. However, conventional error estimating techniques, only account for the error in the last iteration.

4.3. Bryan's Method: a MEM algorithm

We will now sketch Bryan's numerical algorithm to find the optimal spectrum. For a more detailed description, we refer the reader to his paper [49]. We have found his algorithm to be very appropriate for the numerical analytic continuation problem for two reasons: First, due to the exponential nature of the kernel which relates A to the data $\bar{G}$, we typically have $L \gg N_{good}$. Thus, the problem is usually "oversampled." Bryan tailored his numerical algorithm [49] to this type of problem by working in a reduced space whose dimension is determined by singular-value-decomposition of the kernel K and is equal to the largest possible number of good singular values (i.e., numerically significant) which may parametrize the solution. The dimension of this space is usually much less than the number of A_i, and we found the computational advantage over methods that use the entire space determined by the number of A_i to be significant. Second, for the analytic continuation problem, the approximation of setting α equal to its optimal value is questionable because of the wide range of reasonably acceptable values of α. Bryan deals with this by calculating a result which is averaged over $P(\alpha|G,m)$.

4.3.1. *Typical Algorithms*

What distinguishes Bryan's numerical algorithm from its predecessors is the way in which the space of possible solutions is searched. Typical algorithms search for an optimal A by stepping through the entire space of A

$$A \to A + \delta A \tag{193}$$

with

$$\delta A = -(\nabla\nabla Q)^{-1}\nabla Q. \tag{194}$$

The Hessian $(\nabla\nabla Q)^{-1}$ is

$$(\nabla\nabla Q)^{-1} = (\alpha\nabla\nabla S - \nabla\nabla L)^{-1} = \left(\alpha\{A\}^{-1} - \nabla\nabla L\right)^{-1}, \tag{195}$$

where $\{A\}$ is a diagonal matrix with the elements of A along its diagonal. It may conceptually be expanded using the binomial theorem so that $(\nabla\nabla Q)^{-1}$ may be written as a power series in $\{A\}\nabla\nabla L$. Thus, δA may be written as a combination of $\{A\}\nabla Q = \{A\}(\alpha\nabla S - \nabla L)$, and powers of $\{A\}\nabla\nabla L$ acting on $\{A\}\nabla S$ and $\{A\}\nabla L$. Each of these vectors defines a direction in which the search can precede. Typically, between three and ten directions are used; however, these directions are often inappropriate for the problem at hand, because as mentioned earlier, the space of all possible solutions is too large for such oversampled data.

4.3.2. Singular-Space Algorithm

To alleviate this problem, Bryan performs a singular-value decomposition (SVD) of the kernel K, i.e., $K = V\Sigma U^T$ where U and V are orthogonal matrices and Σ is a diagonal matrix, and works in the resulting singular space. To see that this space still contains the solution, we consider

$$\nabla L = \frac{\partial F}{\partial A}\frac{\partial L}{\partial F} = K^T\frac{\partial L}{\partial F} \tag{196}$$

where $F = KA$. We see that ∇L lies in the vector space defined by the columns of K^T. We next perform a SVD on K and assume the diagonal elements of Σ are ordered from largest to smallest. The smallest elements are essentially zero (to the numerical precision of the computer) since the kernel is effectively singular. However, s of the elements are assumed finite. Now the vector space spanned by the columns of K^T is the same as the space spanned by the columns of U associated with the non-singular values. Bryan calls this reduced space the *singular space*. Thus, to the precision that can be represented on the computer, $\{A\}\nabla L$ and all of the search directions formed by acting with $\{A\}\nabla\nabla L$ lie in the singular space spanned by the columns of $\{A\}U_s$, where U_s is the singular space projection of U. The only direction not in this space is $\{A\}\nabla S$. Thus, Bryan's algorithm works in *at most* an $s+1$-dimensional subspace of the N-dimensional space of A.

In this singular space, the condition for an extremum of Q, $\nabla Q = 0$, is

$$\alpha\nabla S - \nabla L = 0 \rightarrow -\alpha\ln(A_i/m_i) = \sum_j K_{ji}\frac{\partial L}{\partial F_j}. \tag{197}$$

Thus, the solution may be represented in terms of a vector u

$$\ln(A/m) = K^T u. \tag{198}$$

Unless K is of full rank, so that $s = N$, the components of u will not be independent. However, since K^T and U share the same vector space and since most of the relevant search directions lie in the singular space, Bryan proposes that the solution be represented in terms of U and u as

$$A_i = m_i \exp\sum_n U_{in}u_n. \tag{199}$$

Thus, to the precision to which it may be represented on the computer and determined by SVD, the space u must contain the solution defined by $\nabla Q = 0$, and the search can be limited to this s-dimensional space.

Bryan's algorithm precedes by first reducing all the relevant matrices to the singular space. With the definitions $K = V\Sigma U^T$ and $\log(A/m) = Uu$, the condition for an extremum becomes

$$-\alpha U u = U\Sigma V^T \frac{\partial L}{\partial F}, \tag{200}$$

or

$$-\alpha u = \Sigma V^T \frac{\partial L}{\partial F} \equiv g, \tag{201}$$

where each of these matrices and vectors has been reduced to the singular space. (u is now a vector of order s, Σ is an $s \times s$ diagonal matrix etc.) Bryan then uses a standard Newton's search to find the solution in the singular space, starting from an arbitrary u. The increment at each iteration is given by

$$J\delta u = -\alpha u - g, \tag{202}$$

where $J = \alpha I + \partial g/\partial u$ is the Jacobian matrix, I the identity matrix, and

$$\frac{\partial g}{\partial u} = \Sigma V^T \frac{\partial^2 L}{\partial F^2} \frac{\partial F}{\partial A} \frac{\partial A}{\partial u}. \tag{203}$$

With the definition $W = \partial^2 L/\partial F^2$ (which is just the diagonal matrix with elements $1/\sigma'^2_l$), $M = \Sigma V^T W V \Sigma$, and $T = U^T A U$. M and T are symmetric $s \times s$ matrices, the Jacobian $J = \alpha I + MT$, and

$$(\alpha I + MT)\,\delta u = -\alpha u - g. \tag{204}$$

At each iteration δu must be restricted in size so that the algorithm remains stable. Thus, another parameter μ (a Marquart-Levenberg parameter) is added

$$[(\alpha + \mu)I + MT]\,\delta u = -\alpha u - g \tag{205}$$

and adjusted to keep the step length $\delta u^T T \delta u$ below some the limit

$$\delta u^T T \delta u \leq \sum_i m_i \tag{206}$$

so the search is within the range of validity of a local quadratic expansion of Q.

This search can be made more efficient if Eq. 205 is diagonalized, so that of order s operations are required for each α μ pair. First, we diagonalize T

$$TP = P\Gamma \tag{207}$$

where P is an orthogonal matrix and Γ is diagonal with finite elements γ_i. Then we define

$$B = \{\gamma^{1/2}\} P^T M P \{\gamma^{1/2}\} \tag{208}$$

and solve the second eigenvalue equation

$$BR = R\Lambda \tag{209}$$

where R is orthogonal and Λ the diagonal matrix with finite elements λ_i. Finally, to diagonalize Eq. 205 we define

$$Y = P\{\gamma^{-1/2}\}R. \tag{210}$$

Then $Y^{-T}Y^{-1} = T$, and $Y^{-1}MY^{-T} = \Lambda$, so that

$$Y^{-1}[(\alpha+\mu)I + MT]\,\delta u = [(\alpha+\mu)I + \Lambda]Y^{-1}\delta u = Y^{-1}[-\alpha u - g] \tag{211}$$

which yields s independent equations for $Y^{-1}\delta u$. Again, as these equations are iterated, μ must be adjusted to keep the step length

$$\delta u^T T\,\delta u = \left|Y^{-1}\delta u\right|^2 \leq \sum_i m_i. \tag{212}$$

4.3.3. Selection of α

The value α is adjusted so that the solution iterates to either a fixed value of χ^2 (for historic MEM) or to a maximum value of $P(\alpha|G,m)$ given by Eq. 181 (for classic MEM). Then, A is obtained from

$$A_i = m_i \exp\left(\sum_{n=1}^{s} U_{in} u_n\right). \tag{213}$$

Alternatively, Bryan suggests that one may start the algorithm with a large α for which $P(\alpha|\bar{G},m)$ is negligibly small, and then iterate to $\alpha \approx 0$ so that the averaged spectrum may be approximated

$$\langle A\rangle = \int_0^\infty d\alpha P(\alpha|G,m)\hat{A}(\alpha) \tag{214}$$

where $\hat{A}(\alpha)$ is the optimal spectrum (that for which $\nabla Q = 0$) for the value of α specified in the argument. This latter step may be necessary when $P(\alpha|G,m)$ is not a sharply peaked distribution. In fact this is usually the case, as may be seen in Fig. 32.

4.3.4. Error Propagation

As discussed in Sec. 4.2.8, it is possible to assign error bars to integrated functions of the spectrum $H = \int d\omega A(\omega)h(\omega)$

$$\left\langle(\delta H)^2\right\rangle = \int\int d\omega d\omega' h(\omega)h(\omega')\left\langle\delta A(\omega)\delta A(\omega')\right\rangle, \tag{215}$$

where

$$\left\langle\delta A(\omega)\delta A(\omega')\right\rangle = -\left(\nabla\nabla Q|_{\hat{A}}\right)^{-1}. \tag{216}$$

This is the inverse of the Hessian of the algorithm discussed above. $\nabla\nabla Q|_{\hat{A}}$ and is easily calculated in terms of singular-space quantities

$$-\nabla\nabla Q|_{\hat{A}} = \{1/A\}UY^{-T}\{\alpha I + \Lambda\}Y^{-1}U^{T}\{1/A\}. \tag{217}$$

Its inverse

$$-\left(\nabla\nabla Q|_{\hat{A}}\right)^{-1} = \{A\}UY\left\{\frac{1}{\alpha+\lambda}\right\}Y^{T}U^{T}\{A\} \tag{218}$$

may be used to calculate the error of H, $\sqrt{\overline{(\delta H)^2}}$ for any α. In principle, one should average the error over $P(\alpha|m,\bar{G})$; however, we find that it is generally adequate to calculate the error of the spectrum at the optimal $\hat{\alpha}$.

We close this section with several practical comments: On a workstation, finding the optimal spectrum by searching in the singular space requires only a few minutes of computer time. This efficiency is in sharp contrast with the amount of computer we needed [50] even on a "supercomputer" for standard Newton algorithms [43] or simulated annealing methods that use the full space of A. We found it essential to use 64 bit arithmetic to obtain stable results. Also, we use LINPACK's [51] singular-value decomposition routine to do the SVD and also to compute any eigenvalues and eigenvectors. The SVD routine in Numerical Recipes [27] and the EISPACK [52] eigenvalue-eigenvector routine RS are not as stable.

4.4. Case Study I

In this section, we will demonstrate that it is possible to extract spectral densities from the quantum Monte Carlo data that are essentially free from artifacts caused by over fitting to the data and have only small and controllable amounts of statistical error. We will use as an example the electronic spectral densities of the infinite-dimensional periodic Anderson model (PAM). We have already qualified the local Greens function data to remove correlations using the procedure discussed in Sec. 4.2.4, so we can begin to process the data to obtain the single-particle density of states spectral function.

For the majority of this section, we will consider particle-hole symmetric data $G(\tau) = G(\beta-\tau)$, and spectra $A(\omega) = A(-\omega)$. This prior information may imposed on the solution by constructing a symmetric kernel and default models. We will use three symmetric default models: two non-informative models — the flat model $m(\omega) =$ constant and a simple Gaussian

$$m(\omega) = \frac{1}{\Gamma\sqrt{\pi}}\exp[-(\omega/\Gamma)^2] \tag{219}$$

and also a third one obtained from second-order perturbation theory in U [53, 54]. The kernel for symmetric Fermionic Green's functions may be modified to reflect the symmetry and the associated integral restricted to positive frequencies

$$G(\tau) = \int_0^{\infty} d\omega A(\omega)\frac{e^{-\tau\omega}+e^{-(\tau-\beta)\omega}}{1+e^{-\beta\omega}}. \tag{220}$$

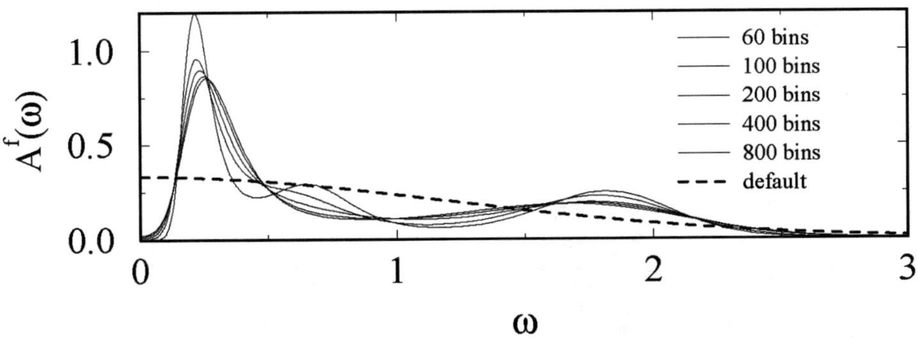

FIGURE 35. A sequence of spectral densities generated with increasingly accurate data. Every time the number of bins of data is doubled, the error is reduced by $1/\sqrt{2}$. A Gaussian default model, the dashed line, was used. N_{bins} increases beyond $2L = 82$, spurious structures are quickly suppressed.

4.4.1. Convergence of the Spectra

To minimize the effects of statistical error, the accuracy of the data needs to be increased until the spectral density has converged. This is demonstrated in Fig. 35, where the accuracy of the data are improved by increasing the number of bins of data. Here, a Gaussian default model is used whose width $\Gamma = 1.6$ (chosen by an optimization procedure to be discussed below). Each time the number of bins of data is doubled, the accuracy of the data increases by 41%. The spectral densities corresponding to smallest number of bins of data have spurious features associated with over fitting. These features are associated with difficulties in calculating the covariance matrix, as discussed in Sec. 4.2.4. As N_{bins} increases beyond $2L = 82$, the spurious structure is quickly suppressed. By the time 800 bins of data have been used, the spectral density appears to be converged to several line widths.

4.4.2. Default Model Selection

One may also test the dependence of the spectral density on the default model by changing its parameters or by using different models. The best model is the one with the largest posterior probability, calculated by assuming that the prior probability of the default model is flat, so that $P(A,\alpha,m|\bar{G}) \propto P(A,\alpha|\bar{G},m)$. Then $P(m|\bar{G})$ is obtained by integrating $P(A,\alpha,m|\bar{G})$ over A and α. The effects of varying the default model parameters are shown in Fig. 36a where the same data set is analytically continued with Gaussian default models whose widths satisfy $1.0 < \Gamma < 2.4$. The posterior probability $P(m|\bar{G})$ of these default models, shown in the inset, is peaked around $\Gamma \approx 1.6$. (We note that the normalization of $P(m|\bar{G})$ is unknown, since the prior probability of the default model and data are unknown). The resulting spectral densities are shown in Fig. 36b and are found to depend only weakly upon the default model. It is also possible to optimize the perturbation theory default model and hence to optimize the corresponding spectral

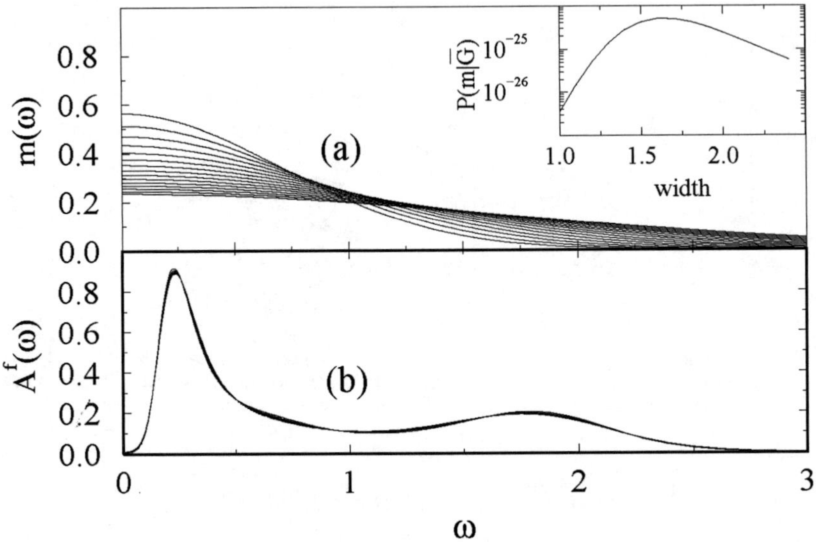

FIGURE 36. Dependence of the spectral density upon the default model. The width Γ of the Gaussian default model (a) is varied, producing a series of spectral densities (b). In the inset to (a) is the posterior probability of the default model $P(m|\bar{G})$, produced by integrating the joint probability $P(A,\alpha,m|\bar{G})$ over α and A, is plotted as a function of Γ. The normalization of $P(m|\bar{G})$ is unknown because it depends upon the probability of the data and the prior probability of the default model which are unknown.

densities. In the optimization of the default for the PAM spectra, the *df*-hybridization V may be treated as a variational parameter.

4.4.3. Error Propagation

In Fig. 37, we compare the optimal spectral densities obtained with the optimal perturbation theory, Gaussian, and flat default models. (The flat default model, with no adjustable parameters, is not optimized.) The posterior probabilities for each result indicate that the perturbation theory default model produces by far the most probable spectral density. However, we note that the qualitative features of the spectral density change little with the default model even though a large variety of default models were used. This independence is one signature of good data!

As a final test of the quality of the spectral density, one can evaluate its error in different intervals of frequency. In Fig. 37, we chose to assign error bars to the integrated spectral density ($h(\omega) = 1$) over different non-overlapping regions. The width of the region centered at each error bar is indicated by the horizontal spread of the error bar, the spectral weight within this region is indicated by the value of the data point, while the estimate of the uncertainty is indicated by the vertical spread. The perturbation theory default model yields the most precise spectra at all frequencies, consistent with the

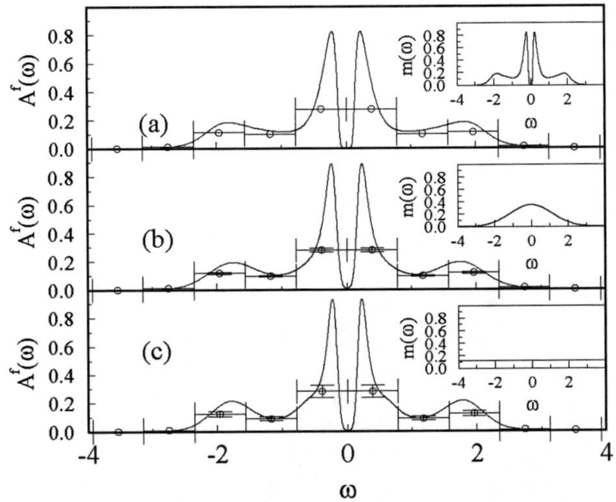

FIGURE 37. The f-electron density of states $A^f(\omega)$ generated using (a) a perturbation theory, (b) a Gaussian, and (c) a flat default model. These models are shown as insets to each graph. The data points indicate the integrated spectral weight within 10 non-overlapping regions of width indicated by the horizontal error bar. The vertical error bar indicates the uncertainty of the integrated weight within each region.

posterior probabilities of the models.

4.4.4. Two-Particle Spectra

There are special difficulties associated with the calculation of spectral densities associated with two-particle Green's functions. These difficulties include noisier and more correlated data and the lack of a good default model. The latter problem stems from the traditional difficulties of performing perturbation theory for two-particle properties.

As an example, we will analytically continue the local f-electron dynamic spin susceptibility $\chi''(\omega)$ of the symmetric PAM. As discussed in Sec. 4.2, the Monte Carlo data $\chi(\tau) = 2\langle S^-(\tau)S^+(0)\rangle$ is related to $\chi''(\omega)$ by

$$\chi(\tau) = \int_0^\infty d\omega \frac{\omega[e^{-\tau\omega} + e^{-(\beta-\tau)\omega}](\chi''(\omega)/\omega)}{1-e^{-\beta\omega}} . \tag{221}$$

To construct a model we will employ an alternative to perturbation theory, and construct a default model from different moments of the spectral function. They will be used as constraints to the principle of maximum entropy. The moments used to generate the default model are

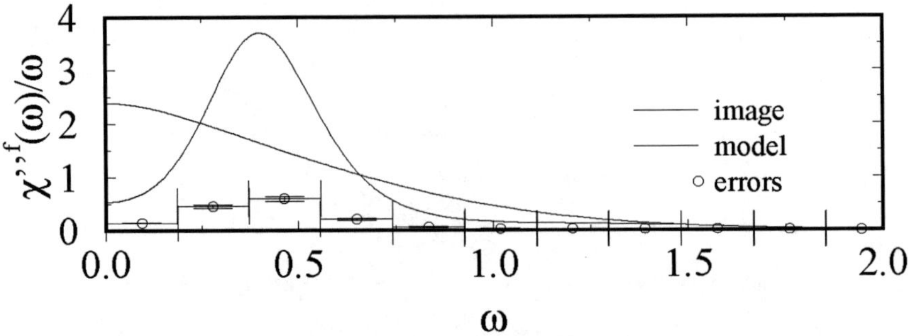

FIGURE 38. $\chi''(\omega)/\omega$ for $V = 0.6$, $U = 2$ and $\beta = 20$ for the PAM generated using a default model defined by two moments of the spectral density. The data points indicate the integrated spectral weight within 10 non-overlapping regions of width indicated by the horizontal error bar. The vertical error bar indicates the uncertainty of the integrated weight within each region.

$$\frac{1}{2}\chi(\omega=0) = \int_0^\infty d\omega(\chi''(\omega)/\omega) \tag{222}$$

$$\chi(\tau=0) = \int_0^\infty d\omega\,(\chi''(\omega)/\omega)\,\omega\coth(\beta\omega/2). \tag{223}$$

The (unnormalized) model is then generated by maximizing the entropy subject to these constraints imposed with Lagrange multipliers λ_0 and λ_1 and is easily found to be

$$m(\omega) = \exp[\lambda_0 + \lambda_1\omega\coth(\beta\omega/2)] \tag{224}$$

where λ_0 and λ_1 are determined by the constraint equations above.

Clearly this procedure may be generalized to utilize an arbitrary number of measured moments and often provides a better default model than perturbation theory. However, as shown in Fig. 38, the final spectral density can differ significantly from the default model when defined in this way. Nevertheless, the error bars indicate that the spectral density is trustworthy.

4.4.5. Annealing Method

Occasionally we have reason to calculate a series of spectra for a variety of temperatures (i.e., for the calculation of transport coefficients). If this set is sufficiently dense, then starting from a perturbation theory default at high temperature, we may use the resulting spectra as a default model for the next lower temperature. As far as we know, this procedure has no Bayesian justification; however, it has significant physical motivation. At sufficiently high temperatures, perturbation theory often becomes exact. Thus, this annealing procedure may be initialized with an essentially exact result. Furthermore, as the temperature is lowered, we expect the high frequency features of many

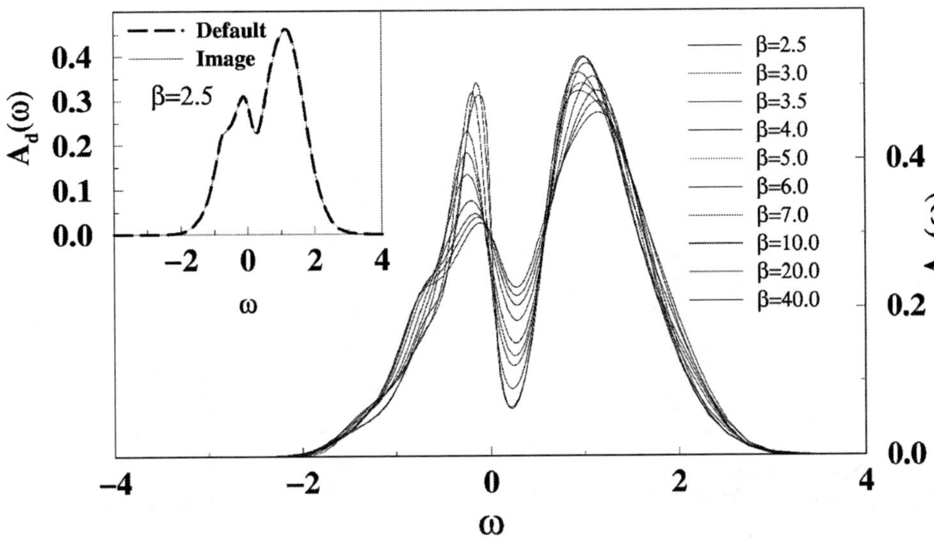

FIGURE 39. *The evolution of the d-electron density of states of the asymmetric PAM when* $U = 1.5$, $V = 0.6$ $n_d = 0.6$, *and* $n_f = 1.0$. *At high temperatures, as shown in the inset, the spectra is in essentially exact agreement with second-order perturbation theory. In addition, the d-electron states far from the Fermi surface are weakly renormalized by the strong electronic correlation on the f-orbitals. Thus, as the temperature is lowered, the low-frequency spectra change continuously, whereas the high frequency features change very little.*

spectra to freeze out (this is an essential assumption behind the numerical renormalization group method). Thus, the QMC is only required to supply information about the low-frequency features. Since QMC is a discrete sampling procedure in Matsubara time, according to Nyquist's theorem QMC only provides information below the Nyquist frequency $\omega_N = \pi/\Delta\tau$. Thus, the perturbation theory provides the high-frequency information, the QMC the low-frequency information, and MEM provides a natural method for combining these information sources.

For example, the evolution of the d-electron density of states of the asymmetric PAM is shown in Fig. 39. At high temperatures, as shown in the inset, the spectra is in essentially exact agreement with second-order perturbation theory. In addition, the d-electron states far from the Fermi surface are weakly renormalized by the strong electronic correlation on the f-orbitals. Thus, as the temperature is lowered, the low-frequency spectra change continuously, whereas the high frequency features change very little.

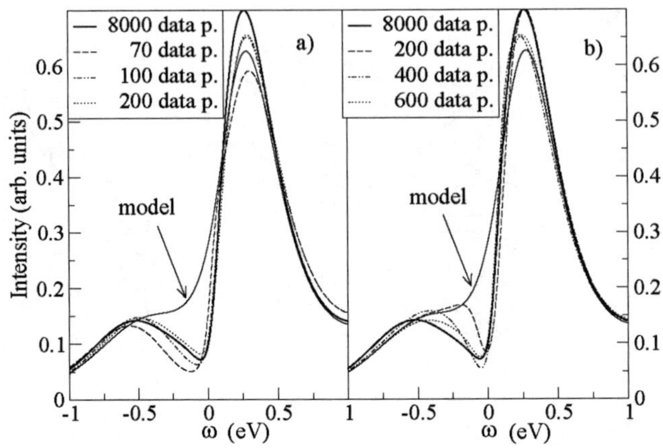

FIGURE 40. One-particle spectra at $K = (\pi, \pi/2)$ calculated with different amounts of data using a) the new method and b) the old method.

4.5. Case Study II

4.5.1. Spectra form data with sign problem

Here we present a comparison between the spectra obtained with the conventional (old) approach which does not consider the sign covariance, and the (new) method described in Sec. 4.2.5. With the old method, Eq. 158 is used to calculate the binned data when the sign is not always one, and Eq. 151 is used to calculate its covariance. In the new method, the sign is incorporated into the data and Eq. 161 is used to calculate the covariance. In this example, we calculate the the one-particle spectrum. At the highest temperature, the model $m(\omega)$ used in the entropy functional is chosen to be a Gaussian function. The annealing procedure is then used to obtain the spectrum at lower temperatures.

In Fig. 40 (a) and (b) we show the one-particle spectra of the Hubbard model at $K = (\pi, \pi/2)$ calculated for different amounts of data with the new and respectively with the old method. In both cases, when a large amount of data is used (8000 data points) the spectrum (thick continuous line) is converged. Moreover the two methods produce the same spectrum. However, notice that with the new method a reasonably good spectrum, i.e., a spectrum close to the converged one, can be obtained with an amount of data as small as 100 data points (see the double-dotted dashed line in Fig. 40 (a)). On the other hand, the old method requires at least 600 data points for a spectrum of comparable quality (see the dotted line in Fig. 40 (b)). Thus, for the case under investigation, we find that the new method reduces the computational cost of calculating the one-particle spectra about six times. For other problems characterized by stronger correlations the improvement is even more significant.

If the sign is not very small, some experience is needed to decide whether to use the old or the new method. Clearly, the old method will fail when the sign is exactly one since the covariance, Eq. 161, will be pathological since the sign has no error. As a rough rule of thumb, we tend to use the old method when the sign is greater than 0.8 and the new method when it falls below this value.

4.6. Summary

The Maximum Entropy Method is a precise and systematic way of analytically continuing Matsubara-time quantum Monte Carlo results to real frequencies. Due to the exponential nature of the kernel which relates the spectra and the data, there are many A with very different character which correspond to the same $\bar{G}$. With the MEM we employ Bayesian statistics to determine which of these is most probable. Bayesian inference is also used to assign error bars to integrals over the spectrum and optimize the default model.

The posterior probability of the spectrum is given by the product of the prior probability and the likelihood function. The entropic nature of the prior insures that the only correlated deviations from the default model which appear in the spectrum are those which are necessary to reproduce the data. The form of the likelihood function is determined by the central limit theorem, assuming that the data are statistically independent and Gaussianly distributed. Insuring these preconditions is the most critical step in the MEM procedure, and requires that the data be systematically rebinned and that the data and the kernel be rotated into the space in which the covariance of the data is diagonal.

Once the data has been properly characterized, we calculate the optimal spectrum using Bryan's algorithm which searches for a solution in the reduced singular space of the kernel. Bryan's method is more efficient than conventional techniques which search the entire spectral space. For any search algorithm three different techniques can be employed to set the Lagrange parameter α which determines the relative weight of the entropy and misfit: the historic, classic or Bryan's averaging technique. With precise uncorrelated data, each returns essentially the same spectrum, but with less-precise uncorrelated data, Bryan's technique yields the best results. Also, as the QMC data are systematically improved, images produced with Bryan's technique appear to converge more quickly than those produced by the other techniques.

While the systematic preparation of the data as described in Sec. 4.2.4 and the qualification of the spectrum described in later subsections is time-consuming, we believe that it is as important to quality of the final result, as is an accurate MEM code.

Together, the techniques discussed in this section provide a powerful, accurate, and systematic approach to the analytic continuation problem. In each case where we have employed these techniques we have been able to produce spectra that are precise at low frequencies, and free from spurious (unjustified) features at all ω.

4.6.1. Steps to ensure a robust MEM calculation.

In this section we have summarized the proper implementation of MEM and given a number of cautionary notes. As a parting discussion, these will be summarized here along with a few other common sense rules of thumb.

1. Rebin your data to remove correlations in QMC time.
2. Generate sufficient bins of data so that $N_{bins} \gtrsim 2L$ where L is the number of time slices used.
3. If DMFA/DCA is used to generate the data, be sure that the error in $\mathscr{G}$ from the previous iteration is negligible.
4. Calculate the covariance of the data making sure that: (1) the eigenvalue spectrum is continuous (if not, increase N_{bins}), and (2) that the covariance matrix is well enough conditioned to allow it to be diagonalized (if not, the data is oversampled in Matsubara time).
5. Diagonalize the covariance and rotate the data and kernel into the diagonal frame.
6. Choose a good default model, hopefully you can use the annealing technique. Always use a non-informative default model unless you are *certain* that the information in the model is exact.
7. When possible, use Bryan's MEM for marginalizing over α.
8. Systematically improve your data until the calculated spectrum converges.
9. When the annealing method is used, if the temperature step appears large (i.e. the spectrum changes abruptly) you may want to introduce data at additional intermediate temperatures.
10. If the annealing method is not used, try different non-informative default models. A reliable result is independent of the model. You may also want to use the model with the highest posterior probability (calculated when Bryan's method is used).
11. If the sign problem is present in the QMC calculations relate the fluctuations in sign to the norm of the spectra and treat the correlations between the sign and the data by including the sign in the covariance matrix as discussed in Sec. 4.2.5.

5. CONCLUSION

We have presented a pedagogical introduction to the dynamical mean field and dynamical cluster approximations, and described two quantum Monte Carlo methods as cluster solvers with the Maximum entropy method used to obtain real frequency spectra. The DMFA is a very powerful method for the study of correlated systems, while DCA incorporates non-local corrections. Both the DMFA and DCA map the lattice onto a self-consistently embedded impurity problem, so that conventional cluster solvers, like Hirsch-Fye or Continuous time QMC, may be used. Both techniques are straightforward, but are limited by the minus sign problem to the study of relatively small clusters at high temperatures. The maximum entropy method is used to affect the Wick rotation of the QMC data from Matsubara time or frequency to real frequency. Although MEM is a reliable and systematic method, great care must be taken in the application of MEM to ensure that the best possible spectra is found devoid of spurious features. Together, DMFA/DCA with QMC and MEM provide the state-of-the-art for the study of correlated model systems.

We would like to acknowledge useful conversations and fruitful collaborations with J. K. Freericks, B. Goodman, C. Groetsch, E. Gull, M. Hettler, H. R. Krishnamurthy, M. Ma, Th. Maier, R. Scalettar, R. N. Silver, D. Sivia, Th. Pruschke, A. N. Tahvildarzadeh. M. Troyer, and F. C. Zhang. This work was supported by the National Science Foundation grants DMR-0706379 and DMR-0312680, and Department of Energy grant DE-FG02-04ER46129. The work of JEG was supported of the U. S. Department of Energy through the LANL/LDRD program.

REFERENCES

1. J. Hubbard, Proc. R. Soc. a**276**, 238(1963); M. C. Gutzwiller, Phys. Rev. Lett. **10**, 159(1963); J. Kanamori, Prog. Theor. Phys. **30**, 257(1963).
2. Proceedings of the international conference on "Itinerant-Electron Magnetism". Physica B+C**91** (1977); for a review about the theory of the Hubbard model see also D. Vollhardt, Rev. Mod. Phys. 56, 99(1984).
3. E. H. Lieb and F. Y. Wu, Phys. Rev. Lett. **20**, 1445 (1968); H. Frahm and V. E. Korepin, Phys. Rev. B**42**, 10533(1990); N. Kawakami and S.-K. Yang, Phys. Rev. Lett. **65**, 2309(1990).
4. G. Czycholl, J. Mag. Mag. Mat. **47,48**, 350 (1985).
5. W. Metzner and D. Vollhardt, Phys. Rev. Lett. **62**, 324 (1989).
6. E. Müller-Hartmann, Z. Phys. B **74**, 507 (1989).
7. P. G. J. van Dongen and D. Vollhardt, Phys. Rev. Lett. **65**, 1663 (1990).
8. M. H. Hettler, A. N. Tahvildar-Zadeh, M. Jarrell, T. Pruschke, and H. R. Krishnamurthy, Phys. Rev. **B 58,** 7475 (1998).
9. M. H. Hettler, M. Mukherjee, M. Jarrell, and H. R. Krishnamurthy, Phys. Rev. B 61, 12739 (2000).
10. M. Jarrell Phys. Rev. Lett. **69**, 168(1992).
11. Y. Kuramoto, Springer Series in Sol. State Science, Vol. 62, 152–162 (1985); C. Kim, Y. Kuramoto and T. Kasuya, J. Phys. Soc. Japan **59**, 2414 (1990).
12. T. Pruschke, M. Jarrell and J .K. Freericks, Adv. in Phys. **42,** 187 (1995).
13. A. Georges, G. Kotliar, W. Krauth and M. J. Rozenberg, Rev. Mod. Phys. **68,** 13 (1996).
14. R. M. Fye and J. E. Hirsch, Phys. Rev. B **38**, 433 (1988).
15. M. Jarrell, Th. Maier, C. Huscroft, and S. Moukouri, Phys. Rev. B 64, 195130/1-23 (2001).
16. M. Jarrell, and J. E. Gubernatis, Physics Reports Vol. **269** #3, (1996).
17. K. Aryanpour, M. H. Hettler and M. Jarrell, Phys. Rev. B 65, 153102/1-4 (2002).
18. V. Zlatić and B. Horvatić, Sol. St. Comm. **75**, 263 (1990).
19. G. Baym and L. P. Kadanoff, Phys. Rev. **124**, 287, (1961).
20. P. Kent, M. Jarrell, T. Maier, and T. Pruschke, Phys. Rev. B 72, 060411 (2005).
21. D. D. Betts, H. Q. Lin, and J. S. Flynn, Can. J. Phys. 77, 353 (1999); D. D. Betts and G. E. Stewart, Can. J. Phys. 75, 47 (1997).
22. A. A. Abrikosov, L. P. Gorkov, I. E. Dzyalishinski, *Methods of Quantum Field Theory in Statistical Physics*, (Dover, New York, 1975).
23. J. E. Hirsch and R. M. Fye, Phys. Rev. Lett. **56**, 2521 (1986).
24. J. E. Hirsch, Phys. Rev. B, **28**, 4059 (1983), and **31**, 4403 (1985).
25. N. Metropolis, A. W. Rosenbluth, M. N. Rosenbluth. A. H. Teller and E. Teller, J. Chem. Phys. **21**, 1087-1092 (1953).
26. J. J. Deisz, D. W. Hess and J. W. Serene, *Recent Progress in Many-Body Theories*, edited by E. Schachinger *et al.* (Plenum, New York, 1995), Vol. 4.
27. W. H. Press, S. A. Teukolsky, W. T. Vetterling, and B. P. Flannery, *Numerical Recipes in Fortran 77*, Second Edition (1992), Cambridge University Press.
28. M. Jarrell, H. Akhlaghpour, and Thomas Pruschke, *Quantum Monte Carlo Methods in Condensed Matter Physics*, Ed. M. Suzuki, (World Scientific, 1993).
29. Transactions on Mathematical Software, Vol. **22**, No. 3, p. 362–371 (September, 1996).
30. A. N. Rubtsov, V. V. Savkin, A. I. Lichtenstein, Phys. Rev. B **72**, 035122 (2005)
31. F. F. Assaad and T. C. Lang, Phys. Rev. B **76**, 035116 (2007).

32. M. Wagner, Phys. Rev. B **44**, 6104 (1991); J. Rammer, H. Smith, Rev. Mod. Phys. **58**, 323 (1986).
33. H.-B. Schüttler and D. J. Scalapino, Phys. Rev. Lett. **55**, 1204 (1985); Phys. Rev. B **34**, 4744 (1986).
34. H. J. Vidberg and J. W. Serene, J. Low Temp. Phys. **29**, 179 (1977).
35. G. Wahba, SIAM Journal on Numerical Analysis **14**, 651 (1977).
36. S. R. White, D.J. Scalapino, R. L. Sugar, and N. E. Bickers, Phys. Rev. Lett. **63**, 1523 (1989).
37. M. Jarrell and O. Biham, Phys. Rev. Lett. **63**, 2504 (1989).
38. Th. Maier, M. Jarrell, Th. Pruschke, and M. Hettler, Rev. Mod. Phys. **77** 1027 (2005).
39. J. Skilling, in *Maximum Entropy and Bayesian Methods* edited by J. Skilling (Kluwer Academic, Dordrecht), p. 45 (1989).
40. S .F. Gull and J. Skilling, IEE Proceedings **131**, 646 (1984).
41. For a discussion of kurtosis and skewness, as well as a discussion of the probability that a distribution is Gaussian, see Ref [27] chap. 14.
42. It is tempting to disregard (i.e., set to 0) the off-diagonal elements of the covariance matrix as an alternative method of alleviating this pathology. Then, the eigenvalues will simply be the well-defined variance of $\bar{G}(\tau)$. However, this procedure neglects the correlations in the error which are clearly visible in Fig. 30 and yeilds an incorrect likelihood function. We have found that this procedure produces unpredictable results, especially when the data quality is marginal.
43. J. Skilling and R. K. Bryan, Mon. Not. R. astr. Soc. **211**, 111 (1984).
44. S. F. Gull, in *Maximum Entropy and Bayesian Methods* edited by J. Skilling (Kluwer, Dordrecht), p. 53 (1989).
45. S. F. Gull and G. J. Daniels, Nature, **272**, 686 (1978).
46. H. Jeffreys, *Theory of Probability*, (Oxford, Claredon Press, 1939); see also E. Jaynes, IEEE Trans. Sys. Sci. Cyb. Vol. SSC-4 (1993).
47. J. Skilling, in *Maximum Entropy and Bayesian Methods*, edited by J. Skilling, Kluwer Academic, Dordrecht, p. 455 (1989).
48. B. Efron, *The Jackknife, the Bootstrap and Other Resampling Plans* (SIAM, Philadelphia, 1982); B. Efron and G. Gong, The American Statistician, **37**, 36 (1983).
49. R .K. Bryan, Eur. Biophys. J. **18**, 165 (1990).
50. M. Jarrell, J .E. Gubernatis, and R. N. Silver. Phys. Rev. B, **44**, 5347-50 (Sept. 1991).
51. J. J. Dongarra, C. B. Moler, J. R. Bunch, and G.W. Stewart, *LINPACK User's Guide*,SIAM, Philadelphia, (1979).
52. B. T. Smith, J. M. Boyle, Y. Ikebe, V. C. Klema, and C. B. Moler, *Matrix Eigensystems Routines — EISPACK Guide* , Springer-Verlag, New York (1976).
53. M. Jarrell, H. Akhlaghpour, and Th. Pruschke, Phys. Rev. Lett. **70**, 1670 (1993); Phys. Rev. B. **51**, 7429-40 (15, March 1995).
54. B. Horvatic, D. Sokcevic, and V. Zlatic, Phys. Rev. B **36**, 675 (1987).
55. S. R. White, D.J. Scalapino, R. L. Sugar, E. Y. Loh, J. E. Gubernatis, and R. T. Scalettar, Phys. Rev. B. **40** 506 (1989); E. Dagotto, Rev. Mod. Phys. **66**, 763 (1994) .

Quantum phase transitions

Hilbert v. Löhneysen* and Peter Wölfle†

*Physikalisches Institut, Universität Karlsruhe (TH), 76128 Karlsruhe, Germany, and
Forschungszentrum Karlsruhe, Institut für Festkörperphysik, 76021 Karlsruhe, Germany
†Institut für Theorie der Kondensierten Materie, Universität Karlsruhe (TH), 76128 Karlsruhe, Germany

Abstract. In these lecture notes, recent theoretical and experimental work on quantum phase transitions, in particular, in metals with strong electronic interactions will be reviewed. After an introduction to classical and quantum phase transitions, we will briefly discuss the standard model of metals, i.e., the Landau Fermi-liquid theory, and routes to non-Fermi-liquid behavior. We will then focus on two different types of quantum phase transitions, magnetic quantum phase transitions in heavy-fermion systems, and metal-insulator transitions in heavily doped semiconductors.

Keywords: Quantum phase transitions, Fermi liquids, non-Fermi-liquid behavior, heavy-fermion compounds, Kondo effect, metal-insulator transitions
PACS: 71.10.Ay, 71.10.Hf, 71.27.+a, 71.30.+h, 72.20.-i, 73.43.Nq, 75.30.Kz, 75.30.Mb

1. INTRODUCTION

Phase transitions are ubiquitous in Nature. The transitions between ice, liquid water and water vapor and the associated enthalpy changes are important in regulating the Earth's climate both globally and regionally. Within the condensed phases (liquids and solids) different phases are possible. In particular, structural transitions occur in many solids as a function of an external parameter such as temperature, or pressure. Of particular interest are transitions that are driven by or directly affect the electronic degrees of freedom, e.g., transitions from a paramagnetic state to a magnetically ordered state, or from a metallic state to an insulating state.

Since some of the most interesting (and least understood) quantum phase transitions occur in metals with strong electronic correlations, we will review the salient features of Landau's Fermi-liquid theory, i.e., the quasiparticle concept and the Fermi-liquid properties in Chapter 2. Of particular importance in the field of strongly correlated materials is the Kondo effect, arising from the coupling between localized magnetic moments (generally arising from impurities) and conduction electrons. The strong-coupling groundstate of such a system can be described - under certain conditions - as a local Fermi liquid. If one has a stoichiometric compound where the "impurities" occupy regular lattice sites, the mass of the quasiparticles may reach several hundreds of the free-electron mass. In the simplest view this "heavy-fermion" behavior arises from a lattice-coherent superposition of local Fermi liquids. Since the mid-80's of the last century, a number of systems have been found which defy a simple Fermi-liquid description. Here the cuprate high-temperature superconductors stand out with an electrical resistivity linear in temperature T over a large temperature range. It became soon clear that non-Fermi-liquid (NFL) behavior, observed since then in many systems,

CP1014, *Lectures on the Physics of Strongly Correlated Systems XII: Twelfth Training Course*, edited by A. Avella and F. Mancini

may arise from distinctly different microscopic origins which will be briefly discussed. An experimental compilation of NFL data was given by Stewart [1].

One of the possible origins is the proximity of the system to a quantum phase transition. Quantum phase transitions (QPT) are phase transitions that occur in a strict sense only at temperature $T = 0$. As opposed to a classical phase transition at finite temperature, quantum effects play a decisive role, as the fluctuations driving the transition follow quantum instead of classical statistical mechanics. Having been considered for a long time as a mere curiosity of theoretical physics, the impact of this field on physics is now emerging with steadily increasing pace. As will be detailed below, quantum critical points (QCPs) which control continuous QPTs are at the heart of violent quantum fluctuations, a direct consequence of Heisenberg's uncertainty principle. They affect the finite-temperature behavior of condensed matter as well, with a multiplicity of new and unexpected phenomena. For instance, the standard model of metals with electronic interactions, the Landau Fermi-liquid picture, breaks down in the vicinity of a magnetic QPT under certain conditions. In addition, since QPTs occur between nearly degenerate phases whose characteristic energies or energy differences are driven to zero, new small energy scales may become important which would be otherwise masked by the primary energy scales. This has the prospect of finding novel phases around QCP. In an even broader sense, research on quantum phase transitions may play an important role also for the development of new concepts in theoretical particle physics where systems often are at essentially zero temperature.

In Chapter 3 we will discuss different types of quantum critical points. The treatment of continuous quantum phase transitons (QPTs) in metallic systems was pionered by Hertz [2] and extended by Millis [3]. Recently, it has become increasingly clear that in a number of experiments, this model does not account for the observations and a number of theoretical alternatives have been advocated, including a breakdown of the Fermi liquid at a QPT. Conceptually, one should distinguish between two quite different statements: (i) observation of NFL behavior, i.e., apparent deviations from the Fermi-liquid phenomenology at low temperature T such as a constant specific-heat coefficient, and (ii) breakdown of Fermi-liquid theory implying that the concept of a FL, e.g., the notion of quasiparticles with a one-to-one correspondence to the noninteracting electron system, becomes invalid. An example of the former is the theory of Hertz [2] for magnetic QPTs in three dimensions, where the relevant low-energy excitations are the usual fermionic quasiparticles (QPs) and magnetic collective excitations that become soft at the QCP, mediating a singular interaction between QPs and thus inducing NFL behavior. Nevertheless, the methods and concepts of FL theory can still be applied - this assumption is in fact the basis of the theroy. What happens in case where FL theory breaks down (in the more radical sense above) and low-energy excitations can not be mapped to fermionic quasiparticles? In some instances, new weakly interacting quasiparticles with different quantum numbers and interactions can be found, e.g., in Luttinger liquid with purely bosonic charge excitations, or the fractional quantum Hall effect with quasiparticles with fractional charge. In other cases, it may turn out that no well-defined quasiparticles exist.

In Chapters 4 and 5 we will review, with emphasis on experimental results, two types of quantum phase transitions involving itinerant degrees of freedom: Fermi-liquid instabilities at magnetic quantum phase transitions, focusing on heavy-fermion systems and itinerant transition-metal magnets, and the metal-insulator transition in heavily doped semiconductors driven either by doping or uniaxial strain.

2. FERMI LIQUIDS

Systems of interacting fermions at low temperature have been of interest early on in the development of condensed matter theory. Landau put forth a phenomenological theory of interacting Fermi systems, the Fermi-liquid theory or Landau theory, which is based on the concept of quasiparticles [5, 6, 7]. It proposed to map the properties of Fermi systems at low temperature T onto a dilute gas of strongly interacting fermionic excitations. To some extent, a microscopic justification of this picture was given by Landau and others, although a rigorous general mathematical proof is not available. Recent studies of this problem have used the renormalization group (RG) method [8, 9] which can be used to establish rigorous mathematical bounds on the stability of the FL state [8]. In the following we review the salient assumptions and results of FL theory for later reference. We closely follow our previous review [4].

2.1 The quasiparticle concept

Let us start by considering the non-interacting system first. In the ground state all single-particle states $|\mathbf{k}\sigma\rangle$ with momentum $\mathbf{k}$ less than the Fermi momentum k_F and spin orientation σ are occupied and all other states are empty, i.e., the occupation numbers are given by

$$n_{\mathbf{k}\sigma}^{T=0} = \theta(k_F - k) \tag{1}$$

where $\theta(x)$ is the step function. The Fermi momentum is determined by the total number of particles in the given volume, i.e., the density n, by

$$n = \sum_{\mathbf{k}\sigma} n_{\mathbf{k}\sigma}^{T=0} = \frac{k_F^3}{3\pi^2} \tag{2}$$

Let us now imagine that the interaction between the particles is turned on adiabatically. If the single-particle energy spectrum of the interacting system is in one-to-one correspondence with the Fermi-gas spectrum and if the ground state retains the full symmetry of the Hamiltonian the system is termed "normal" or more explicitly, a "normal Fermi liquid". Note that the interaction will lead to the appearance of collective modes. However, these bosonic excitations occupy a negligible fraction of phase space in the limit of low temperatures and therefore do not spoil the principal one-to-one correspondence of single-particle states. In an ordered state this one-to-one correspondence is lost. Therefore, the state of the Fermi liquid may again be described by the distribution function $n_{\mathbf{k}\sigma}$ of single-particle excitations. These single-particle excitations are called "quasiparticles". In particular, the ground state of the system is characterized by the distribution function $n_{\mathbf{k}\sigma}$ defined in (1).

The energy of a quasiparticle, $\varepsilon_{\mathbf{k}\sigma}$, is defined as the amount of energy by which the total energy E of the system increases, if a quasiparticle is added to the unoccupied state

$|\mathbf{k}\sigma\rangle$,

$$\varepsilon_{\mathbf{k}\sigma} = \frac{\partial E}{\partial n_{\mathbf{k}\sigma}} \tag{3}$$

where $\partial n_{\mathbf{k}\sigma}$ is the corresponding change of the distribution function. As a consequence of the interaction, the single-particle energies depend on the state of the system, $\varepsilon_{\mathbf{k}\sigma} = \varepsilon_{\mathbf{k}\sigma}\{n_{\mathbf{k}'\sigma'}\}$. The energy of a single low-energy quasiparticle added to the ground state may be parametrized as

$$\varepsilon_{\mathbf{k}\sigma}\{n^{T=0}_{\mathbf{k}'\sigma'}\} = v_F(k - k_F) \tag{4}$$

for an isotropic system at small energies, with

$$v_F = \frac{k_F}{m^*} \tag{5}$$

being the Fermi velocity. Note that we measure the single-particle energies relative to the chemical potential μ ($= \varepsilon_F$ at $T = 0$). The effective mass m^* determines the density of states for one spin projection at the Fermi level

$$N_0 = \frac{m^* k_F}{2\pi^2} \tag{6}$$

(here and in the following we use units where $\hbar = k_B = 1$, unless explicitly stated). The effect of interactions with other excited quasiparticles on the energy of a specific quasiparticle may be expressed in terms of an effective two-particle interaction function or "Fermi-liquid" interaction $f_{\mathbf{k}\sigma\mathbf{k}'\sigma'}$

$$\delta\varepsilon_{\mathbf{k}\sigma} = \sum_{\mathbf{k}'\sigma'} f_{\mathbf{k}\sigma\mathbf{k}'\sigma'}\,\delta n_{\mathbf{k}'\sigma'} \tag{7}$$

where $\delta n_{\mathbf{k}\sigma} = n_{\mathbf{k}\sigma} - n^0_{\mathbf{k}\sigma}$. To simplify the notation, we have assumed here that the spin quantization axis of all quasiparticle states is the z-axis. This precludes the discussion of transverse spin excitations.

For isotropic systems with short-range interaction the FL interaction function only depends on the angle between the momenta $\mathbf{k}$ and $\mathbf{k}'$ and on the relative spin orientation of σ and σ', and hence may be parametrized as

$$f_{\mathbf{k}\sigma\mathbf{k}'\sigma'} = \frac{1}{2N_0}\sum_{\ell=0}^{\infty} P_\ell(\hat{\mathbf{k}}\cdot\hat{\mathbf{k}}')\left[F^s_\ell + F^a_\ell\sigma\sigma'\right]. \tag{8}$$

Here $\hat{\mathbf{k}} = \mathbf{k}/|\mathbf{k}|$, $\sigma = \pm 1$, $P_\ell(x)$ are the Legendre polynomials, and F^s_ℓ and F^a_ℓ are the dimensionless spin-symmetric and spin-antisymmetric "Landau parameters", which characterize the effect of the interaction on the quasiparticle energy spectrum. For Galileian invariant systems the Landau parameter F^s_1 and the effective mass m^* are related through $m^*/m = 1 + F^s_1/3$.

In a crystal lattice, the symmetry of the system is reduced to discrete rotations/reflections (the elements of the point group of the lattice), and rotations in spin space, or in the case of strong spin-orbit coupling, to discrete orbital-spin-rotations only. As a consequence, the band structure $\varepsilon_{\mathbf{k}}$ and the FL interaction $f_{\mathbf{k}\sigma\mathbf{k}'\sigma'}$ may be strongly anisotropic. The parametrization of $\varepsilon_{\mathbf{k}}$ and of $f_{\mathbf{k}\sigma\mathbf{k}'\sigma'}$ then requires additional parameters, which weakens the predictive power of FL theory. In applications of FL theory to metals, it is frequently assumed that an isotropic approximation in 3D or quasi-2D systems can give a reasonable account of the FL properties. A further aspect not included in the original formulation of FL theory is the effect of disorder, in the form of impurities, lattice defects, vacancies, etc. We will discuss several ways in which disorder changes the properties of metals later.

2.2 Thermodynamic properties

The equilibrium distribution function $n^0_{\mathbf{k}\sigma}$ at finite temperature T follows from the assumed one-to-one correspondence:

$$n^0_{\mathbf{k}\sigma} = n_F(\varepsilon_{\mathbf{k}\sigma}) \equiv \frac{1}{e^{\varepsilon_{\mathbf{k}\sigma}/T}+1}. \tag{9}$$

This is a complicated implicit equation for $n^0_{\mathbf{k}\sigma}$ due to the dependence of $\varepsilon_{\mathbf{k}\sigma}$ on $\{n^0_{\mathbf{k}'\sigma'}\}$.

We are now in a position to calculate specific heat, compressibility and spin susceptibility. The derivative of the internal energy with respect to temperature yields the specific heat at constant volume. The leading term at $T \ll T_F$ ($T_F = \varepsilon_F$ is the Fermi temperature) is linear in T, as for the free Fermi gas, and given by the (renormalized) density of states

$$C_V = \frac{2\pi^2}{3} N_0 T = \gamma T\,. \tag{10}$$

The spin susceptibility χ at $T \ll T_F$ follows as

$$\chi = \frac{2\mu_m^2 N_0}{1+F_0^a} \tag{11}$$

where μ_m is the magnetic moment of electrons. χ is seen to be given by the Pauli susceptibility of a non-interacting Fermi gas of particles with mass m^*, "screened" by a polarization field proportional to the Landau parameter F_0^a. The density response to a change in chemical potential is obtained in complete analogy as

$$\frac{dn}{d\mu} = \frac{2N_0}{1+F_0^s}. \tag{12}$$

2.3 Instabilities within a Fermi-liquid description

Thermodynamic stability requires that the susceptibilities χ and $dn/d\mu$ be positive, which leads to the requirements $F_0^{a,s} > -1$. A general analysis of the stability of the system with respect to any variation of $n_{\mathbf{k}\sigma}$ results in the stability conditions [10]

$$F_\ell^{a,s} > -(2\ell+1)\;,\;\; \ell = 0,1,\ldots \tag{13}$$

In the spin-symmetric isotropic case the compressibility $dn/d\mu$ diverges when $F_0^s \to -1$, which is an indication of phase separation into a dense and a dilute phase. More common is the case of ferromagnetism, which appears when $F_0^a \to -1$. In the case of an instability at $\ell > 0$ the corresponding susceptibility of an anisotropic density excitation in **k**-space diverges, which is termed Pomeranchuk instability. It may lead to an anisotropic deformation of the Fermi surface. While spatially uniform Fermi surface deformations can be captured by Fermi-liquid theory, this is more difficult in the case of instabilities (charge and spin density waves) at finite momentum, as the full momentum dependence of $f_{kk'}(q)$ becomes important. The critical behavior of the FL properties on approaching a Pomeranchuk instability is discussed by Quintanilla and Schofield [11].

A different class of instabilities is signaled by a singularity in the quasiparticle scattering amplitude at zero total momentum. It usually leads to the formation of pair-correlated ordered states, i.e., unconventional superconductors, some features of which will be briefly discussed in Sec. 4.3.

2.4 Finite-temperature corrections to Fermi-liquid theory

The leading corrections to Fermi-liquid theory at low temperatures $T \ll T_F$ have been the subject of extensive theoretical and experimental study. While for a Fermi gas these corrections are of relative magnitude $(T/T_F)^2$, collective effects in an interacting Fermi system generally lead to much larger corrections. Thus one finds that the specific-heat coefficient varies with temperature as

$$\gamma(T) - \gamma(0) = \begin{cases} -g_3 T^2 \ell n(T) & \text{for} \quad d = 3 \\ -g_2 T & \text{for} \quad d = 2 \end{cases} \tag{14}$$

The coefficients g_2 and g_3 have been calculated exactly in lowest order perturbation theory (for a review and references see, [12]). Sub-leading corrections for a Fermi gas with weak repulsion have been derived within an RG approach by [13]. The leading corrections to the spin susceptibility have been found as

$$\chi(T) - \chi(0) = \begin{cases} -c_3 T^2 & \text{for} \quad d = 3 \\ -c_2 T & \text{for} \quad d = 2 \end{cases} \tag{15}$$

In contrast to the specific heat, in $d = 3$ a non-analytic contribution is absent [12, 14]. There is, however, a non-analytic dependence on the wavevector and on magnetic field. In two dimensions the leading T-power of the correction is again reduced from T^2 to T by singular interaction processes [15]. Experimentally, the clearest evidence for the above finite-temperature corrections has been reported in ^{3}He, where a $T^2 \ell n(T)$ contribution in $\gamma(T)$ could be identified [16].

2.5 Electrical resistivity

As an example of a transport coefficient we consider the electrical conductivity $\sigma = \rho^{-1}$ (where ρ is the resistivity), defined as the response of the electric current density $\mathbf{j}$ to the (screened) electric field $\mathbf{E}$, $\mathbf{j} = \overleftrightarrow{\sigma}\mathbf{E}$. In terms of the linearized distribution function $\delta n_{\mathbf{k}} = \lim_{\omega\to 0}\lim_{\mathbf{q}\to 0}\delta n_{\mathbf{k}}(q,\omega)$ the conductivity is given by

$$\sigma_{ij} = e\sum_{\mathbf{k}\sigma} v_{ki}\delta n_{\mathbf{k}}/E_j\,. \tag{16}$$

Here $\delta n_{\mathbf{k}}$ satisfies the Boltzmann equation

$$e\mathbf{v}_{\mathbf{k}}\cdot\mathbf{E}\left(\frac{\partial n_k^0}{\partial\varepsilon_{\mathbf{k}}}\right) = \delta I\{\delta n_{\mathbf{k}}^{\ell}\} \tag{17}$$

where $\delta n_{\mathbf{k}}^{\ell}$ is the deviation of n_k from local equilibrium. The collision integral δI describes the effect of inter-quasiparticle collisions and any other collision processes. For a translation-invariant system the quasiparticle collisions are momentum conserving and the resistivity would be zero. The resistivity is finite in a real solid if umklapp scattering is possible. The most important source of momentum dissipation at low temperature is, however, impurity scattering.

In lowest approximation the collision integral may be modelled as

$$\delta I = -\hat{C}_{e-e}\delta n_k^{\ell} - \hat{C}_i\delta n_k^{\ell} \tag{18}$$

where $\hat{C}_{e-e}$ and $\hat{C}_i$ are the linear integral operators describing electron–electron and electron–impurity collisions, respectively. Taking into account that $\hat{C}_{e-e}\sim\tau^{-1}\sim T^2$ at low temperatures, one finds for the resistivity

$$\rho(T) = \rho_0 + AT^2 + \ldots \tag{19}$$

Here ρ_0 is the residual $T=0$ resistivity arising from impurity scattering. The coefficient A is given by a weighted angular average of the squared quasiparticle scattering amplitudes $A_{0,1}(\theta,\phi) = 2N_0 a(1,2;3,4)$, which depends on the anisotropy of the scattering and the band structure in a sensitive way. Here, $a(1,2;3,4)$ is the transition amplitude for a quasiparticle in state $\mid 1\rangle = \mid \mathbf{k}_1\sigma_1\rangle$ scattering off a quasiparticle in state $\mid 2\rangle$, the two particles ending up in final states $\mid 3\rangle$ and $\mid 4\rangle$. The resulting transport scattering rate τ_{tr}^{-1} is in general different from the single-particle relaxation rate τ^{-1} (except for isotropic scattering), as only finite-angle scattering affects transport.

Provided that the transition amplitudes $a(1,2;3,4)$ are weakly material dependent, i.e., when they are governed by local physics, the ratio of the resistivity coefficient and the square of the specific-heat coefficient, A/γ^2, may be expected to be material-independent since $A\propto N_0^2$ and $\gamma\propto N_0$. This is indeed observed for a large number of heavy-fermion systems [17], and A/γ^2 is termed the Kadowaki-Woods ratio. A corresponding dimensionless quantity may be defined as

$$R_{\text{KW}} = \frac{\rho(T)-\rho_0}{\rho_0}\frac{n^2}{(\gamma T)^2}\,. \tag{20}$$

2.6 Kondo effect – concept of a local Fermi liquid

The Kondo problem [18] goes back to the discovery of a resistance minimum at low temperatures in metals with dilute magnetic impurities. The minimum and low-T increase of the resistance were successfully explained by Kondo [19] within a perturbative calculation. Within the so-called Kondo (or s-d) model, a magnetic impurity is described by a local spin $\mathbf{S}$ (assumed to be $S=\frac{1}{2}$ located at $\mathbf{r}=0$) exchange-coupled to the local conduction-electron spin density $\mathbf{s}_0$,

$$H_{sd} = \sum_{\mathbf{k},\sigma} \varepsilon_k c^\dagger_{\mathbf{k}\sigma} c_{\mathbf{k}\sigma} + J\mathbf{S}\cdot\mathbf{s}_0 \tag{21}$$

with

$$\mathbf{s}_0 = \frac{1}{2}\sum_{\mathbf{k},\mathbf{k}'}\sum_{\sigma\sigma'} c^\dagger_{\mathbf{k}\sigma}\tau_{\sigma\sigma'}c_{\mathbf{k}'\sigma'}\,. \tag{22}$$

Here $\tau_{\sigma\sigma'}$ is the vector of Pauli matrices, and J is the exchange coupling constant. Kondo found that the electrical resistivity ρ due to scattering of conduction electrons off the impurity acquires a logarithmic dependence on temperature in third order in J

$$\rho = \rho_B\left[1 + 2N_0J\,\ell n\frac{D}{T} + \ldots\right] \tag{23}$$

where $\rho_B \propto J^2$ is the usual Born-approximation result, N_0 is the local conduction-electron density of states at the Fermi level for one spin projection, and D is the half-width of the conduction band. The reason for the temperature dependence in ρ (or energy dependence in the scattering cross section) lies in the internal degrees of freedom of the magnetic impurity. As seen from the result (23), perturbation theory breaks down below the Kondo temperature

$$T_{\mathrm{K}} = D\sqrt{N_0J}\exp\left(-\frac{1}{N_0J}\right), \tag{24}$$

when the first-order correction term becomes comparable to the Born approximation. (The prefactor in (24) requires a more detailed analysis, it is determined by the condition that the term $1/\ell n^2(T/T_{\mathrm{K}})$ in the high-temperature expansion of the magnetic susceptibility $\chi_{\mathrm{imp}}(T)$ is absent.) In the limit $J \to 0$ the Kondo temperature T_{K} depends on the coupling J in a non-analytic fashion.

As pointed out by Anderson and collaborators, the ground state of the Kondo model (21) is non-degenerate. The $S=\frac{1}{2}$ impurity spin is fully compensated by a "screening cloud" of conduction-electron spins containing in total one electron spin, bound to the impurity in a singlet state. The impurity complex formed in this way acts like a potential scatterer. The low-energy physics of this system has been formulated by Nozières [20] in terms of a local Fermi-liquid picture.

It is useful to introduce a model of a magnetic impurity in which the electronic structure of the impurity is displayed directly. This Anderson impurity model [21] consists of an impurity orbital (a d or f orbital of an incompletely filled inner atomic shell) hybridizing with a conduction band,

$$H = \sum_{\mathbf{k},\sigma} \varepsilon_{\mathbf{k}} c^{\dagger}_{\mathbf{k}\sigma} c_{\mathbf{k}\sigma} + \varepsilon_f \sum_{\sigma} f^{\dagger}_{\sigma} f_{\sigma} + U n_{f\uparrow} n_{f\downarrow} + V \sum_{\mathbf{k},\sigma} (c^{\dagger}_{\mathbf{k}\sigma} f_{\sigma} + h.c.) \tag{25}$$

where $f^{\dagger}_{\sigma}$ creates an electron with spin projection σ in the f orbital, $n_{f\sigma} = f^{\dagger}_{\sigma} f_{\sigma}$, and V is the hybridization matrix element. The decisive feature of the model is the Coulomb interaction U between electrons in the f orbital, which is usually strongly repulsive and in many cases may be taken to infinity. The electron correlations induced by U lead to local-moment formation: if the impurity level lies below the Fermi energy, $\varepsilon_f < 0$, while $\varepsilon_f + U > 0$ and the bare hybridization width $\Gamma = \pi N_0 V^2$ is small, $\Gamma \ll |\varepsilon_f|, \varepsilon_f + U$, the impurity level is mainly occupied by a single electron (rather than being empty or doubly occupied) and thus represents a local moment of spin $\frac{1}{2}$.

By projecting the Anderson Hamiltonian (25) onto the subspace of singly occupied impurity states [22] one is led to the Kondo model. The exchange coupling constant of the projected model is

$$J = 2V^2 \left(\frac{1}{|\varepsilon_f|} + \frac{1}{\varepsilon_f + U} \right). \tag{26}$$

J is positive, indicating antiferromagnetic coupling.

We note that Kondo physics is not restricted to the screening of *magnetic* degrees of freedom, it can occur in any situation where transitions between a multi-level impurity are induced through the interaction with a bath of fermionic particles. The particular example of the quadrupolar Kondo effect will be briefly discussed in Sec. 2.8.

We will now present a more detailed analysis of the low-temperature FL regime of the exactly screened Anderson model. This can be expected to be continuously connected to the non-interacting limit of the Anderson model, and consequently perturbation theory in powers of the Coulomb interaction U (assuming a non-degenerate ground state) is appropriate [18]. To capture the crossover from high to low temperature, the concept of scaling and of the renormalization group (RG) has turned out to be the principal tool. In Anderson's "poor man's scaling" approach [23] applied to the Kondo model, the Hamiltonian is projected onto a smaller Hilbert space where the conduction electron bandwidth D_0 has been reduced by an infinitesimal amount dD. The effective Hamiltonian still takes the form of the Kondo model, with a modified coupling constant $J(D)$. The flow of the dimensionless coupling $j = N_0 J$ is described by the so-called β function, which can be calculated in a power series in j:

$$\beta(j) = -\frac{dj}{d\ell nD} = j^2 - \frac{1}{2} j^3 + \ldots \tag{27}$$

The radius of convergence of this series is not known a priori. However, from (27) we can infer that for small antiferromagnetic $j > 0$ the effective coupling grows under the reduction of the bandwidth. In contrast, for ferromagnetic coupling ($j < 0$), j scales to zero for $D \to 0$, leaving a free spin and a free conduction band. This solves the problem for ferromagnetic coupling; the finite-temperature thermodynamics displays logarithmic corrections to the free-spin behavior due to the flow of j to zero. For the antiferromagnetic case, which is the usual situation, there are obviously two possibilities: (i) the coupling tends to a finite value as $D \to 0$, (ii) the coupling grows indefinitely, $j \to \infty$. It was recognized early on by Anderson and coworkers [23, 24] that the latter scenario is realized in the usual (single-channel) Kondo situation.

Using the fact that the fixed point of $j \to \infty$ is indeed stable [20], one may now discuss the local properties of conduction electrons at the impurity, taking into account (i) the energy shift provided by the scattering off the static potential of the fully screened impurity, (ii) the interaction between conduction electrons at the impurity site induced by virtual excitations of the impurity complex. This is elegantly done in terms of a phenomenological description following the Landau FL theory [20]. Let us consider an additional quasiparticle near the Fermi energy. Its effect on the phase shift $\eta_\sigma(\varepsilon)$ will be (i) to change the chemical potential by $\delta\mu$, (ii) to shift the scattering phase due to its interaction with other particles.

The above phenomenological considerations provide a correct and useful picture of the Kondo behavior at small temperatures and fields. In order to describe the crossover from high-temperature local-moment physics to the low-temperature Kondo-screened state, more elaborate methods are required. The thermodynamic properties of the Kondo model (21) and the Anderson model (25) may be calculated exactly by using the Bethe ansatz method [25, 26, 27]. The dependence of the free energy and its derivatives on temperature and magnetic field is found to exhibit single parameter scaling behavior in $(T/T_{\rm K})$ and $(B/T_{\rm K})$ from the low-energy regime below $T_{\rm K}$ up to high energies $\gg T_{\rm K}$, but sufficiently below the conduction electron bandwidth. Most importantly, there is no phase transition upon variation of T or B for the metallic single-impurity Kondo problem. Likewise, the electrical resistivity for the Anderson model was found to obey single-parameter scaling behavior in T/T_K [28].

2.7 Anderson and Kondo lattice models

So far we have considered the properties of a single quantum impurity in a host metal. We now turn to a discussion of materials where "quantum impurity ions" are put on a lattice. Generalizing the single-channel impurity Anderson model (25) to a lattice of localized orbitals, f_i, one obtains the so-called periodic Anderson model (PAM),

$$\begin{aligned} H &= \sum_{\mathbf{k}\sigma} \varepsilon_k c^\dagger_{\mathbf{k}\sigma} c_{\mathbf{k}\sigma} + \varepsilon_f \sum_{i\sigma} f^\dagger_{i\sigma} f_{i\sigma} + \\ &+ U \sum_i n^f_{i\uparrow} n^f_{i\downarrow} + V \sum_{i\sigma} \left(f^\dagger_{i\sigma} c_{i\sigma} + h.c. \right). \end{aligned} \tag{28}$$

Both direct hopping of and direct exchange between f electrons are neglected here. In a situation of large U and negative ε_f local moments on the f sites become well defined. Employing a Schrieffer-Wolff transformation as in the single-impurity case, the PAM maps onto the Kondo lattice model

$$H = \sum_{\mathbf{k}\sigma} \varepsilon_k c^\dagger_{\mathbf{k}\sigma} c_{\mathbf{k}\sigma} + J \sum_i \mathbf{S}_i \cdot \mathbf{s}_i \tag{29}$$

where the Kondo coupling J is related to the parameters of the Anderson model through (26).

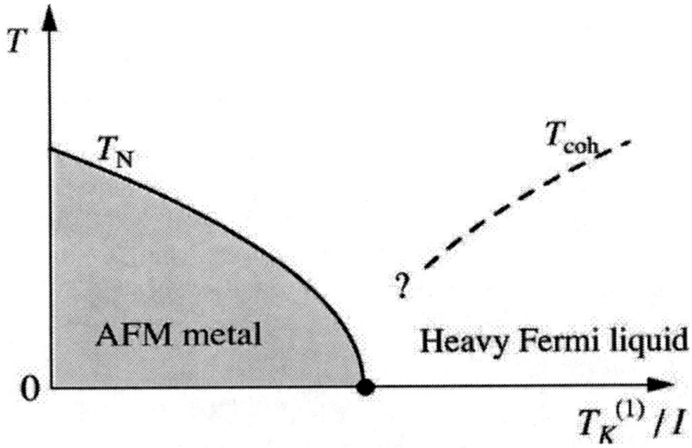

FIGURE 1. Doniach's phase diagram of the Kondo lattice, as function of the ratio of Kondo and inter-moment exchange energies. The heavy Fermi liquid is formed below the coherence temperature T_{coh}. The behavior of T_{coh} across the phase transition will be discussed in Sec. 3.8.

We now discuss the qualitative features of the Anderson and Kondo lattice models. For small interaction U, the periodic Anderson model (28) can be expected to describe a Fermi liquid with two bands. As detailed below, a Fermi liquid at lowest temperatures can also survive in the large-U limit and thus in the Kondo-lattice case – this requires the local moments to be screened by a lattice generalization of the Kondo effect. The resulting Fermi liquid, formed below a coherence temperature T_{coh}, will have a Fermi volume containing both conduction (c) electrons and local moments (dubbed "large Fermi volume"), consistent with the Luttinger theorem [29]. The resistivity will follow the usual quadratic T dependence (19). It is interesting to discuss the evolution of this state with temperature: For $T \gg T_{coh}$ (but still assuming well-formed local moments) the system can be described as c fermions with a "small" Fermi volume interacting weakly with a paramagnetic system of localized spins. The resistivity is often rather low in this situation. In the crossover region $T \sim T_{coh}$ the Fermi surface fluctuates strongly, giving rise to a very high resistivity of the order of the unitarity limit – experimentally, this resistivity maximum is often used to define T_{coh}. The screening of the local moments, required for FL behavior in the Kondo lattice, competes with interactions between the local moments. Such interactions can be due to direct hopping or exchange between the f orbitals, but are also generated due to the polarization of the conduction electrons. This indirect Ruderman-Kittel-Kasuya-Yosida (RKKY) interaction is given in lowest (quadratic) order in J by

$$H_{RKKY} = \sum_{i,j} I_{ij} \; \mathbf{S}_i \cdot \mathbf{S}_j, \tag{30}$$

$$I_{ij} = N_0 J^2 F(k_F R_{ij}) \tag{31}$$

where $F(x) = (x\cos x - \sin x)/x^4$ and R_{ij} is the distance between lattice sites i,j. In the Kondo-screened state I_{ij} is expected to be renormalized, in particular at long distances, but a reliable determination of I_{ij} is not available at present. The competition between the Kondo coupling and the inter-moment interaction will govern the phase diagram of the Kondo lattice [30]. The most natural competitor of the Fermi liquid is a magnetically ordered metal, Fig. 1, but in the presence of strong quantum effects and geometric frustration spin-glass and spin-liquid states may also occur. The fact that the Doniach phase diagram leads to a magnetic instability was pointed out by Continentino [31, 32].

We now present an analysis of the FL phase of the periodic Anderson model – here non-local interaction effects are believed to be unimportant. A viable method is perturbation theory in U [18]. Since the interaction U is assumed to act only between electrons in the f level, there is only one self-energy $\Sigma_\sigma(\mathbf{k},\omega)$. Expanding Σ near the Fermi energy $\omega = 0$ and near the Fermi surface $k = k_F$ (assuming an isotropic band structure) and using the fact that $\mathrm{Im}\Sigma \sim \omega^2$ in this region, one may define a quasiparticle component of the single-particle Green's function $G^{ff}(\mathbf{k},\omega)$ of weight

$$Z = \left[1 - \frac{\partial \Sigma(k_F,\omega)}{\partial \omega}\right]^{-1}_{\omega=0} \tag{32}$$

located at energy

$$\tilde{\varepsilon}_{f,k} = Z\left[\varepsilon_f + \Sigma(k_F,0) + (\mathbf{k}-\mathbf{k}_F)\cdot\nabla_k\Sigma\right] \tag{33}$$

and hybridizing with the conduction band with renormalized strength $\tilde{V} = Z^{1/2}V$. One finds two quasiparticle bands with dispersion

$$\tilde{\varepsilon}_k^{\pm} = \frac{1}{2}\left\{\tilde{\varepsilon}_{f,k} + \varepsilon_k \pm \left[(\tilde{\varepsilon}_{f,k} - \varepsilon_k)^2 + 4\tilde{V}^2\right]^{1/2}\right\}. \tag{34}$$

These are the same bands as in the non-interacting case ($U = 0$) except that the bare energies ε_f and V are replaced by $\tilde{\varepsilon}_{f,k}$ and $\tilde{V}$, respectively. Neglecting the self-energy's k dependence, which is expected to be weak compared to the ω dependence, one finds for the quasiparticle effective mass at the Fermi energy

$$\frac{m^*}{m} = 1 + \frac{V^2}{\left[\varepsilon_f + \Sigma(k_F,0)\right]^2}\frac{1}{Z} \tag{35}$$

The specific-heat coefficient and the spin susceptibility are found as

$$\gamma = \frac{2\pi^2}{3}\tilde{N}(0), \quad \chi = \frac{2\mu_m^2\tilde{N}(0)}{1+F_0^a} \tag{36}$$

where $\tilde{N}(0)$ is the renormalized total density of states at the Fermi level

$$\tilde{N}(0) = N_c^{(0)}(0) + \frac{1}{Z}N_f^{(0)}(0) \tag{37}$$

with $N_{c,f}^{(0)}(0)$ the densities of states of conduction electrons and f electrons, respectively, in the limit $U = 0$. The factor $R = (1+F_0^a)^{-1}$, often called the generalized Wilson ratio, in χ (36) expresses the effect of quasiparticle interactions in terms of the Landau parameter F_0^a. Experimentally, one indeed finds a close correlation between γ and χ for various systems see Fig. 2 [33], with $R \approx 2-5$ corresponding to $F_a \approx -0.5$ to -0.8.

There are various approximation schemes available, allowing to estimate the parameters Z, $\tilde{\varepsilon}_f$ in the above semi-phenomenological quasiparticle theory. The simplest one uses slave-boson mean-field theory, in the limit $U \to \infty$ [34]. One finds an approximate mapping to a model of two non-interacting, hybridizing fermion bands, with a quasiparticle dispersion given in (34). Assuming a flat conduction-band density of states $N_c^{(0)}(\varepsilon) = 1/(2D)$, and for $D \gg \varepsilon_f, V$, the renormalized f level $\tilde{\varepsilon}_f$ is found as a solution of

$$\varepsilon_f - \tilde{\varepsilon}_f = \frac{2}{\pi}\Gamma \ell n(\tilde{\varepsilon}_f/D) \tag{38}$$

which defines a characteristic energy scale, $T^* = \tilde{\varepsilon}_f$, i.e., the distance of the *renormalized* f level to the Fermi energy ($\Gamma = \pi N_c^{(0)}(0)V^2$ is the bare hybridization width). The renormalized f-electron density of states can be expressed as

$$\frac{1}{Z}N_f^{(0)}(0) = \frac{2}{\pi}\frac{\Gamma}{\tilde{\varepsilon}_f}\frac{1}{T^*} \tag{39}$$

and the chemical potential μ follows from the total particle number per site as

$$n_{\text{tot}} = \frac{2}{\pi}\left(\frac{\Gamma}{\tilde{\varepsilon}_f} + \frac{\pi}{2}\right). \tag{40}$$

The specific heat and the spin susceptibility are given by (36), with the Landau parameter $F_0^a = 0$. A finite value of F_0^a can be obtained from the contribution of fluctuations about the mean field [35]. If the f states are sufficiently far below the Fermi level, the characteristic temperature assumes within this approximation the same functional dependence as the Kondo temperature of the single impurity problem:

$$T^* \sim D\exp\left(-\frac{\pi|\varepsilon_f|}{2\Gamma}\right) \tag{41}$$

where $T^* \ll D$, implying an exponentially small quasiparticle weight factor $Z \sim T^*/\Gamma \ll 1$. This is the regime of heavy-fermion metals, with large effective mass ratio $m^*/m \gg 1$.

A similar mean-field approach can be taken to the Kondo lattice model [36], showing that (at least) *two* energy scales are relevant for the Kondo lattice problem. The onset of Kondo screening upon lowering T happens around $T_K^{(1)}$, the (single-impurity) Kondo temperature, whereas the Fermi liquid is only established below T_{coh}. Typically, $T_{\text{coh}} < T_K^{(1)}$, leading to a crossover regime which is wider than in the single-impurity case (dubbed protracted screening). In the weak-coupling limit, both $T_K^{(1)}$ and T_{coh} can be obtained analytically, with the result:

$$\begin{aligned} T_K^{(1)} &= D\exp\left(-\frac{1}{N_0 J}\right)F_K(n_c), \\ T_{\text{coh}} &= D\exp\left(-\frac{1}{N_0 J}\right)F_{\text{coh}}(n_c) \end{aligned} \tag{42}$$

where F_K and $F_{\rm coh}$ are functions of filling (and shape) of the conduction band, and N_0 is the conduction-band density of states for $J = 0$. In this approximation the ratio $T_{\rm coh}/T_{\rm K}^{(1)}$ is a function of the conduction-band properties only, but is independent of the Kondo coupling J.

Beyond slave bosons, the local correlation physics of Anderson and Kondo lattice model has been studied using the Dynamical Mean-Field Theory (DMFT). The DMFT makes use of the limit of infinite spatial dimensions [37], in which the self-energy $\Sigma(\mathbf{k},\omega)$ becomes independent of momentum $\mathbf{k}$. This corresponds to a mapping of the Anderson lattice model to an effective Anderson impurity model with energy-dependent hybridization function $\Delta(\omega)$ [38]. The effective Anderson model has been solved by Quantum Monte Carlo (QMC, see [38]) and Numerical Renormalization Group (NRG, see [39]) techniques. The most accurate study using NRG [40] shows that near half-filling, $n_c = 1$, the coherence temperature is actually larger than the single-impurity Kondo temperature, their ratio depending on the Kondo coupling J. More importantly, at lower fillings $n_c \leq 0.8$, the ratio $T_{\rm coh}/T_{\rm K}^{(1)}$ was numerically found to be *independent* of J, and in the small-n_c limit was proportional to n_c.

The results for the J dependence of the coherence scale of [36] and [40], namely $T_{\rm coh} \propto T_{\rm K}^{(1)}$, indicate that the so-called exhaustion scenario of [42], predicting $T_{\rm coh} \propto (T_{\rm K}^{(1)})^2/D$, is incorrect. As detailed by [41], the original exhaustion argument [42] is too simplistic, e.g., it does not correctly account for the flow of the Kondo coupling. We also note that the exhaustion scenario has been falsified experimentally: In $Ce_xLa_{1-x}Pb_3$ Kondo behavior of single-impurity type has been observed down to lowest temperatures for x up to 80%, with $T_{\rm K} = 3.3$ K essentially independent of x [43], whereas the exhaustion arguments would predict a suppression of $T_{\rm K}$ for $x > 0.1\%$.

2.8 Multi-channel Kondo effect

Nozières and Blandin [44] realized the importance of the proper matching of the degrees of freedom of impurity and environment for local non-Fermi-liquid behavior in Kondo models. This can be discussed using a generalized Kondo model, where an impurity spin of size S is coupled to several identical conduction bands (labeled α, $\alpha = 1,....,M$, and called "channels"):

$$H = \sum_{\mathbf{k},\sigma,\alpha} \varepsilon_k c^\dagger_{\mathbf{k}\sigma\alpha} c_{\mathbf{k}\sigma\alpha} + J\mathbf{S}\cdot\sum_\alpha \mathbf{s}_\alpha \tag{43}$$

where $\mathbf{s}_\alpha$ is the local conduction electron spin at the impurity site contributed by conduction band α. The Hamiltonian (43) has a high degree of symmetry: in addition to rotation invariance in spin space there is separately invariance against unitary transformations in channel space.

For the low-energy behavior of (43) with $J > 0$, three cases have to be distinguished: (i) $S = M/2$, here the impurity spin is exactly screened by M electrons of spin $1/2$, yielding local FL behavior; (ii) $S > M/2$ yields an effective uncompensated spin of size $S - M/2$ – this underscreened Kondo effect leads to non-analytic corrections to local FL behavior; (iii) $S < M/2$ results in overscreening, associated with true local NFL behavior – this will be discussed in the following.

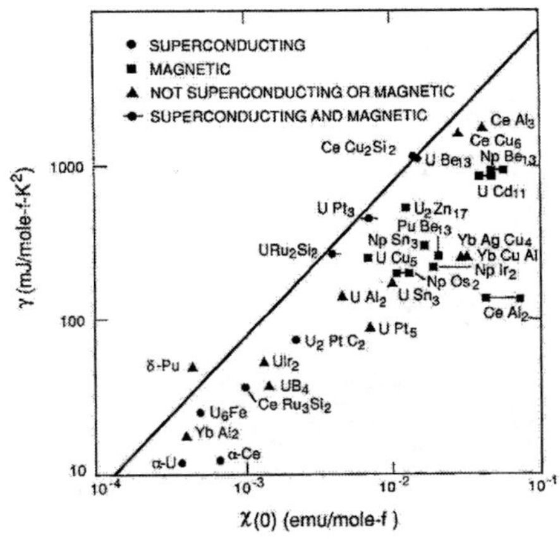

FIGURE 2. Linear specific-heat coefficient γ vs. magnetic susceptibility fo several heavy-fermion systems on a log-log plot. Straight line indicates the free-electron behavior. From [33].

The occurrence of a new type of ground state for $S < M/2$ can be understood as follows [44]. Suppose $S = \frac{1}{2}$, and $M = 2$; if the effective exchange coupling j would scale to infinity as in the single-channel case considered in the last section, the impurity would bind one electron with opposite spin orientation in each of the two channels. The net spin projection of the overscreened impurity site would be $S_z' = -S_z$ if S_z is the momentary impurity spin projection. Now, virtual hopping processes of neighboring electrons (with spin projection S_z due to the Pauli principle) induce an exchange interaction between the effective spin $\mathbf{S}'$ and the neighboring electron spins, which is again antiferromagnetic. According to the initial assumption, this coupling scales to infinity; this process repeats itself and thus does not converge. One concludes that the fixed point at infinite coupling is not stable, i.e., there must exist a fixed point at some finite coupling strength. In the limit of large channel number M the new fixed point is accessible by "poor man's scaling" [44]. One finds that the β function for the scale-dependent coupling j now has a weak-coupling expansion:

$$\beta(j) = j^2 - \frac{M}{2} j^3 + \ldots \tag{44}$$

replacing (27). Thus there exists a fixed point ($\beta = 0$) at $j^* = N_0 J^* = 2/M \ll 1$, within the range of validity of the expansion.

At the new fixed point the impurity spin is not exactly screened, which destroys the Fermi liquid found in the exactly screened case, and leads to local non-Fermi-liquid behavior. Exact treatments based upon the Bethe ansatz method [45, 46, 47], bosonization and conformal field theory [48, 49] have been used to determine the limiting low-temperature behavior.

For the M-channel, spin $S = 1/2$, Kondo model (43) the most striking result is a finite impurity entropy $S(T=0)$ indicating some kind of fractional degeneracy of the ground state [45],

$$S(T=0)=\ell n\left[2\cos\left(\frac{\pi}{M+2}\right)\right]. \qquad (45)$$

The low-temperature specific heat and spin susceptibility in zero magnetic field follow power-law behavior [45, 46]

$$\frac{C_v^{\text{imp}}}{T}\sim\chi_{\text{imp}}(T)\sim T^{\frac{4}{2+M}-1},\ M>2 \qquad (46)$$

and a logarithmic law for $M=2$ [47]. The leading low-temperature power-law correction of the resistivity is found as

$$\rho(T)-\rho_0\sim T^{\alpha_\rho} \qquad (47)$$

where

$$\alpha_\rho=\frac{2}{2+M};\ S=\frac{1}{2},\ M\geq 2. \qquad (48)$$

As noted by [44], perfect channel symmetry is rather unlikely due to anisotropies in realistic crystalline electric fields. The presence of channel asymmetry will always cause a flow to the single-channel Kondo fixed point. Nonetheless, there is the possibility that the symmetry-breaking fields are accidentally small, and the system would be governed by the multi-channel fixed point over a sizeable energy/temperature range, before branching off to a single-channel fixed point at some low energy scale.

An interesting route to two-channel Kondo behavior is the so-called quadrupolar Kondo effect. Here, the impurity has a doubly degenerate *non-magnetic* ground state, and the role of the two internal states is taken by orbital degrees of freedom. Then, the conduction electron spin directions provide two independent screening channels, with channel symmetry being protected by spin symmetry.

To date clear experimental realizations of multi-channel or quadrupolar Kondo physics in Kondo alloys have been elusive, although one of the first reports on NFL behavior, i.e., on $U_xY_{1-x}Pd_3$, suggested this scenario [50]. For a comprehensive discussion of experimental candidate systems and a presentation of theoretical results available, see [51].

2.9 Non-Fermi-liquid behavior from disorder

The influence of disorder on metallic systems has a wide range of aspects, ranging from diffusive transport to Anderson localization, as reviewed by [52]. As discussed above, the (perfectly screened) Kondo effect leads to a local Fermi liquid around a magnetic impurity. Usually, in a random dilute solution of impurities, the microscopic parameters J and N_0 which determine the coupling between magnetic impurity and conduction electrons, acquire well defined values yielding a unique Kondo temperature for the system under consideration. However, in disordered systems, a distribution of Kondo temperatures may arise from statistically fluctuating J and/or N_0. This might occur in metals near the metal–insulator transition as exemplified by heavily doped semiconductors [53, 54, 55, 310, 56] and will be discussed in detail in Chapter 5. In alloys different non-magnetic constituents may lead to different local environments. A similar situation arises in disordered heavy-fermion systems, where the competition of

Kondo screening and interaction of magnetic moments takes place in a spatially non-uniform environment, often leading to non-Fermi-liquid behavior [1].

The fact that disorder can give rise to phases with anomalous behavior has been known for some time [57]. The behavior is not confined to a critical point in the phase diagram but may persist over an entire region. These so-called Griffiths-McCoy phases were discussed originally in disordered classical systems [58]. Much stronger singularities are found in quantum models as has been recently established theoretically. The subject of NFL behavior driven by disorder has been reviewed recently [59].

3. FERMI-LIQUID INSTABILITIES AT QUANTUM PHASE TRANSITIONS: THEORY

A quantum system can undergo a continuous phase transition at $T = 0$ upon variation of some non-thermal control parameter [60, 61, 62]. Near the critical point of such a quantum phase transition in the itinerant electron systems of interest here, the finite-temperature behavior is characterized by scaling laws with temperature exponents different, in general, from those of Fermi-liquid theory. This may be considered as a breakdown of the Fermi-liquid state induced by quantum fluctuations near the critical point. As is the case with usual classical continuous phase transitions, the different systems fall into universality classes, depending on the symmetry properties of the phase transition. In contrast to classical phase transitions, also affects the dynamics critical thermodynamic properties, and therefore a larger number of different universality classes can be expected, e.g., depending on the presence or absence of an efficient coupling between the order parameter and fermionic quasiparticles.

3.1 Classical vs. quantum phase transitions

Phase transitions at $T = 0$ are dominated by quantum effects, in contrast to classical phase transitions at $T > 0$, even though both may occur in the same physical system. Any continuous finite-temperature phase transition is classical in the following sense: continuous phase transitions are accompanied by a divergence of both correlation length and time. This means that the order parameter ϕ (magnetization, staggered magnetization, etc.) fluctuates coherently over increasing distances and time scales as one approaches the transition. The latter implies that there exists a characteristic frequency, ω_ϕ, for order-parameter fluctuations, which tends to zero at the transition. The system behaves classically (even if quantum effects are important at short length scales) if the transition temperature T_c satisfies $k_B T_c \gg \hbar\omega_\phi$. Any finite-$T_c$ transition will follow this inequality for $T \to T_c$. This argument shows at once that quantum phase transitions, for which $T_c = 0$, are qualitatively different: their critical fluctuations require a *quantum-statistical* description.

We start the discussion with the free energy $F = -(1/\beta)\ell n Z$ of the system with Hamiltonian H, where Z is the partition function

$$Z = \mathrm{Tr}\, e^{-\beta H} \tag{49}$$

and $\beta = 1/k_BT$ is the inverse temperature (we do not distinguish canonical and grand canonical ensembles for this qualitative discussion). The trace is taken over a complete basis set of states. For the purpose of discussing quantum critical phenomena it is useful to reformulate the problem in a way as closely related to a classical system as possible. For any classical continuous phase transition, the partition function may be represented in terms of the relevant time-independent order-parameter field $\phi(\mathbf{r})$ as a functional integral over all configurations of ϕ [63]

$$Z_{cl} = Z_{cl}^0 \int \mathscr{D}\phi(\mathbf{r}) \exp(-\beta F_{cl}\{\phi\}) \tag{50}$$

where F_{cl} is the Landau-Ginzburg-Wilson (LGW) free energy functional

$$\begin{aligned} F_{cl}\{\phi\} &= \int d^d r \mathscr{F}_{cl}[\phi(\mathbf{r})] \\ &= \varepsilon_0 \int d^d r \Big[\phi(\mathbf{r})(\delta_0 + \xi_0^2 \nabla^2)\phi(\mathbf{r}) + u_0 \phi^4(\mathbf{r}) + \ldots\Big] \end{aligned} \tag{51}$$

and Z_{cl}^0 is the partition function of the non-critical degrees of freedom, and ε_0 is a microscopic energy scale. The parameter δ_0 is related to the temperature and tunes the system through the phase transition.

For the quantum system we can represent the partition function in a somewhat similar form, the difference being that the quantum nature of the order-parameter field requires that one keeps track of the order in which the ϕ operators appear in the power series expansion of $\exp(-\beta F)$, as they do not commute with each other. Realizing that $\exp(-\beta H)$ is a time evolution operator on the imaginary time axis, the partition function $Z(\beta)$ may be represented as a path integral over all configurations $\phi(\mathbf{r},\tau)$, starting at $\phi(\mathbf{r},0)$ and ending at $\phi(\mathbf{r},\hbar\beta) = \phi(\mathbf{r},0)$:

$$Z(\beta) = Z_0 \int \mathscr{D}\phi(\mathbf{r},\tau) \exp(-S\{\phi\}) \tag{52}$$

where S is the Euclidian action,

$$S\{\phi\} = \int_0^{\hbar\beta} d\tau \int d^d r \mathscr{L}\Big\{\phi(\mathbf{r},\tau)\Big\} \tag{53}$$

with the Lagrange density $\mathscr{L}$ given by the LGW free-energy expression amended by the kinetic energy of quantum fluctuations of the order-parameter field, $\mathscr{L}_{kin}$,

$$\mathscr{L} = \mathscr{L}_{kin} + \mathscr{F}_{cl}\Big[\phi(\mathbf{r},\tau)\Big] \; . \tag{54}$$

The actual form of $\mathscr{L}_{kin}$ depends on the dynamics of the system. We note that these general arguments, however, do not exclude the possibility that $\mathscr{L}$ is a highly non-local, non-linear object, such that an efficient description of the critical behavior in terms of an order-parameter field $\phi(\mathbf{r},\tau)$ is no longer useful. We will defer a discussion of these complications, which in fact do arise in ferromagnetic metals to Sec. 3.6.

If the dependence of $\phi(\mathbf{r},\tau)$ on τ can be neglected, as is the case at a finite-temperature phase transition, the contribution from quantum fluctuations $\mathscr{L}_{kin} \to 0$, and the τ-integral in (53) gives back the factor of β in the expression (50) for the classical partition function. This can be seen more clearly in the Fourier representation, where S, e.g., in the case of an insulating magnet, takes the form

$$S = \frac{1}{\beta V} \sum_{\mathbf{k}\omega_n} \phi_{-\mathbf{k},-\omega_n} \left[\varepsilon_0(\delta_0 + \xi_0^2 k^2) + \omega_n^2 \right] \phi_{\mathbf{k},\omega_n} + S_4 \tag{55}$$

with $\omega_n = 2\pi n k_B T$ the (imaginary) frequency and $\mathbf{k}$ the wavevector of the order-parameter fluctuation. The term S_4 represents the fourth-order term in the LGW function. The non-thermal control parameter δ_0 acquires a (temperature-dependent) renormalization from the interactions; the transition occurs at $\delta_0 = \delta_c$ where the renormalized δ vanishes, with $\delta_c \to 0$ for $u_0 \to 0$. In the following we will denote the distance to the quantum critical point (QCP) by

$$r = \delta_0 - \delta_c(T=0), \tag{56}$$

which can be tuned by varying pressure [then $r = (p - p_c)/p_c$], magnetic field, or chemical composition.

Upon approaching a *finite-temperature* phase transition, the energy of the characteristic order-parameter fluctuations ω_ϕ (proportional to the renormalized value of δ) becomes eventually smaller than $k_B T$. Then only the $\omega_n = 0$ term in (63) contributes to the critical behavior, which is thus governed by spatial order-parameter fluctuations only, Eq. (51). In contrast, at $T = 0$ temporal fluctuations are not negligible, but appear to be intimately intertwined with spatial fluctuations. The representation of the partition function suggests that the system behaves like a $(d+1)$-dimensional classical system, which is anisotropic since the "gradient energy" in the time direction may be of different order (e.g. first order, see below) than in the three spatial directions (usually second order), and the time axis is restricted to the interval $\hbar\beta$. The anisotropy of the fictitious classical system may be characterized by the so-called dynamical exponent z, defined by the scaling of frequency with wavevector $\omega \sim k^z$. For the LGW function, ω scales as k^1 and hence the bare value of $z = 1$ in this case. For metallic magnets we will have $z > 1$.

The effective dimensionality for a system near a quantum phase transition is thus $d_{\text{eff}} = d + z$. In many cases d_{eff} is equal to or larger than the upper critical dimension, d_c^+, of the respective field theory; $d_c^+ = 4$ for the magnetic transitions to be discussed below. While a phase transition with $d_{\text{eff}} < d_c^+$ is controlled by an interacting fixed point and usually obeys strong hyperscaling properties, a transition with $d_{\text{eff}} \geq d_c^+$ can be described within mean-field theory, and hyperscaling is violated due to the presence of dangerously irrelevant operators.

3.2 Scaling properties near quantum phase transitions

The functional integral formulation of the partition function allows one to employ the well-established picture of scaling near a continuous phase transition. As such a

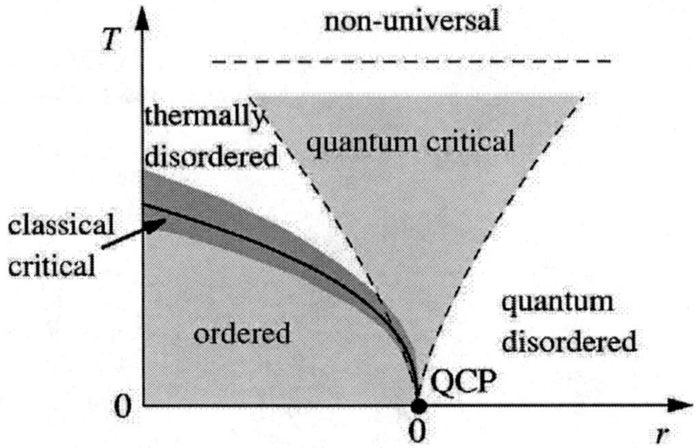

FIGURE 3. Generic phase diagram in the vicinity of a continuous quantum phase transition. The horizontal axis represents the control parameter r used to tune the system through the QPT, the vertical axis is the temperature T. The dashed lines indicate the boundaries of the quantum critical region. The lower crossover lines are given by $T \propto |r|^{\nu z}$; the high-temperature crossover to non-universal (lattice) physics occurs when the correlation length is no longer large to microscopic length scales. The solid line marks the finite-temperature boundary between the ordered and disordered phases. Close to this line, the critical behavior is classical.

transition is approached, both the order-parameter correlation length ξ and correlation time ξ_τ (i.e., the correlation length along the imaginary time axis) diverge:

$$\xi \sim |r|^{-\nu}, \quad \xi_\tau \sim \xi^z, \tag{57}$$

where r measures the distance to the QCP. We note that the correlation-length exponent ν of the quantum transition is different from the one of a possible finite-temperature transition at T_c in the same system (which describes the divergence $\xi \sim |T - T_c|^{-\nu_{cl}}$). At finite temperature, the $(d+z)$-dimensional quantum system has a finite length in the time direction, $L_\tau = \hbar\beta$. Its properties can then be deduced from finite-size scaling [64]. We are now in the position to discuss the properties of a system near a quantum critical point, located at $T = 0$, $r = 0$ (Fig. 3). The QCP is usually the endpoint of a line of continuous finite-temperature transitions. (Exceptions are low-dimensional systems where order at finite T is prohibited by the Mermin-Wagner theorem, or systems where no order parameter can be defined for $T > 0$, as is the case for metal–insulator transitions or transitions in the topology of the Fermi surface.) In general, the boundary of the ordered phase follows $T_c \propto (-r)^\psi$ where ψ is the so-called shift exponent. In the immediate vicinity of this boundary there is a region of classical non-Gaussian criticality. The disordered phase of the system at finite T can be divided into distinct regimes: For low temperatures and $r > 0$ thermal effects are negligible ($L_\tau \gg \xi_\tau$, equivalently $T \ll r^{\nu z}$), and the critical singularity is cut off by the deviation of the control parameter r from criticality. This regime is dubbed "quantum disordered" and is characterized by well-defined quasiparticle excitations; for a magnetic transition in a metallic system this will be the usual Fermi-liquid regime. For $r < 0$ and $T > T_c$, but still $L_\tau \gg \xi_\tau$, we are in the "thermally disordered" regime; here the order is destroyed by thermal fluctuations of the ordered state (yet quasiparticles are still well defined on intermediate scales). A completely different regime is the high-temperature regime above the QCP where $\xi_\tau \gg L_\tau$. In this "quantum critical" regime, bounded by crossover lines $T \sim |r|^{\nu z}$, the critical singularity is cutoff by the finite temperature. The properties are determined by the unconventional excitation spectrum of the quantum critical ground state, where the quasiparticles of the stable phases are replaced by a critical continuum of excitations. In the quantum critical regime, this continuum is thermally excited, resulting in unconventional power-law temperature dependencies of physical observables.

Assuming that the critical behavior is governed by ξ and ξ_τ, the critical contribution to the free energy density, $f_{\text{cr}} = f - f_{\text{reg}}$, should follow the homogeneity law

$$f_{\text{cr}}(r,T) = b^{-(d+z)} f_{\text{cr}}(r b^{1/\nu}, T b^{z}) \tag{58}$$

where b is an arbitrary scale factor. Note that this "naive scaling" (equivalent to hyperscaling) is valid only below the upper critical dimension, $d_{\text{eff}} < d_c^+$, and we will comment on deviations later on. Choosing $b = \xi$, Eq. (58) can be cast into the scaling form $f_{\text{cr}} = \xi^{-(d+z)} \phi_1(\xi_\tau / L_\tau)$, or, equivalently, the ansatz:

$$\begin{aligned} f_{\text{cr}} &= \rho_0 r^{\nu(d+z)} \phi_2\left(\frac{T}{r^{\nu z}}\right) \\ &= \rho_0 T^{(d+z)/z} \phi_3\left(\frac{r}{T^{1/(\nu z)}}\right), \end{aligned} \tag{59}$$

where T is measured in units of T_0, T_0 and ρ_0 being non-universal constants, while $\phi_{1,2,3}(x)$ are *universal* scaling functions.

From Eq. (59) we can immediately deduce the critical contribution to the specific heat, $C = T\partial S/\partial T$, as

$$C_{\text{cr}}(r=0,T) \propto T^{d/z} \tag{60}$$

in the quantum critical regime. If the quantum disordered regime of Fig. 3 is a Fermi liquid, then Eq. (59) yields for its specific-heat coefficient $C_{\text{cr}}/T(T \to 0) \propto r^{\nu(d-z)}$.

As is clear from Fig. 3, a quantum critical point can be generically approached in two different ways: $r \to 0$ at $T = 0$ or $T \to 0$ at $r = 0$. The power-law behavior of physical observables in both cases can often be related. We discuss this idea by looking at the entropy S. It goes to zero at the QCP, but its derivatives are singular. The specific heat C will show power-law behavior, as does the observable $B = \partial S/\partial r$. At a pressure-tuned phase transition, $r = (p - p_c)/p_c$, B measures the thermal expansion,

$$\alpha = \frac{1}{V}\frac{\partial V}{\partial T}\bigg|_p = -\frac{1}{V}\frac{\partial S}{\partial p}\bigg|_T . \tag{61}$$

The quantity B/C is equivalent to the Grüneisen parameter Γ,

$$\Gamma = \frac{\alpha}{C_p} = -\frac{1}{V_m T}\frac{(\partial S/\partial p)_T}{(\partial S/\partial T)_p} \tag{62}$$

where $V_m = V/N$ the molar volume. Taking the ratio of the *singular parts* of B and C one observes that the scaling dimensions of T and S cancel, and therefore B/C scales as the inverse of the tuning parameter r. Thus, one obtains a *universal* divergence in the low-temperature limit [65]

$$\Gamma_{\rm cr}(T=0,r) = \frac{B_{\rm cr}}{C_{\rm cr}} = G_r|r|^{-1} \tag{63}$$

similarly,

$$\Gamma_{\rm cr}(T,r=0) = G_T T^{-1/(\nu z)} \ . \tag{64}$$

With the help of the scaling ansatz (59) the full scaling form of Γ can be determined, for details see [65]. Remarkably, in the $T \to 0$ limit, even the prefactor G_r is universal and given by a combination of critical exponents. Further we note that Γ does not diverge at a finite-temperature phase transition, thus a divergence of Γ is a unique signature of a continuous QPT.

If the control parameter of the QPT is not pressure but an external magnetic field H, the quantity B is the T-derivative of the magnetization M, and the role of the Grüneisen ratio is played by

$$\Gamma_H = -\frac{(\partial M/\partial T)_H}{c_H} = -\frac{1}{T}\frac{(\partial S/\partial H)_T}{(\partial S/\partial T)_H} = \frac{1}{T}\left.\frac{\partial T}{\partial H}\right|_S . \tag{65}$$

It can be determined directly from the magnetocaloric effect by measuring the change of temperature in response to an adiabatic ($S=$ const) change of H.

We finally turn to dynamical scaling. Any physical quantity depending on **r** and t (or equivalently **k** and ω) in the critical region close to the QPT (but sufficiently far from the associated finite temperature transition) should depend on space and time only through the scaled variables $k\xi$ and $\omega\xi_\tau$, since ξ is the only length scale and ξ_τ is the only time scale in that regime. The Fourier components of a physical quantity X affected by the transition are thus expected to exhibit the following scaling behavior

$$X(k,\omega;r,T) = \xi^{d_x} F_x(k\xi, \omega\xi_\tau, \xi_\tau/L_\tau) \tag{66}$$

where d_x is the scaling dimension of the observable X. In analogy to Eq. (59) this can be cast into

$$X(k,\omega;r,T) = T^{-d_x/z} \tilde{F}_x\left(\frac{k^z}{T}, \frac{\omega}{T}, \frac{T}{r^{\nu z}}\right) \tag{67}$$

Exactly at the quantum critical point this reduces to

$$X(k,\omega;r{=}0,T{=}0) = k^{-d_x} F_x^*(k^z/\omega) \tag{68}$$

We again note that all scaling relations are only expected to be valid if the critical point satisfies hyperscaling properties, which is true below the upper critical dimension d_c^+.

3.3 Itinerant fermion systems

Quantum phase transitions in itinerant electron systems were first studied in a pioneering paper by Hertz [2]. He pointed out that near a phase transition at $T = 0$ static and dynamic properties are inextricably mixed and applied a renormalization group (RG) treatment to model systems of this type. This work was later reconsidered and extended by [3]. We will now discuss the main features of this model.

In the context of strongly correlated electron systems one is mainly interested in magnetic phase transitions involving the conduction electrons. As prototypes we will consider ferromagnetic (FM) and antiferromagnetic (AFM) phase transitions. We will assume the collective behavior near the transition to be characterized by a real N-component order-parameter field ϕ, representing the magnetization (for the ferromagnet) or the staggered magnetization (for the antiferromagnet). A number of simplifications occur in the limit $N \to \infty$, although the actual number of components is $N \leq 3$. The effective action may be derived from the Hamiltonian either by introducing the collective field in functional integral representation and integrating out the electron degrees of freedom [2], or by more conventional techniques [66]. *Assuming* that the resulting action $S\{\phi\}$ can be expanded in powers of ϕ with spatially local coefficients, one arrives at the Hertz model,

$$S = S_2 + S_4 + S_h + \ldots . \tag{69}$$

Here the second-order term is given by

$$S_2 = \frac{1}{\beta V} \sum_{\mathbf{k},\omega_n} \varepsilon_0 \left[\delta_0 + \xi_0^2 k^2 + \frac{|\omega_n|}{\gamma(k)} \right] \phi_{\mathbf{k},\omega_n} \cdot \phi_{-\mathbf{k},-\omega_n} \tag{70}$$

where the prefactor of ϕ^2 is nothing but the inverse of the dynamical spin susceptibility $\chi^{-1}(\mathbf{k},\omega_n)$. In this case the microscopic correlation length ξ_0 is $\sim k_F^{-1}$, where k_F is a Fermi wavevector, and ε_0 is the microscopic energy scale, given by the Fermi energy ε_F. The momentum summation extends up to a (bare) cutoff Λ_0.

The dynamic contribution $|\omega_n|/\gamma(k)$ accounts for damping of the spin fluctuations $\phi_{\mathbf{k},\omega_n}$ in the electron system induced by excitation of particle–hole pairs across the Fermi level ("Landau damping"). The linear increase of the phase space for particle–hole excitations with excitation energy ω is reflected in the form of this contribution. For a ferromagnetic transition (or other transitions with a $\mathbf{Q} = 0$ order parameter), $\gamma(k) = v_F k$ as $k \to 0$, i.e., the damping rate diverges due to the abundance of particle–hole pairs with small momentum. This results in a theory with (bare) dynamical exponent $z = 3$. For an antiferromagnetic transition $\gamma(k) \sim \gamma_0$, independent of k, yielding $z = 2$. These forms of $\gamma(k)$ hold if the wavevector of the spin mode in either case is well inside the particle-hole continuum, i.e., if the ordering wavevector $\mathbf{Q}$ connects points on a $(d-2)$-dimensional manifold of points on the Fermi surface. For an antiferromagnetic system with a small Fermi volume and a large ordering vector $Q > 2k_F$, the particle–hole pairs decouple from the spin fluctuations and ω enters quadratically as in (55). The crossover from linear to quadratic ω dependence has been discussed in detail by [67, 61]. A special situation arises when an antiferromagnetic mode is tied to wavevector $2k_F$ at the edge of the particle-hole continuum ("nesting") [3, 4].

The fourth-order term S_4 of the action accounts for the self-interaction of spin excitations,

$$S_4 = u_0 \int d\tau \int d^d r \, [\phi(r,\tau)^2]^2 \tag{71}$$

with u_0 denoting the strength of the interaction. The last term in the action (69), S_h, contains the effect of a uniform (staggered) magnetic field $\mathbf{h}$ of wavevector $\mathbf{k}$ in the ferromagnetic (antiferromagnetic) case,

$$S_h = \frac{1}{\beta V} \sum_{\mathbf{k},\omega_n} \mathbf{h}_{\mathbf{k},\omega_n} \cdot \phi_{-\mathbf{k},-\omega_n} \, . \tag{72}$$

Let us point out here that the damping term in the Hertz theory has been derived under the *assumption* of Fermi-liquid behavior of the electronic quasiparticles. This needs to be justified a posteriori. We also note that in the ordered phase, i.e., $r < 0$, $T < T_c$, the action does *not* apply: the form of the damping term will be modified due to the appearance of a gap in the electronic band structure.

3.4 Scaling equations and thermodynamic properties

The Hertz model defined by the action (69) and(70) has been studied near its critical point [2, 3] using the perturbative RG. To define a RG transformation one investigates how a change of the cutoff (either in momentum space or frequency space or both) and their subsequent rescaling can be absorbed in a redefinition of the coupling constants, using for example the perturbative expressions for the free energy [3]. A detailed derivation of the RG scaling equations can be found in [3, 4].

The RG equations have a Gaussian fixed point at $T = u = \delta = 0$, which is unstable with respect to the tuning parameter δ. Of particular interest is the differential equation for the quartic coupling u: if $d+z > 4$, u scales to zero, i.e., the upper critical dimension is given by $d_c^+ = 4-z$ rather than $d_c^+ = 4$ as for classical critical phenomena. Thus in many cases of interest one can expect the critical behavior to be that of the Gaussian model. For example, in the cases considered above, $z = 2$ (antiferromagnetic metal) or $z = 3$ (ferromagnetic metal), and in dimensions $d \geq 2$ the Gaussian model applies.

The case $d = z = 2$ is marginal and needs special consideration. Here the expansion in the quantum critical regime (see Fig. 3) is not very revealing as it is only valid if $\ell n \ell n(\bar{c}/T) \gg 1$. As pointed out by [68], this condition will never be satisfied in practice. Thus, the $d = z = 2$ quantum critical theory is in general *not* in a weak-coupling regime at any $T > 0$. Instead, a strongly coupled effective classical model emerges that can be used to determine the critical dynamics, for details see [68]. Abanov and Chubukov [69] have shown that in $d = 2$ the derivation of the LGW theory breaks down completely. Somewhat similar to the FM case to be discussed in Sec. 3.6, the low-energy modes of the Fermi liquid lead to long-range order-parameter interactions. Formally, the coefficients of the high-order interactions in the LGW functional diverge, leading to an infinite number of marginal operators. An analysis of the resulting theory is difficult and again requires a treatment of a coupled field theory of order-parameter fluctuations and fermions – this has not been done until now. It was concluded that the 2D metallic AFM

TABLE 1. Results for thermal expansion α (61), specific heat C, and Grüneisen ratio Γ (62), at a QCP in the quantum disordered (Fermi-liquid) regime $T \ll r^{z/2}$. The left columns ($z=2$) are for a metallic AFM, the right ones ($z=3$) for a metallic FM. Non-universal prefactors of $\alpha_{\rm cr}$ and $C_{\rm cr}$ are not shown. The prefactors of $\Gamma_{\rm cr}$ and $\Gamma_{H,{\rm cr}}$ are (up to the logarithmic correction for $d=z$) universal; these quantities are given by $\Gamma_{\rm cr} = (dr/dp)\Gamma_{r,{\rm cr}}/V_m$ for pressure-tuned QCP with $r=(p-p_c)/p_c$, and $\Gamma_{H,{\rm cr}} = (dr/dH)\Gamma_{r,{\rm cr}}$ for field-tuned QCP with $r=(H-H_c)/H_c$. Note that for $d=3$, $z=2$ the specific heat is dominated by a non-critical fermionic contribution $C \sim T$ [65].

	$d=2$ $z=2$	$d=3$ $z=2$	$d=2$ $z=3$	$d=3$ $z=3$
$\alpha_{\rm cr} \sim$	Tr^{-1}	$Tr^{-1/2}$	$Tr^{-3/2}$	Tr^{-1}
$C_{\rm cr} \sim$	$T\ell n\frac{1}{r}$	$-Tr^{1/2}$	$Tr^{-1/2}$	$T\ell n\frac{1}{r}$
$\Gamma_{r,{\rm cr}} =$	$\left(r\ell n\frac{1}{r}\right)^{-1}$	$-(2r)^{-1}$	$(2r)^{-1}$	$\left(r\ell n\frac{1}{r}\right)^{-1}$

shows a continuous transition with non-trivial exponents [69], but concrete predictions, e.g., for transport are lacking.

As mentioned above, in the cases where $d+z>4$ the system scales to the non-interacting (Gaussian) fixed point. Although the quartic coupling u is formally irrelevant, it still affects a number of physical quantities, like the location of the critical line or the order-parameter susceptibility at $r=0, T>0$. To describe this within a scaling approach, as in Sec. 3.2, one has to include explicitly the quartic coupling u in the scaling ansatz for the free energy

$$f_{\rm cr} = \rho_0 T^{(d+z)/z}\phi_4\left(\frac{r}{T^{1/(\nu z)}}, uT^{(d+z-4)/z}\right) \tag{73}$$

which replaces Eq. (59). While u is irrelevant, it has to be kept as ϕ_4 can become a singular function of u. In such a case the "naive" scaling relations derived from Eq. (59) are modified, and u is called a "dangerously irrelevant" variable. The free energy in the theory of Hertz is obtained by integrating the RG equation for the free-energy density up to the scale where $\delta(b_0)=1$ and adding the Gaussian free energy at this scale, see [4] for details.

Shown in Tables 1 and 2 are the results for the critical contributions to the specific heat C, the thermal expansion α, Eq. (61), and the resulting Grüneisen ratio Γ, Eq. (62), derived from the free energy $F = F_G + F_{SC}$, in both the quantum disordered and the quantum critical regimes. Up to logarithmic corrections the results agree with those derived from scaling in Sec. 3.2. Note that for $d=z$ the prefactor in (63) vanishes. The $1/r$ dependence of $\alpha_{\rm cr}$ for $d=z$ arises from a $T^2\ell n(1/r)$ correction to $F_{\rm cr}$ not captured by scaling. For the quantum critical regime in $d=1/\nu=2$ the thermal expansion is logarithmic. The argument of the logarithm is a power of T for $d+z>4$ and is itself logarithmically dependent on T for $d+z=4$; these features reflect the dangerously irrelevant or marginal nature of the quartic coupling u_0.

In addition to the critical contributions, the measured quantities also contain non-critical background components. We list here the full results for the purpose of comparisons with experiments in heavy-fermion compounds undergoing an antiferromagnetic transition ($z = 2$). Consider first $d = 3$. At the QCP ($r = 0$)

$$\alpha = a_1 T^{1/2} + a_2 T , \tag{74}$$

where the a_2 term comes from the (fermionic) background contribution. However, approaching the QCP in the Fermi-liquid regime gives:

$$\alpha = (a_1/r^{1/2} + a_2)T . \tag{75}$$

For $d = 2$ and $z = 2$, we have at the QCP ($r = 0$)

$$\alpha = a_1 \ell n[b \ell n \frac{T_0}{T}] + a_2 T , \tag{76}$$

and in the Fermi-liquid regime approaching the QCP:

$$\alpha = (a_1/r + a_2)T . \tag{77}$$

In two dimensions, the thermal expansion coefficient at $r = 0$ diverges in the zero-temperature limit, Eq. (76), in sharp contrast to the textbook statement that $\alpha(T \to 0) = 0$.

Finally, we turn to other thermodynamic quantities. The static susceptibility is not critical in the case of an antiferromagnet and therefore the calculation of χ is delicate (for the ferromagnet $\chi \sim \xi^2$). Ioffe and Millis [70] performed a careful calculation of the various contributions and concluded that in $d = 2$ (with $z = 2$), the leading low-T dependence of the susceptibility is given by $\chi \approx \chi_0 - DT$ with non-universal constants χ_0 and D (probably positive for typical band structures). In analogy, we expect $\chi \approx \chi_0 - D'T^{3/2}$ in $d = 3$.

For a QPT driven by a magnetic field B, however, the susceptibility, $\chi = \partial M/\partial B = -\partial^2 F/\partial B^2$, is more singular since the uniform magnetic field is a relevant perturbation acting as the control parameter, $r \propto B - B_c$. Below the upper critical dimension one obtains from scaling $\chi \propto T^{(d+z-2/\nu)/z}$ in the quantum critical regime. For an itinerant

TABLE 2. Results for LGW QCP in the quantum critical regime $T \gg |r|^{z/2}$ (cf. Table 1).

	$d=2$, $z=2$	$d=3$, $z=2$	$d=2$, $z=3$	$d=3$, $z=3$
$\alpha_{cr} \sim$	$\ell n \ell n \frac{1}{T}$	$T^{1/2}$	$\ell n \frac{1}{T}$	$T^{1/3}$
$C_{cr} \sim$	$T \ell n \frac{1}{T}$	$-T^{3/2}$	$T^{2/3}$	$T \ell n \frac{1}{T}$
$\Gamma_{r,cr} \sim$	$\frac{\ell n \ell n \frac{1}{T}}{T \ell n \frac{1}{T}}$	$-T^{-1}$	$T^{-2/3} \ell n \frac{1}{T}$	$(T^{2/3} \ell n \frac{1}{T})^{-1}$

magnet, $d = 3, z = 2$, Fischer and Rosch [71] found in contrast that the critical contribution χ_c of χ is a singular function of the irrelevant coupling u, $\chi_c \sim \xi(T)T \sim T^{1/4}/\sqrt{u}$. It therefore strongly violates scaling.

The above discussion was based on a renormalization group analysis following the work of Hertz [2]. However, many of the main results, e.g., for the phase diagram and susceptibility of nearly ferromagnetic [71] or antiferromagnetic metals [72] in three dimensions, have been obtained earlier by Moriya and coworkers. Their so-called SCR (self-consistently renormalized) formalism, described in detail in [66], is a self-consistent one-loop approximation for the scattering of spin fluctuations. Within the RG approach one can understand the enormous success of SCR theory and where it fails. The SCR formalism works above the upper critical dimension, especially for $d = 3$ and $z = 2$ or 3; these approximations reproduce correctly the leading behavior of the relevant physical quantities including prefactors. The SCR formalism fails upon approaching the classical transition as it is blind for the Ginzburg criterion. A recent extension of the SCR method [73] has been used to obtain the correct behavior at the upper critical dimension, $d + z = 4$.

3.5 Transport properties

For quantum critical systems above the upper critical dimension, the low-T thermodynamics can be calculated reliably using standard methods like RG (see Sec. 3.4). Much less is known about transport properties. Earlier predictions of the resistivity close to an antiferromagnetic QCP (see [66] and references therein), for example, are not valid for perfectly clean [74] or weakly disordered samples [75, 76]. The difficulty arises because scattering from AFM spin waves is extremely anisotropic and affects only a small fraction of the Fermi surface. Therefore the transport properties depend both qualitatively and quantitatively on how *other* scattering mechanisms re-distribute quasiparticles and scatter them into these small regions.

In $d = 3$ and for small static electric fields, it is possible to treat transport within a simple Boltzmann approach, as the QCP is above its upper critical dimension, spin–spin interactions are irrelevant in the RG sense, and because furthermore the concept of Fermi quasiparticles is still (marginally) valid in three dimensions. In $d = 2$, a quasiparticle description is not possible at the QCP (see Sec. 3.6) and a transport theory is more complicated [77]. In the linear response regime in $d = 3$, for the case of a uniform scattering of FM QCP or if the scattering is dominated by (short-range) impurities, i.e. for the quantum critical region, one obtains [78, 66, 79]

$$\Delta\rho \sim \begin{cases} T^{3/2} & \text{AFM disorder-dominated, } \Delta\rho \ll \rho_0 \\ T^{5/3} & \text{FM} \end{cases} \tag{78}$$

This treatment fails completely in the case of an AFM QCP in a weakly disordered metal where quasiparticles scatter strongly close to lines on the Fermi surface with $\varepsilon_{\mathbf{k}} = \varepsilon_{\mathbf{k}+\mathbf{Q}}$, the so-called "hot lines". This has been first realized by Hlubina and Rice [74], who argued that in a clean metal close to an AFM QCP, the resistivity is dominated by quasiparticles from regions of the Fermi surface far away from the hot lines, where scattering rates are proportional to T^2. Accordingly, $\rho \propto T^2$ is expected in an ultra-clean metal close to the AFM QCP. This effect can be understood in a simple relaxation-time approximation, where the resistivity is calculated from a Fermi-surface average of the $\mathbf{k}$-dependent scattering time, $\rho \propto 1/\langle \tau_{\mathbf{k}} \rangle_{FS}$. Clearly, the longest scattering times dominate the conductivity and short-circuit contributions from the hot lines where τ is small. This effect is missed when $1/\langle \tau_{\mathbf{k}} \rangle_{FS}$ is effectively replaced by $\langle 1/\tau_{\mathbf{k}} \rangle_{FS}$. However, the predicted T^2 dependence of ρ is unobservable in real systems, as tiny amounts of disorder change the picture qualitatively – this has been pointed out by Rosch [75, 76]. He solved the Boltzmann equation numerically and analytically (in a certain scaling limit). For sufficiently small magnetic fields, weak disorder, and close to an AFM QCP in the paramagnetic phase in $d > 2$, the temperature-dependent part of the resistivity, $\Delta\rho = \rho(T) - \rho_0$, obeys a scaling form as function of magnetic field B, residual resistivity ρ_0, and distance from the QCP, here denoted as $r > 0$:

$$\Delta\rho(T) \sim T^{d/2} f\left[\frac{T^{(d-1)/2}}{\rho_0}, \frac{r^{(d-1)/2}}{\rho_0}, \frac{B}{\rho_0\sqrt{T}}\right]. \tag{79}$$

Most interesting is the case $B = 0, r = 0$ where the scaling function crosses over from $f(\alpha,0,0) \sim$ const for lowest temperature, i.e. $\alpha \to 0$, to $f(\alpha,0,0) \sim \alpha^{-(4-d)/(5-d)}$ for temperatures above a crossover scale proportional to $\rho_0^{2/(d-1)}$. Various crossovers in $d = 3$ are discussed below in Eq. (80) for $B = 0$.

The magnetoresistivity and the sensitivity to disorder can be used to decide whether an observed NFL behavior arises only from some "hot lines" on the Fermi surface typical for a spin-density wave transition or from a breakdown of the Fermi liquid on the whole Fermi surface which is expected in some other scenarios. Sensitivity to weak disorder and large non-linear effects in the magnetoresistivity are characteristic for the Hertz scenario.

For $B = 0$ and $d = 3$, one obtains for clean samples, characterized by a large residual resistivity ratio [RRR, defined as $\rho(T = 300\,\mathrm{K})/\rho_0$].

$$\frac{\Delta\rho(T)}{\rho(T_0)} \sim \begin{cases} t\sqrt{x} & , \ \max[x, \sqrt{rx}] < t < \sqrt{x} \\ t^{3/2} & , \ r < t < x \\ t^2/\sqrt{r} & , \ t < \min[r, \sqrt{xr}] \end{cases} \tag{80}$$

The dimensionless temperature, disorder and distance from the critical point are defined by $t \sim T/T_0$, $x \sim \rho_0/\rho(T_0) \sim 1/\mathrm{RRR}$, $r \sim 1/(k_F\xi)^2 \propto p - p_c$ (note $z\nu = 1$ here). T_0 is a characteristic temperature scale, e.g., an effective Kondo temperature of a heavy-fermion system. In a sufficiently clean sample with a RRR of the order of 100, the resistivity at the QCP is *linear* in temperature over a large regime $T_0/\mathrm{RRR} \leq T \leq T_0/\sqrt{\mathrm{RRR}}$ due to the interplay of weak impurity scattering all over the Fermi surface and strong inelastic scattering close to the hot lines. Even at some distance from the QCP, one has to go to rather low temperatures to recover Fermi-liquid behavior. In a very clean sample with a large RRR, the crossover temperature $t \sim \sqrt{rx}$ can be considerably lower than the characteristic scale ($t \sim r$) below which Fermi-liquid behavior is recovered in thermodynamic measurements.

The picture developed by Rosch [75, 76], fits well with some experimental trends. In a large number of systems, the resistivity at the AFM QCP seems to rise with $T^{1.5}$, e.g., in $CeCu_2Si_2$ [80], $CeNi_2Ga_2$ [81], $CeCu_{6-x}Ag_x$ [82] $CeNi_2Ge_2$ [83], $CePd_2Si_2$, or $CeIn_3$ [84]. All of these systems are "dirty" in the sense that $\Delta\rho \ll \rho_0$ in the temperature range where the above exponents have been fitted. In a few systems, the Lonzarich group [84, 85, 83] succeeded in preparing high-quality samples with RRR values of the order of 100 and the resistivity seems to rise with exponents smaller than 1.5. In the cleanest samples the resistivity is almost linear in T over a substantial temperature range. Especially, the sensitivity to weak disorder might be interpreted as a signature that the systems can be described within the theory sketched above. Also, resistivity measurements experiments in U_2Pt_2In and $U_3Ni_3Sn_4$ [86, 87] have been fitted to the theory. A number of systems do not fit into the scenario above described, an example being $CeCu_{6-x}Au_x$ where $\Delta\rho$ is linear in T in a regime with $\Delta\rho \ll \rho_0$. Remarkably, one obtains a linear T dependencewhen one assumes that 3D electrons scatter from 2D spin fluctuations [88] – see Sec. 4.1 for a more detailed discussion.

3.6 Multiple dynamical exponents at ferromagnetic quantum critical points

The metallic ferromagnet is an example where the LGW approach of Hertz fails due to the presence of fermionic modes in the system. The idea can be discussed in terms of time scales: In a nearly-critical quantum system, the length scale ξ may induce several diverging time scales. The order parameter fluctuates on the time scale $\xi_\tau \propto \xi^{z_{OP}}$, with, e.g., $z_{OP} = 3$ in a clean FM according to the theory of Hertz. A different time scale is induced by the fermions. In a clean system, electrons cross an ordered domain of size ξ ballistically in the much shorter time $t_B \propto \xi^{z_B}$, $z_B = 1$. In a disordered system, charge or spin (if conserved) diffuse over a distance of ξ in the time $t_D \propto \xi^{z_D}$, $z_D = 2$.

Vojta et al. [89] (see also [90, 91, 93]) have shown that for itinerant quantum critical ferromagnets these other slow modes are indeed important (for AFM the effect is less severe.) The problem becomes apparent when deriving the LGW functional from a microscopic theory in perturbation theory. A detailed analysis shows that a description of the QPT in terms of the critical modes *alone* is not possible for the ferromagnet [93]. A generalized critical theory has been set up and analyzed for the ferromagnetic QPT in a diffusive metal [90, 91, 92] – this theory involves both the ferromagnetic order-parameter field ϕ and the diffusive modes of the disordered metal. Remarkably, the Gaussian critical theory was found to be correct only up to logarithms, as suggested by the scaling with $z = z_B = 1$ or $z = z_D = 2$, respectively.

In the *clean* quantum critical itinerant ferromagnet, the logarithmic corrections to the actions in S_2^c and S_4^c lead probably to an instability of the ferromagnetic second-order transition: the ϕ^4 term changes sign at an exponentially small temperature which induces a weak first-order transition [94]. Such a first-order transition has, e.g., been reported in $ZrZn_2$ as we discuss in Sec. 4.2. This generic scenario has been discussed by Belitz et al. [94], and the resulting phase diagram is shown in Fig. 4. The $-q^2 \ell n 1/q$ correction in S_2^c can also drive the transition to an AFM QCP (see [95]). At some distance from the first-order or AFM transition, the predictions of the Hertz approach for the 3D ferrmomagnet regarding resistivity and specific heat are likely to hold (up to logarithmic corrections), with a resistivity $\propto T^{5/3}$ at the QCP and logarithmically diverging specific-heat coefficient C/T. The exponents, e.g., for the pressure dependence of the Néel

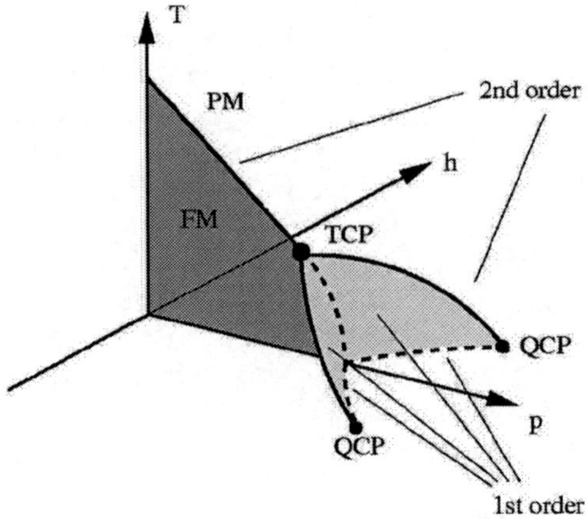

FIGURE 4. Generic phase diagram of an itinerant ferromagnet, as function of temperature T, tuning parameter p, and magnetic field h. PM (FM) denote the paramagnetic (ferromagnetic) phases, and TCP is a tricritical point. From [94].

temperature [3] close to the QCP (but not too close to the first-order transition) require more careful considerations – outside the scope of this review – as they involve directly the scaling dimension of the ϕ^4 term. Chubukov et al. [96] have critically studied whether self-energy effects and vertex corrections wash out the singularities that lead to the first-order transition. Interestingly, they found that such corrections can in principle modify critical exponents, but in the case of the ferromagnet the transition remains of first order.

In the *disordered* diffusive ferromagnet, the Gaussian fixed point is stable, but the exponents deviate strongly from mean-field behavior [90]. The quantum critical behavior shows up, e.g., in the resistivity. The $\sqrt{T}$ cusp in the resistivity of a dirty Fermi liquid is modified and a $T^{1/3}$ temperature dependence with logarithmic corrections is expected.

3.7 Pseudogaps close to antiferromagnetic critical points

In the antiferromagnetically ordered phase of a metal, gaps open in parts of the Fermi surface with $\varepsilon_{\mathbf{k}} \approx \varepsilon_{\mathbf{k}\pm\mathbf{Q}} \approx 0$, provided that the Fermi volume is sufficiently large. This feature is not captured by the LGW order-parameter theory of Hertz – this theory is not valid in the ordered phase, e.g., it wrongly predicts damped Goldstone modes. It is therefore important to ask whether precursors of these gaps will show up already in the paramagnetic phase close to the quantum critical point. In this regime, the behavior of the system is dominated by large antiferromagnetic domains of size ξ, slowly fluctuating on the time scale $\tau_\xi \sim \xi^{z_{\mathrm{OP}}}$ where z_{OP} is the dynamical critical exponent of the order parameter. As ξ is diverging when the QCP is approached, it is suggestive [97] to assume that the electrons will adjust their wave functions adiabatically to the local antiferromagnetic background and will therefore show a similar behavior as in the AFM ordered phase. Will precursors of this effect show up and induce pseudogaps in the paramagnetic phase for sufficiently large ξ? This would imply a breakdown of the LGW approach of Hertz. Pseudogaps play an important role in the physics of underdoped cuprates [98] and it has been speculated that they are indeed precursors of gaps in either superconducting, antiferromagnetic, flux or striped phases (not discussed in this review). Within the approach of Hertz, $z_{OP} = 2$ and the critical dimension for pseudogap formation is therefore $d_c = 2$, i.e., no pseudogaps are expected from this argument in $d = 3$ as long as interactions are not too strong [99].

This analysis is based on the assumption that *amplitude* fluctuations of the staggered order parameter exist and destroy the pseudogap. Electrons can adjust their wave functions much better adiabatically to angular fluctuations of the direction of the staggered magnetization than to fluctuations of its size. Schrieffer [97] has argued that pseudogap behavior will occur always sufficiently close to an AFM QCP. He considered models without amplitude fluctuations. Within the theory of Hertz, however, amplitude fluctuations are present close to the AFM QCP in $d = 3$ because the system is above its upper critical dimension. Furthermore, strong statistical interactions of the electrons with the magnetic excitations [99] might destroy pseudogaps even in the absence of amplitude fluctuations.

3.8 Breakdown of the Kondo effect in heavy-fermion metals

For heavy-fermion systems (HFS) it is generally accepted that the magnetic transition is driven by a competition of the lattice Kondo effect, which favors a paramagnetic ground state, and a magnetic RKKY or superexchange interaction between the local f moments [30]. In the heavy Fermi-liquid state the local moments contribute to the Fermi volume leading to a "large" Fermi volume, see Sec. 2.7 – the f electrons are usually termed "delocalized" in this situation.

Before discussing scenarios for the magnetic transition we have to think about the nature of the ordered phase in HFS (the state on the left-hand side of Fig. 1). Two distinct types of magnetically ordered metals appear possible. (i) Magnetism can arise from a spin-density wave instability of the parent heavy FL state. Here, Kondo screening is essentially intact, with a weak polarization of the local moments which are still "delocalized". We will refer to such a state as the SDW metal. (ii) A different kind of magnetic metal is possible where the localized moments order due to RKKY exchange interactions, and do not participate in the Fermi volume, i.e., Kondo screening is absent. We will denote this second state as a "local-moment magnetic (LMM) metal". The distinction between these two kinds of magnetic states can be drawn sharply, if the Fermi surfaces have different topologies (albeit the same volume modulo that of the Brillouin zone of the ordered state), such that they cannot be smoothly connected to one another.

Returning to the transition from the paramagnet to the antiferromagnet, one possibility is that the Fermi liquid undergoes a transition to a SDW metal – this QCP is well described by the LGW approach of Hertz, (69). In this situation the local moments remain screened accross the phase transition, i.e., a suitably defined lattice Kondo temperature stays finite at the QCP. The anomalous behavior close to AFM QCP in heavy-fermion systems like $CeCu_{6-x}Au_x$ and $YbRh_2Si_2$ (discussed in detail in Sec. 4.1 – inconsistent with the Hertz scenario – has stimulated speculations about a different type of transition [101, 102, 103, 104, 105, 106, 107, 108, 109, 110, 111]: If the ordered state is a LMM metal, the transition to be considered now involves the breakdown of Kondo screening (due to competing magnetic fluctuations), accompanied by an abrupt change (i.e., collapse) of the Fermi surface. Our present theoretical understanding of such transitions is limited, and we will describe a few theoretical approaches below. An obvious question then is: Can there be a continuous transition where the Kondo screening disappears concomitantly with the appearance of magnetic long-range order? Parenthetically, we note that some materials show a first-order volume collapse transition at finite temperature [112] – in contrast, we are interested here in a continuous zero-temperature transition.

A zero-temperature transition involving the breakdown of Kondo screening ("Kondo transition") has interesting implications: The collapse of the Fermi surface may be detected via photoemission or de Haas-van Alphen measurements, and transport properties like the Hall conductivity will show a jump upon crossing the transition at lowest T [104, 109, 111]. (At a SDW transition the Fermi surface evolves continuously, and the Hall coefficient displays a kink, but no jump; only at a magnetic-field driven transition does the derivative of the Hall current with respect to the magnetic field jump.) As the Kondo transition is not associated with a single critical (fermionic) wavevector, one may expect critical fluctuations in an extended range of the reciprocal space. These qualitative theoretical considerations fit remarkably well some recent experiments: In $CeCu_{6-x}Au_x$ (Sec. 4.1) the susceptibility at the AFM QCP was found to obey $\chi_{\mathbf{q}}(\omega) \sim 1/(f_{\mathbf{q}}+(T+ai\omega)^{\alpha})$ with an anomalous exponent $\alpha \sim 0.8$, obtained from fits to susceptibility measurements and inelastic neutron scattering data at various positions in momentum space including $\mathbf{q}=0$ [101, 102]. Momentum and frequency dependence "separate", this favors an interpretation in terms of a Kondo transition. For $YbRh_2Si_2$ (Sec. 4.1) no neutron scattering data are available to date. However, a recent magneto-transport measurement may indicate a jump in the Hall coefficient at the magnetic QCP [113]; in addition, magnetostriction data of Gegenwart et al. [114] showed the vanisching of several energy scales at the same QCP of $YbRh_2Si_2$.

Away from the critical point, a characteristic signature of a breakdown of Kondo screening may be a shift of the maximum in the resistivity to lower temperatures upon approaching the QCP (as this signals the crossover from "small" to "large" Fermi volume), however, concrete calculations for the transport crossover are lacking.

A first approach designed to capture the breakdown of the lattice Kondo effect due to magnetic bulk fluctuations employs an extension of the dynamical mean-field theory (DMFT) and has been worked out by Si and coworkers [104, 117, 105, 106]. It led to the proposal of a "local" QCP, based on the idea that the breakdown of Kondo screening is a spatially local phenomenon, i.e., it affects every spin of the underlying Kondo lattice independently.

The starting point is the Kondo lattice model, where localized spins $\mathbf{S}_i$ couple to the spin density of conduction electrons at lattice site i,

$$H=\sum_{k\sigma}\varepsilon_k c^{\dagger}_{k\sigma}c_{k\sigma}+J\sum_i \mathbf{S}_i\cdot\mathbf{s}_i+\sum_{i,j}I_{i,j}\mathbf{S}_i\cdot\mathbf{S}_j \tag{81}$$

where a direct spin-spin exchange term (I) has been added to the usual Kondo lattice model. While the usual DMFT maps the lattice problem to a single impurity in a fermionic bath, the extended DMFT (EDMFT) uses a mapping to a so-called Bose-Fermi Kondo model with both a fermionic and bosonic baths.

Within EDMFT, it is possible to describe situations where collective magnetic fluctuations destroy the Kondo effect. The Bose-Fermi Kondo model is known to have a continuous QPT, due to the competition of the two baths, between a phase with Kondo screening and one with universal local-moment fluctuations – for a more detailed discussion. The QCP of the lattice model is thus mapped – via EDMFT – onto an impurity QCP, where the magnetic instability of the lattice drives the Kondo effect critical. At this "local" QCP all self-energies are momentum-independent, and the non-local dynamics of the magnetic fluctuations is Gaussian. A qualitative analysis of the EDMFT equations in $d = 2$ [106] shows that a power-law behavior of $M(\omega)$ is possible at the QCP. A numerical solution of a simplified EDMFT (without fermionic self-consistency and with Ising magnetic symmetry) has confirmed this result [118, 119]. The impurity critical point has been shown to feature ω/T scaling in the susceptibility, similar to what has been found in experiments [102]. The collapse of the Kondo scale necessarily leads to a jump in the Fermi volume [103] that should affect the resistivity close to the QCP. Concrete theoretical predictions for transport are, however, lacking.

A different approach to the breakdown of Kondo screening, without any assumptions about spatial locality, starts by identifying the zero-temperature phase which arises when Kondo screening breaks down *without* the simultaneous onset of magnetic order (or other types of symmetry breaking). As has been detailed by Senthil et al. [107], the resulting state is a paramagnet where the conduction electrons form well-defined quasiparticles on their own and the local moments are in a fractionalized spin-liquid state – this phase has been termed "fractionalized Fermi liquid" (FL*). The spin-liquid component may be gapped or gapless, and may feature a secondary instability to an ordered state.

The QPT from FL* to FL involves a change from a small Fermi volume, containing only conduction electrons, to a large Fermi volume including all Kondo spins. Close to the transition the Fermi surface consists of two sheets with primarily c and f character, respectively. Approaching the QPT from the FL side, the quasiparticle weight on an entire sheet of the Fermi surface vanishes continuously – this illustrates how a discontinuous jump in the Fermi volume can happen at a continuous transition. Clearly, the transition is not associated with a specific critical (fermionic) wavevector, but it is also *not* spatially local, as all self-energies retain their momentum dependence.The Fermi liquid coherence temperature vanishes as the transition is approached from the FL side as $T_{\rm coh} \propto |r|$. The specific heat acquires a singular contribution from the gauge-field fluctuations and behaves as $C/T \sim \ell n(1/T)$ in $d = 3$, resembling the experimental result on $CeCu_{6-x}Au_x$. A preliminary transport calculation [108] led to a resistivity $\rho \sim 1/\ell n(1/T)$, inconsistent with experiments.

So far, magnetism is not involved in this scenario. Clearly, the spin liquid in the FL* phase is potentially unstable towards magnetic order at low temperature – the resulting state will be a LMM metal (Fig. 5). A particularly appealing scenario is that this instability arises as a secondary one, i.e., driven by an operator that is irrelevant at the critical point. This naturally leads to a Landau-forbidden transition within the concept of deconfined criticality, to be discussed below.

A related scenario for the breakdown of Kondo screening has been proposed [110]

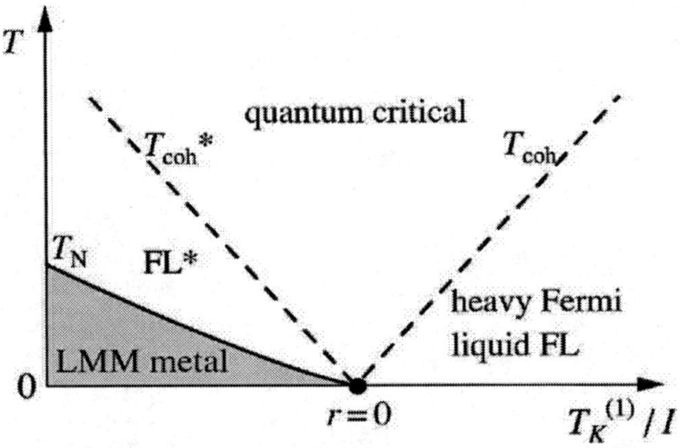

FIGURE 5. Phase diagram near the conjectured FL–LMM quantum transition of a heavy-fermion metal (see also Fig. 1). The primary phase transition, characterized by the breakdown of Kondo screening, is between the heavy Fermi liquid (FL) and fractionalized Fermi liquid (FL*). FL* is unstable at low energies towards local-moment magnetism. Two distinct energy scales are present on the l.h.s., manifested by the differing exponents by which T_N and T_{coh} (or T^*_{coh}) approach the quantum critical point. From [120].

based on the idea that the heavy quasiparticle fractionalizes into a spinon and a spinless fermion at the QCP [109]. Formally, the Kondo interaction of the Kondo lattice model is decoupled here with a fermionic field (in contrast to the slave boson in standard mean-field theory for the Kondo effect), and the dynamics of this fermion is key for the critical behavior. As above, this approach implies a jump in the Fermi volume. Building on a number of phenomenological assumptions on the dynamics and dispersion of the spinless-fermion mode, it is possible to describe various properties of $YbRh_2Si_2$, like a $T^{-1/3}$ upturn of the specific-heat coefficient at low temperatures which is not reflected in transport measurements. However, the theory of [110] is not able to describe the heavy Fermi-liquid state itself, as the spinless fermion cannot condense.

For all the above approaches, two possibilities arise. Generically, the breakdown of Kondo screening will not occur at the same point as the magnetic QCP which is associated with the formation of long-range order. Therefore the Hertz theory for an antiferromagnetic QPT seems to remain valid. The situation is different when the two QCP coincide (or are in close proximity).

Within the EDMFT model, this happens in the case of *two-dimensional* magnetic fluctuations: the local susceptibility at the magnetic QCP diverges which drives the Kondo effect critical, leading to a power-law behavior of $M(\omega)$. (In $d = 3$ the magnetic transition precedes a possible breakdown of Kondo screening.) Interestingly, the momentum dependence of the critical fluctuations in $CeCu_{6-x}Au_x$ appears to be two-dimensional [121], which also implies that the self-energy of the electrons is weakly momentum-dependent [88]. Thus, the EDMFT model reproduces a number of important aspects of the experimental data on $CeCu_{6-x}Au_x$, but it remains to understand why the magnetic fluctuations are two-dimensional (no obvious structural reason for this behavior is evident), and whether this fact is generic.

4. FERMI-LIQUID INSTABILITIES AT QUANTUM PHASE TRANSITIONS: EXPERIMENT

The main focus of this section wĄill be on experiments on magnetic QPT. Rather than trying to be exhaustive in treating every system for which NFL behavior near a magnetic QPT has been claimed, we will discuss mostly systems that have been characterized thoroughly. These comprise heavy-fermion systems, e.g., $CeCu_{6-x}Au_x$, $YbRh_2Si_2$, and the so-called Ce-115 compounds, as well as transition metal compounds such as MnSi and $ZrZn_2$. Superconductivity will be discussed in so far as it is relevant to quantum criticality. The issue of (one or several) QPT in cuprate high-temperature superconductors and other transition-metal oxides is beyond the scope of this review. Of course, the linear resistivity $\rho(T)$ in cuprate superconductors at optimal doping constitutes an early manifestation of NFL behavior (cf. the marginal Fermi-liquid phenomenology of Varma et al. [122]).

In discussing QPT, two generic cases have to be distinguished. The prototype is a continuous, e.g., second-order transition between two ground states tuned by a non-thermal parameter. If the transition is discontinuous (first order), the evolution of the length and time scales fluctuations will be limited. An interesting case arises when a line of first-order transitions terminates in a critical endpoint and if this critical point can be driven to zero temperature by another tuning parameter, leading to a quantum critical end point. This might be the case for $Sr_3Ru_2O_7$ where tuning parameters are magnetic field and the field orientation [123, 124].

This chapter will focus on intermetallic rare-earth and transition-metal alloys and compounds near a QPT. Actinide systems, with their $5f$ electrons being intermediate between localized (cf. $4f$ states) and itinerant (cf. $3d$ states) will only be discussed in reference to rare-earth or transition-metal systems.

4.1 Quantum-critical behavior in heavy-fermion systems

In general, Fermi-liquid theory has been very successful in describing the low-temperature behavior of metals with strong electronic correlations, including many heavy-fermion systems (HFS) as mentioned in Sec. 2.7. These materials often comprise a regular sublattice of $4f$ or $5f$ atoms, notably Ce, Yb, or U. Under certain conditions, a crossover of the magnetic behavior occurs with decreasing temperature T, from that of a collection of "free" localized $4f$ or $5f$ magnetic moments (subjected to the crystalline electric field and to spin-orbit interaction) coupled weakly to the conduction electrons, to low-T local singlets where the localized moment is screened completely by the conduction electrons by virtue of the Kondo effect, similar to the single-impurity case discussed in detail in Sec. 2.6.

The energy gain of singlet formation $T_K \sim D\exp(-1/N_0J)$ sets the temperature scale where this crossover occurs, i.e., the Kondo temperature T_K. Here N_0 is the (unrenormalized) conduction-electron (c) density of states at the Fermi level and J is the exchange constant between c and f electrons. The onset of a coherent state leading to a Fermi liquid at usually still lower temperatures is mostly signaled by a maximum in the electrical resistivity $\rho(T)$. Although there exists a lot of discussion on the onset of coherence at a coherence temperature $T_{\rm coh}$ thus defined, its relation to $T_{\rm K}$ remains under debate, see Sec. 2.7. Here we use for simplicity $T_{\rm K}$ as a measure of the energy gain due to singlet formation, keeping in mind that $T_{\rm K}$ in HFS might be modified by interactions with respect to the single-ion Kondo temperature in dilute magnetic alloys. Experimentally, this $T_{\rm K}$ is often determined through fits of the intermediate-temperature data for $C(T)$ or $\chi(T)$ to single-impurity results. (Frequently the symbol T^* is used for the characteristic temperature of a Kondo lattice, instead of $T_{\rm K}$.)

At sufficiently low $T \ll T_{\rm K}$, Fermi-liquid (FL) properties are observed in many HFS with a very large effective mass m^* derived from the huge linear specific-heat coefficient $\gamma = C/T$ and a correspondingly large Pauli susceptibility, both being only weakly dependent on T. The electrical resistivity of a FL exhibits a contribution $\Delta\rho = AT^2$ arising from electron–electron collisions, see Sec. 2.5. The phenomenological correlations $\gamma \sim \chi$ [33] and $A \sim \gamma^2$, see Sec. 2.7, observed approximately for different HFS, do suggest the validity of the FL description. The Wilson ratio $R = (\chi/\gamma)(\pi^2 k_B^2/\mu_0\mu_{\rm eff}^2)$ deviates from the free-electron value $R = 1$, see Fig. 2 [33]. The observed values of $R \sim$ 2 to 5 can be accounted for in the frame of FL theory by a negative Landau parameter F_0^a of the order of -0.5 to -0.8, see Sec. 2.7, Eq. (36). The problem of proving or disproving FL behavior in HFS lies in the low energy scale set by the Kondo temperature $T_{\rm K} \sim$ 10 to 100 K compared to conventional metals where $T_F \sim 10^4$ to 10^5 K. The first heavy-fermion system discovered was $CeAl_3$, with a γ coefficient of ~ 1.5 J/molK^2 [125]. The interest in these materials increased tremendously when Steglich et al. [126] reported superconductivity in $CeCu_2Si_2$, with the heavy quasiparticles being responsible for the superconductivity. The early work on heavy-fermion systems has been reviewed [127].

The competition between on-site Kondo interaction, quenching the $4f$ or $5f$ localized magnetic moments, and intersite RKKY interaction between these moments via the conduction electrons allows for non-magnetic or magnetically ordered ground states in HFS. In the Doniach picture [30], this competition is governed by a single parameter, namely the effective exchange constant J between conduction electrons and local moments, which enters the characteristic energy scales $T_{\rm K}$ and $T_{\rm RKKY} \sim J^2N_0)$ for Kondo and RKKY interactions, respectively (see Sec. 2.7). The strength of the exchange interaction is usually tuned by composition or pressure. In addition, a large magnetic field can suppress Kondo screening. Owing to the extremely strong dependence of the Kondo energy scale on the interatomic distance d, which arises from the exponential dependence of $T_{\rm K}$ on J, volume changes are often the dominant effect in producing the magnetic–non-magnetic transition if isoelectronic constituents are substituted against each other. This is also reflected in the very large Grüneisen parameter in these materials [128]. Often the coupling constant $g = N_0J$ is used as the driving parameter for the magnetic–non-magnetic transition. However, in view of the argument just given, J is the quantity of preponderant influence.

Many HFS show some kind of static magnetic order with often very tiny ordered moments (of the order of or smaller than $10^{-2}\,\mu_B$). A particularly intriguing example is $CeAl_3$, long considered the archetypal HFS without magnetic order. However, $CeAl_3$ has been shown to order antiferromagnetically when produced in single-crystalline form [129]. This is probably due to very subtle structural differences, e.g., strains, between polycrystals and single crystals. An inhomogeneous distribution of Kondo temperatures was inferred from NMR measurements in this material [130].

This example shows that disorder can have a decisive influence on the low-temperature properties even in nominally stoichiometric HFS samples. For non-stoichiometric substitutional alloys one can distinguish between substitution of the $4f$ or $5f$ site and on the ligand site. In both cases, a pronounced effect may be anticipated. In the former case, replacement of Ce or U by a non-magnetic atom in a otherwise stoichiometric HFS might lead to the formation of a Kondo hole. In the latter case, replacement of the ligand atom around a given Ce or U site might change the hybridization and hence of the local T_K. In general, predictions are difficult as to which effect will be stronger in a given system. Although much work on QPT that were tuned by composition of substitutional alloys has been done, stoichiometric compounds avoiding disorder are preferable. Alternatively, different tuning parameters should be employed to check the role of disorder, as has been done for $CeCu_{6-x}Au_x$.

4.1.1 $CeCu_{6-x}Au_x$ and $CeCu_{6-x}Ag_x$

$CeCu_6$ has been established as a HFS showing no long-range magnetic order down to the range of ~ 20 mK [131, 132]. $CeCu_6$ crystallizes in the orthorhombic Pnma structure and undergoes an orthorhombic–monoclinic distortion around 200 K [133]. The change of the orthorhombic angle is only small ($\sim 1.5°$). In order to avoid confusion, we always use the orthorhombic notation for the direction of the lattice vectors. $CeCu_6$ exhibits a pronounced magnetic anisotropy with the magnetization ratios along the three axes $M_c : M_a : M_b \approx 10 : 2 : 1$ at low T [131]. The measurements of the specific heat C were extended down to 10 mK and of the magnetic susceptibility χ to even below 1 mK [134]. Their analysis of χ at very low T (after subtraction of an impurity contribution attributed to Gd) suggests magnetic order around 5 mK. This is backed by NQR measurements which likewise hint at (possibly nuclear) magnetic order [135]. Direct evidence for magnetic order below 2 mK was found in the *ac* magnetic susceptibility and thermal expansion [136]. μSR measurements have put an upper limit for a static moment of 10^{-2} to 10^{-3} μ_B/Ce-atom (depending on the assumption of long-range magnetic vs. spin-glass order) above 40 mK [137].

Although $CeCu_6$ does not order magnetically above 5 mK, the expectation $C/T \approx$ const for a FL is not met very well (see Fig.6). The single-ion Kondo model with $T_K =$ 6.2 K does not fit the data below ~ 0.4 K [138]. Instead C/T increases slightly towards low T which might be a precursor of the 5-mK order. On the other hand, the T^2 dependence of the electrical resistivity is rather well obeyed between 40 and 200 mK [131]. Antiferromagnetic fluctuations were observed in inelastic neutron scattering by peaks in the dynamic structure factor $S(\mathbf{Q}, \omega)$ for energy transfer $\hbar\omega = 0.3$ meV at $\mathbf{Q}$ = (1 0 0) and (0 1$\pm$0.15 0) [139]. The rather large widths of these peaks correspond to correlation lengths extending roughly only to the nearest Ce neighbors. These correlations vanish at a field of ≈ 2 T applied along the easy c direction, also associated with a shallow maximum at 2 T at very low T in the differential magnetic susceptibility dM/dB [140]. This maximum has been identified with the "metamagnetic transition" in loose analogy to the metamagnetic transition in strongly anisotropic antiferromagnets.

Upon alloying with Au the $CeCu_6$ lattice expands while retaining – in fact: stabilizing – the orthorhombic (at room temperature) Pnma structure [143]. Thus the hybridization between Ce $4f$ electrons and conduction electrons, and hence J, decrease, leading to a

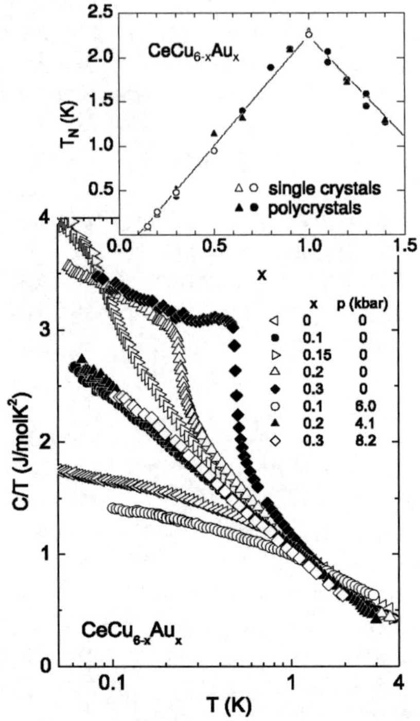

FIGURE 6. Specific heat C of $CeCu_{6-x}Au_x$ in the vicinity of the QPT plotted as C/T vs. temperature T (logarithmic scale). Application of hydrostatic pressure at the respective critical value p_c shifts C/T of the antiferromagnetic samples $x = 0.2$ and 0.3 towards NFL behavior for $x = 0.1$ at ambient pressure [153, 155]. The inset shows the Néel temperature T_N of $CeCu_{6-x}Au_x$ vs. Au concentration x as determined from specific heat (triangles) and magnetic susceptibility (circles). From [143].

stabilization of localized magnetic moments which can now interact via the RKKY interaction. The result is incommensurate antiferromagnetic order in $CeCu_{6-x}Au_x$ beyond a threshold concentration $x_c \approx 0.1$. This was first inferred from sharp maxima in the specific heat $C(T)$ and magnetization $M(T)$ [144] and confirmed by neutron scattering [145, 146, 147].

For $0.1 \leq x \leq 1$ where Au occupies exclusively the Cu(2) position in the $CuCu_6$ structure, the Néel temperature T_N varies linearly with x (Fig.6). For the stoichiometric compound $CeCu_5Au$ a complex magnetic phase diagram has been mapped out [148]. Beyond $x = 1$, T_N decreases again. This is due to a subtle change within the orthorhombic structure: for $x < 1$ the lattice parameters a and c increase while b decreases with growing Au content, whereas for $x > 1$ all three lattice parameters a, b, and c increase [143]. The orthorhombic–monoclinic transition is quickly suppressed with increasing x and vanishes around $x = 0.14$ [149].

Fig. 6 shows specific-heat data for concentrations in the vicinity of the critical concentration $x_c \approx 0.1$ plotted as C/T. The Néel temperature T_N manifests itself as a sharp kink in C/T which becomes less pronounced as T_N decreases. It is, however, still clearly visible, for $x = 0.15$ where $T_N = 0.080\,K$ as confirmed by a maximum in the *ac* and, quite recently, *dc* susceptibility [150]. For $x = 0.1$ we observe the non-Fermi-liquid behavior $C/T = a\,\ell n(T_0/T)$ between 0.06 K and ~ 2.5 K, i.e., over almost two decades in T, with $a = 0.578$ J/mol and $T_0 = 6.2$ K. These data represent one of the best examples of a logarithmic divergence of C/T for $T \to 0$. (The positive deviations above 2.5 K can be attributed to phonon and crystal-field contributions to C.) Concerning the critical NFL behavior at x_c it is important to verify that it does not arise from some inhomogeneity of the alloys, i.e., a distribution of magnetic ordering temperatures. A μSR study has shown that there is no ordered magnetic moment in a $x = 0.1$ sample, with the detection limit $\mu < 10^{-3}\,\mu_B$/Ce-atom [151]. Likewise, a wide distribution of Kondo temperatures appears to be unlikely [152].

The specific heat data can be integrated over temperature to obtain the entropy, where the dominant low-temperature contribution arises from the Ce moments. As a free moment (with effective spin 1/2) would give an entroy of $R\ln 2$, it is convenient to define a temperature $T_{1/2}$ where the entropy reaches $0.5R\ln 2$ – this defines a scale that is a measure of the (Kondo) temperature, below which the moments are screened. The x dependence of $T_{1/2}$ of $CeCu_{6-x}Au_x$is shown in Fig. 7. It is important to note that $T_{1/2}$ does not vanish at x_c. Its decrease with x can be essentially understood from the weakening of the Kondo effect because of the lattice expansion upon alloying Au: the increase of $T_{1/2}$ toward $x = 1$ is likely due to the quenching of the moments by the ordered antiferromagnetism.

As expected from the correlation between molar volume and T_N discussed above, T_N of $CeCu_{6-x}Au_x$ decreases under hydrostatic pressure p [141, 154, 155]. The Néel temperature (again as determined from the inflection point of $C(T)$ above the maximum) decreases linearly with increasing p for $x = 0.3$. For $x = 0.2$ a linear $T_N(p)$ decrease is also compatible with the data. $T_N \approx 0$ is reached for 7 - 8 kbar and 3.2 - 4 kbar for $x = 0.3$ and 0.2, respectively. At these pressures both alloys exhibit NFL behavior with, surprisingly, the same coefficients a and T_0 for both, coinciding with the NFL alloy $x = 0.1$ and $p = 0$ (Fig. 6). At 6.9 kbar for $x = 0.2$ the clear suppression of the low-T increase of C/T towards the data for $CeCu_6$ indicates restoration of the FL (not shown), i.e., one can pressure-tune $CeCu_{5.8}Au_{0.2}$ all the way from antiferromagnetic order through the QCP to FL. Likewise, application of $p = 6.0$ kbar for $x = 0.1$ drives this alloy even further towards FL behavior (Fig. 6), as C/T at low T now falls even below the $p = 0$ data of $CeCu_6$.

The magnetization $M(T)$ measured in a field $B = 0.1$ T for $x = 0.1$ (not shown) exhibits a cusp for $T \to 0$ which can be modeled as $\chi \approx M/B = \chi_0 - \alpha\sqrt{T}$ between 80 mK and 3 K. Roughly the same T dependence of M/B is found upon reduction of the field to 0.01 T with a slightly stronger upturn towards low T below 0.3 K [142]. The 0.1 T data can be very well described also by a different functional dependence, i.e., $\chi(T)^{-1} - \chi(0)^{-1} = aT^{\alpha}$ with $\alpha = 0.8$ [101], see below.

Fig. 8 shows $\rho(T)$ for different $CeCu_{6-x}Au_x$ alloys for current parallel to the orthorhombic a-axis. For $x < x_c \approx 0.1$, $\rho(T)$ increases at low T as $\rho(T) = \rho_0 + AT^2$ which is expected for a FL with dominant quasiparticle–quasiparticle scattering for $T \to 0$. This has been observed before for $CeCu_6$ [131]. For $x = 0.1$ a linear T dependence of ρ is observed between 20 mK and 0.6 K (see Fig. 8), signaling NFL behavior. The anisotropic

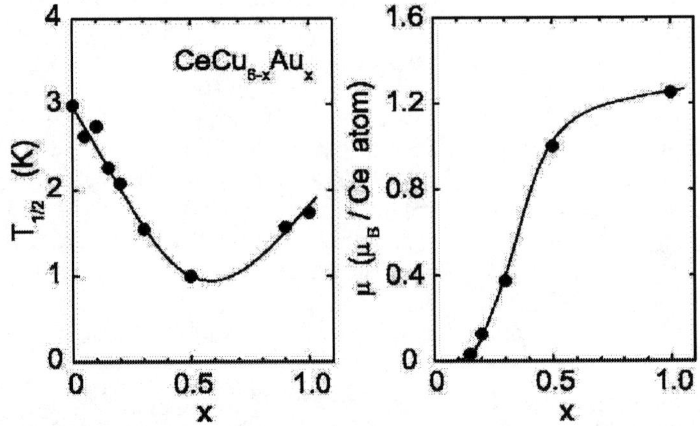

FIGURE 7. Temperature $T_{1/2}$ and staggered moment μ for $CeCu_{6-x}Au_x$Left: Concentration dependence of the temperature $T_{1/2}$ where the magnetic entroy reaches $0.5R \ln 2$, obtained from a temperature integration of the specific-heat data. The decrease in $T_{1/2}$ with x for $x < 0.5$ might reflect the decrease in T_K inferred in an analysis of the high-field specific heat in the frame of a single-ion Kondo model [153] while the larger $T_{1/2}$ toward $x = 1$ likely results from the strong AF order. Right: x dependence of the staggered moment per Ce ion extracted from elastic neutron scattering. From [147]

$\rho(T)$ dependence of the magnetically ordered alloys can be qualitatively interpreted in terms of the observed magnetic order: $\rho(T)$ for all alloys except $x = 1$ increases below T_N for current directions with a non-zero projection of the magnetic ordering vector Q determined from elastic neutron scattering [147]. An increase of $\rho(T)$ below T_N has been observed before in other HFS, for example, in $CeRu_{2-x}Rh_xSi_2$ as will be discussed below [154].

The abundance of low-energy magnetic excitations as $T_N \to 0$ has been suggested to cause the NFL behavior at the magnetic instability [155]. This is supported by the recovery of FL behavior in high magnetic fields B, Fig. 9 [155, 156]. A negative deviation from the $C/T \sim \ell n(T_0/T)$ divergence is seen for all fields $B \geq 0.2\,T$. Defining arbitrarily the temperature where the C/T data in a certain field cross the $B = 0$ straight line in Fig. 8 as a crossover temperature T_{cr}, we find roughly $T_{cr} \sim B$. This recovery of FL behavior of a quantum critical system upon application of a magnetic field has been observed in many other systems since. We add that the high-field specific heat of all $CeCu_{6-x}Au_x$ alloys including $x = 0.1$ can be reasonably well described [138, 141] within a single-ion Kondo model.

The $\ell n(T_0/T)$ dependence of C/T and the linear T dependence of ρ in $CeCu_{6-x}Au_x$ at the magnetic instability have constituted a major puzzle ever since they were first reported. The LGW theories for 3D itinerant fermion systems predict $C/T = \gamma_0 - \beta\sqrt{T}$ and $\Delta\rho \sim T^{3/2}$ for antiferromagnets ($z = 2$), while $C/T = \ell n(T_0/T)$ and $\Delta\rho \sim T^{5/3}$ are expected for ferromagnets ($z = 3$), (see Secs. 3.5 and 3.6). In addition, T_N should depend on the control parameter $r_x = x - x_c$ or $r_p = p - p_c$ as $T_N \sim |r|^\psi$ with $\psi = z/(d+z-2) = z/(z+1)$, for $d = 3$, while for $CeCu_{6-x}Au_x$ $\psi = 1$ for both r_x and r_p. Rosch et al. [88] showed in an analysis similar in spirit to that of Millis [3] that 2D

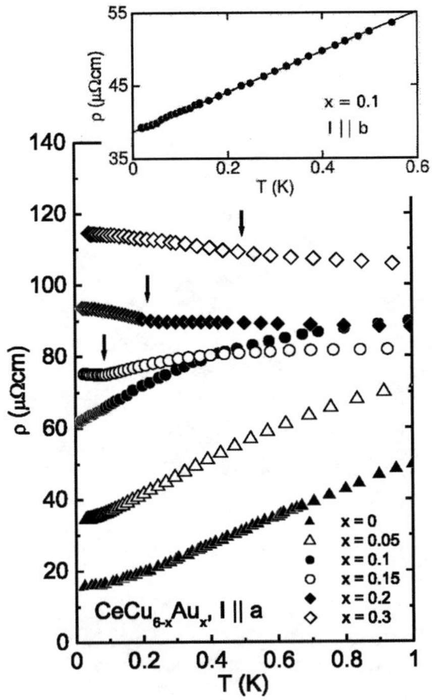

FIGURE 8. Electrical resistivity ρ of $CeCu_{6-x}Au_x$ vs. temperature T, with current applied to the a direction. Arrows indicate the Néel temperature. Inset shows data for $x = 0$ along the b direction. For all directions, $\rho = \rho_0 + A'T$ is observed. From [142].

critical fluctuations coupled to quasiparticles with 3D dynamics lead to the observed behavior $C/T \sim \ell n(T_0/T)$, $\Delta\rho \sim T$ and $T_N \sim |r|$.

Let us discuss the question of 2D vs. 3D magnetism in $CeCu_{6-x}Au_x$. $CeCu_{6-x}Au_x$ does exhibit 3D antiferromagnetic ordering, and the anisotropy of the electrical resistivity along different crystallographic directions does not exceed a factor of 2. Therefore $CeCu_{6-x}Au_x$ looks like a 3D antiferromagnetic metal. The magnetic structure of $CeCu_{6-x}Au_x$ $(0.15 \leq x \leq 1)$ has been investigated with elastic neutron scattering [147]. An example of resolution-limited magnetic Bragg reflections for $x = 0.2$ is shown in Fig. 10. The magnetic ordering vector is $\mathbf{Q}$ = (0.625 0 0.253) for $x = 0.2$ and remains almost constant up to $x = 0.4$. For larger x it jumps onto the a^* axis, $\mathbf{Q}$ = (0.56 0 0) for $x = 0.5$ and (0.59 0 0) for $x = 1$. A detailed investigation of the critical fluctuations at the critical concentration $x = 0.1$ using inelastic neutron scattering [121] showed that the critical fluctuations are strongly anisotropic and extend into the a^*c^* plane. This is inferred from a large number of l scans in the a^*c^* plane, some of which are shown in Fig. 11. Hence the dynamical structure factor $S(\mathbf{q}, \hbar\omega = 0.15\,\text{meV})$ has the form of rods (see Fig. 10). Since a quasi-1D feature in reciprocal space corresponds to quasi-2D fluctuations in real space, the 2D LGW scenario [88] appears to be applicable. The width of

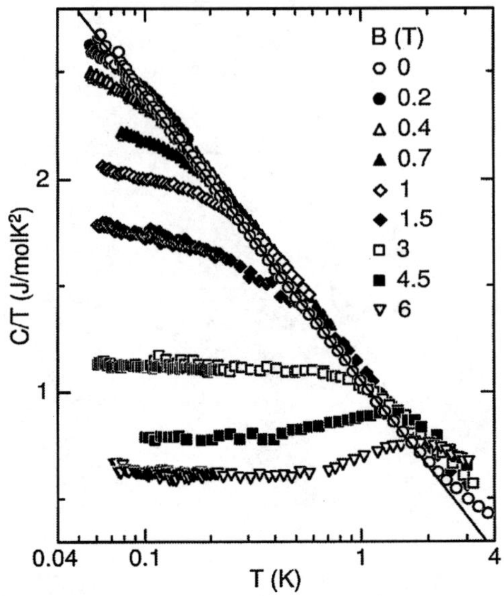

FIGURE 9. Specific heat of $CeCu_{5.9}Au_{0.1}$ plotted as C/T vs. temperature T (logarithmic scale) for magnetic field applied to the c direction. A nuclear Zeeman contribution was subtracted from the data. Solid line represents $C/T = a\ln(T_0/T)$ with $a = 0.578$ J/molK^2 and $T_0 = 6.2$ K. From [156].

$S(\mathbf{q},\omega)$ perpendicular to the rods is roughly a factor of five smaller than along the rods. It is an issue of current debate whether this anisotropy of the correlation length is enough to qualify the fluctuations as being 2D. The 3D ordering peaks for $x = 0.2$ and 0.3 fall on the rods for $x = 0.1$ which therefore can be viewed as a precursor to 3D ordering (Fig. 10). Fig. 11 demonstrates the essentially similar, albeit broader $S(\mathbf{q},\hbar\omega = \text{const})$ dependence for samples away from the critical concentration, i.e., for $x = 0$ and 0.2 [115].

The dynamic structure factor $S(\mathbf{q}=\text{const},\hbar\omega)$ of $CeCu_{6-x}Au_x$was investigated around $\mathbf{Q}$ = (0.8 0 0), i.e., on the rods (Fig. 10), by Schröder et al. [101]. They found a scaling of the dynamical susceptibility of the form

$$\chi^{-1}(\mathbf{q},E,T) = c^{-1}(f(\mathbf{q}) + (-iE + aT)^{\alpha}) \tag{82}$$

with an anomalous scaling exponent $\alpha = 0.74$ (a Lorentzian fluctuation spectrum would be described by $\alpha = 1$). This translates to

$$\chi''(E,T) = T^{-\alpha} g(E/k_B T) \tag{83}$$

with

$$g(y) = c \sin\left[\alpha \tan^{-1}(y)\right] / (y^2 + 1)^{\alpha/2}. \tag{84}$$

Fig. 12 shows the scaling obtained by [101], which has been confirmed by newer data [102]. The exponent $\alpha \approx 0.8$ implies $z = 2/\alpha = 2.5$. It is interesting to note that Eq. (82)

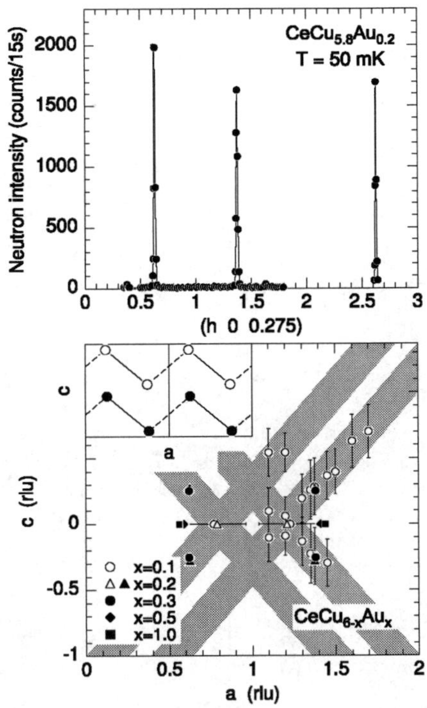

FIGURE 10. Upper panel: Resolution-limited magnetic Bragg reflection for $CeCu_{5.8}Au_{0.2}$ corresponding to an incommensurate magnetic ordering wavevector $\mathbf{Q} = (0.625\ 0\ 0.275)$ [147]. Lower panel: Position of the dynamic correlations ($x = 0.1, \hbar\omega = 0.1$ meV, $T < 100$ mK) and magnetic Bragg peaks ($0.2 \leq x \leq 1.0$) in the a^*c^* plane in $CeCu_{6-x}Au_x$. Closed symbols for $x = 0.2$ represents short-range order peaks. The vertical and horizontal bars indicate the Lorentzian linewidths for $x = 0.1$. The four shaded rods are related by the orthorhombic symmetry (we ignore the small monoclinic distortion). The inset shows a schematic projection of the $CeCu_{6-x}Au_x$ structure onto the *ac* plane where only the Ce atoms are shown. The bars in reciprocal space correspond to planes in real space spanned by *b* and lines in the inset. From [121].

implies for the static uniform susceptibility ($E = 0$, $\mathbf{q} = 0$)

$$[\chi'(T)^{-1} - \chi'(0)^{-1}]^{-1} = c(aT)^{\alpha}. \tag{85}$$

Schröder et al. [101] showed that indeed the static uniform susceptibility can be described by Eq. 73 with $\alpha = 0.8$ between 0.08 and 8 K in agreement with $\alpha = 0.74$ from neutron scattering. The fact that this quantum critical scaling persists to temperatures well above T_K appears surprising.

The two sets of experiments by [121] and [101] are complementary in that the former focuses on the $\mathbf{q}$ dependence and the latter on the E dependence of $\chi(\mathbf{q}, E)$ of the critical fluctuations at the QCP in $CeCu_{6-x}Au_x$. The data agree qualitatively in the range where they overlap. An open question is the origin of the fluctuation spectrum being effectively two-dimensional, and whether this fact may be generic for certain Ce-based materials.

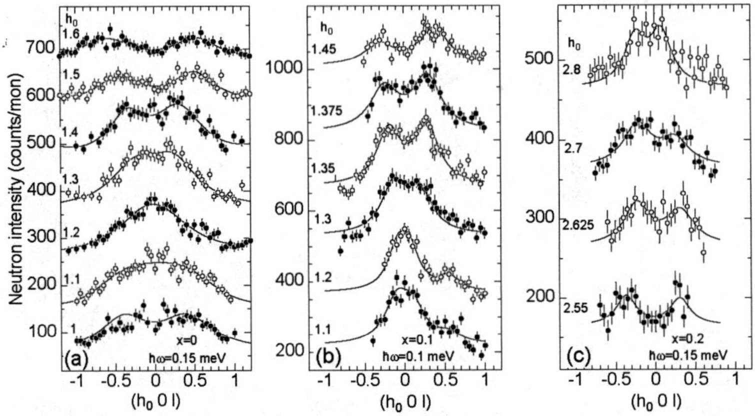

FIGURE 11. Inelastic neutron scattering scans in the reciprocal a^*c^* plane of $CeCu_{6-x}Au_x$ for (a) $x = 0$ with an energy transfer $\hbar\omega = 0.15$ meV, (b) $x = 0.1$, $\hbar\omega = 0.10$ meV, and (c) $x = 0.2$, $\hbar\omega = 0.15$ meV. Data were taken at the triple axis spectrometers IN12 ($x = 0$) and IN14 ($x = 0.1; 0.2$) at the ILL Grenoble. The individual scans are shifted by 100 counts ($x = 0; 0.2$) for 150 counts ($x = 0.1$) with respect to each other. Scans along $(2.8 - \xi 0 l)$ for $x = 0.2$ are symmetry equivalent to $(1.2\ \xi\ 0\ l)$ scans ($x = 0; 0.1$). From [115].

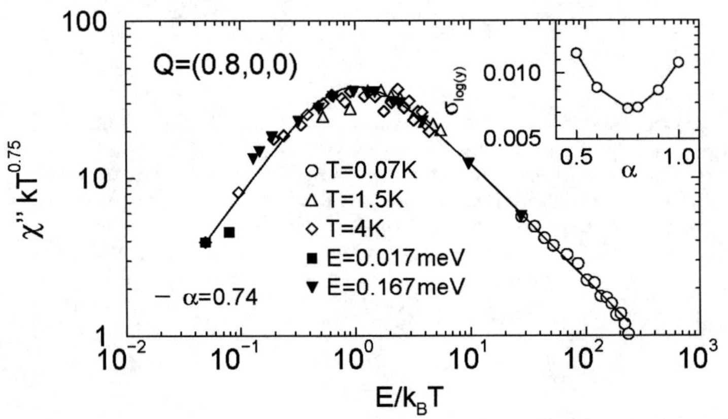

FIGURE 12. Scaling plot of inelastic neutron scattering data for $CeCu_{5.8}Au_{0.2}$ at q = (0.8 0 0) vs. E/k_BT. Solid line corresponds to a fit of the scaling function Eq. (83 with $\alpha = 0.74$. Inset shows how the quality of the scaling collapse varies with α. From [101].

It is clear that the 2D planes are spanned by the b axis and the connection between next-nearest neighbor Ce atoms (see Fig. 10). Only more detailed investigations can establish if, perhaps, the low-dimensionality arises from a strong spatial anisotropy or frustration of the RKKY interaction or the c–f hybridization.

The unusual type of scaling in the dynamical susceptibility of $CeCu_{6-x}Au_x$ at the QPT is incompatible with the LGW model – there, E/T scaling is only expected below the upper critical dimension (which is $d = 2$ for the metallic antiferromagnet). The experiments therefore prompted new theoretical concepts, based on the idea that at the QCP the (lattice) Kondo effect breaks down. Specific proposals include a "local" quantum critical point [105, 106], spin-charge separation [109, 110], or fractionalization of the Fermi liquid [107, 108]. (Note that the scenario of Si [105] requires the existence of two-dimensional critical spin fluctuations.) As discussed in Sec. 3.8, the breakdown of the Kondo effect should be connected with an additional energy scale (associated with Fermi surface fluctuations) vanishing at the QCP, apart from that of magnetic ordering. Experimentally, this has not been verified to date. In $CeCu_{6-x}Au_x$ essentially all experimental results indicate that some scale of the order of a few Kelvin stays finite at the QCP for a doping of $x = 0.1$, as can be seen from the existence of a maximum in the resistivity [142, 115], or the temperature dependence of $T_{1/2}$ (see Fig. 7).

Despite these open questions it should be stressed that $CeCu_{6-x}Au_x$is one of the best characterized heavy-fermion systems exhibiting NFL behavior. It is rewarding that the unusual behavior of the thermodynamic and transport quantities at the QCP can be consistently traced back to an unusual low-dimensional fluctuation spectrum.

Measurements of the ^{63}Cu nuclear spin-lattice relaxation time T_1 on one of the five inequivalent Cu sites in $CeCu_{5.9}Au_{0.1}$ down to 0.1 K, employing the 6.25 MHz ^{63}Cu NQR line, revealed a non-exponential nuclear magnetization recovery which was attributed to different Cu-neighbor configurations in the alloy [157]. More importantly, T_1^{-1} was found to vary as $T^{0.75}$ at low T. Since T_1^{-1} is proportional to the weighted squared average $\langle A(\mathbf{q})^2\rangle$ of transverse hyperfine couplings over the Brillouin zone, the "agreement" with the exponent α found in the T and E dependence of $\chi(\mathbf{q},\omega)$ (Eq. 83) is entirely accidental.

Although $CeCu_6$ alloyed with Au is the most thoroughly studied system, other alloys derived from $CeCu_6$ have been widely studied as well. Magnetic ordering in the system $CeCu_{6-x}Ag_x$ was discovered simultaneously with $CeCu_{6-x}Au_x$[158, 144, 159]. $CeCu_{6-x}Ag_x$ displays critical behavior around $x_c = 0.2$, with thermodynamic properties being very similar to $CeCu_{5.9}Au_{0.1}$, e.g., a logarithmic divergence of C/T. Interestingly, thermal expansion measurements have uncovered that the critical behavior is incompatible with both the 3D and 2D LGW models: the Grüneisen parameter Γ, Eq. (64), is expected to diverge in the quantum critical regime for $T \to 0$ as $\Gamma \sim T^{-1/(z\nu)}$ with $z\nu = 1$. However, the experiment shows a much weaker (logarithmic) divergence of Γ [160].

For field-tuned transitions the behavior appears to be different: NFL behavior in a polycrystalline $CeCu_{4.8}Ag_{1.2}$ alloy, subjected to a magnetic field, was reported, i.e., approximately $C/T \sim \ell n(T_0/T)$ between 0.35 and 2.5 K [161]. However, in the light of the strong magnetic anisotropy of this system, with B affecting crystallites of different orientations quite differently, this result should be viewed with caution. Subsequently, specific-heat data down to 0.07 K on a $CeCu_{5.2}Ag_{0.8}$ single crystal with $T_N = 0.7$ K were reported [161]. For a critical magnetic field $B_c = 2.3$ T applied along the easy direction, C/T varies logarithmically from $\sim$ 1.5 down to 0.2 K and then levels off. Furthermore, the resistivity $\rho(T)$ of $CeCu_{5.2}Ag_{0.8}$ can be described by a $T^{1.5}$ dependence at B_c. The authors interpret the data within the LGW scenario of [79], with $d = 3$ and $z = 2$.

A detailed comparison of pressure-tuned and field-tuned QCP on the *same* system

$CeCu_{5.8}Au_{0.2}$ [162] demonstrated that field, as opposed to pressure, drives the system to a 3D LGW quantum critical point. The data $\rho(T)$ and C/T for $CeCu_{5.2}Ag_{0.8}$ for B_c look quite similar to those for $CeCu_{5.8}Au_{0.2}$ for $B = 0.3$ or 0.5 T, i.e., in the region of $B_c \approx 0.4$ T as determined from neutron scattering. This difference to concentration and – very likely – also to pressure tuning was recently directly confirmed by inelastic neutron scattering where the observed $\omega/T^{1.5}$ scaling corresponding to the LGW scenario is much superior to ω/T scaling [163]. It is open whether the apparent similarity to the spin-fluctuation theory represents cross-over phenomena or the approach to a new QCP.

A few experiments have been performed on further $CeCu_6$-derived alloys. For $CeCu_{6-x}Pd_x$ and $CeCu_{6-x}Pt_x$ the NFL behavior was found in the specific heat at the critical concentration for the onset of magnetic order ($x_c \approx 0.05$ for Pd and ≈ 0.1 for Pt), while for $CeCu_{6-x}Ni_x$ a non-magnetic ground state is stabilized for small x [164]. Notably, the C/T data for the Pd and Pt alloys at their respective x_c agree well with those for $CeCu_{6-x}Au_x$ at the quantum critical point, suggesting universality for all QPT in $CeCu_6$ alloys in zero magnetic field. Note, however, that recent data on a $CeCu_{6-x}Ag_x$ polycrystal with $x = 0.2$, believed to be close to the quantum-critical concentration, exhibit a somewhat steeper slope of C/T vs. $ln(T_0/T)$, with, however, a gentle leveling off toward low T, leading in effect to a C/T larger by 20 % at 0.1 K compared to $CeCu_{5.9}Au_{0.1}$ [160].

4.1.2 $Ce_{1-x}La_xRu_2Si_2$ and $Ce(Rh_{1-x}Ru_x)_2Si_2$

$CeRu_2Si_2$ is, like $CeCu_6$, a canonical heavy-fermion system without apparent magnetic order. It crystallizes in the tetragonal I/4mmm structure. The ground state doublet of Ce^{3+} in $CeRu_2Si_2$ exhibits Ising character [165]. Static magnetism with very small ordered moment was detected by muon-spin-rotation experiments [166]. The magnetic order develops below about 2 K and the ordered moment reaches a value of $\sim 10^{-3}$ μ_B/Ce-atom at low temperatures ($T < 0.1$ K). The Kondo coupling between Ce $4f$ electrons and conduction electrons is stronger than in $CeCu_6$, this is also suggested by the rather low (for a heavy-fermion system) γ value of 0.36 J/molK^2 [167]. From specific-heat data T_K is found to be 20 K. Magnetic intersite correlations are also much stronger in $CeRu_2Si_2$ than in $CeCu_6$. This has been shown by inelastic neutron scattering where q-dependent correlations are peaked at $\mathbf{k}_1$ = (0.3 0 0) and $\mathbf{k}_2$ = (0.3 0.3 0) [139]. In a recent inelastic neutron scattering study, the spin fluctuations were interpreted within the LGW model [168].

A metamagnetic transition occurs in a field of $B_M \approx 8$ T [169, 170], which manifests itself as a steep jump in the magnetization curve $M(B)$. The corresponding maximum in dM/dB at B_M = 8 T is much more pronounced and much sharper than in $CeCu_6$ at $B_M = 2$ T. It is interesting to note in the present context that all the way through the metamagnetic transition, $CeRu_2Si_2$ exhibits well-defined FL behavior as evidenced by the prevalence of de Haas-van Alphen (dHvA) oscillations very close to the transition [171]. Some of the branches could be detected to within 0.005 T of B_M = 7.8 T. However, the branch of the heavy quasiparticles is strongly affected by the metamagnetic transition: its effective mass changes from $m^* = 12\,m_0$ at 6 T to $4\,m_0$ at 12.5 T.

Alloying $CeRu_2Si_2$ on the Ce site, i.e., *diluting* with non-magnetic La, leads to long-range antiferromagnetic order in $Ce_{1-x}La_xRu_2Si_2$. This might at first sight be even more

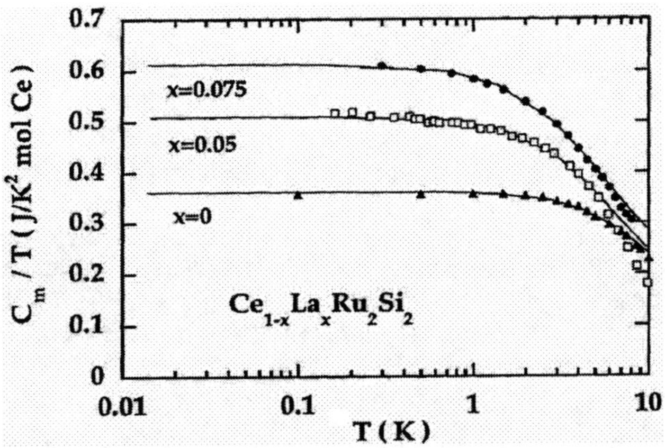

FIGURE 13. Magnetic contribution to the specific heat divided by T, C_m/T, vs. temperature T (logarithmic scale) for $Ce_{1-x}La_xRu_2Si_2$ for $x = 0, 0.05$ and 0.075. Solid lines are fits to the LGW model. From [177].

surprising than the observation of antiferromagnetism induced by Au doping of $CeCu_6$. It may, however, be simply attributed to a lattice expansion caused by replacing Ce by La (the famous lanthanide contraction backwards), leading to a strong reduction of T_K which outweighs the reduction of the number of Ce $4f$ moments for small x. This is nicely supported by measurements on $Ce_{1-x}(La_{0.63}Y_{0.37})_xRu_2Si_2$ which is isochoric to $CeRu_2Si_2$ and does not show magnetic order [172].

The long-range antiferromagnetic order for $x \geq 0.1$ manifests itself by maxima in the specific heat and the magnetic susceptibility [167] and in an increase of the electrical resistivity below the ordering temperature [173]. Upon application of hydrostatic pressure, T_N decreases as shown for $x = 0.2$ [174] where $T_N \rightarrow 0$ for $p_c \approx 7$ kbar. At the same time, the ordered magnetic moment as derived from elastic neutron scattering vanishes [174]. However, the pressure dependence $T_N \sim |p - p_c|^{\psi}$ is clearly sublinear. This matches the sublinear $T_N(x)$ dependence where T_N vanishes for a critical concentration $x_c \approx 0.075$ [175].

Upon approaching $x_c = 0.075$ where $T_N \rightarrow 0$, the specific-heat coefficient increases towards low T [176]. Additional experiments for $x = 0.075$ clearly show the strong enhancement of C/T [177]. They describe are interpreted consistently in terms of the LGW model as applied to heavy-fermion systems by [79]. In contrast to $CeCu_{6-x}Au_x$, the data can be quantitatively interpreted within the latter model for a 3D antiferromagnet, i.e., the specific-heat coefficient γ crosses over from a logarithmic T dependence in an intermediate T range to $C/T = \gamma_0 - \beta\sqrt{T}$ for $T \rightarrow 0$ (Fig. 13), which is the asymptotic behavior within the LGW model, Sec. 3.4. Likewise, the electrical resistivity at the critical concentration shows the LGW dependence for a disordered antiferromagnet, $\Delta\rho \sim T^{3/2}$ [177].

Inelastic neutron-scattering experiments on $Ce_{0.925}La_{0.075}Ru_2Si_2$ showed three-dimensional correlations for this system [178, 179]. At the wavevector $\mathbf{Q}_1 = (0.69\ 0\ 1)$

which corresponds to antiferromagnetic order, an approximate scaling of the form

$$\chi''(E,T) = T^{-1} g(E/(k_B T)^{0.8}) \quad (86)$$

with $g(y) = y/(1+y)^2$ was observed, however, only for $T \geq 5$ K. This apparent scaling is probably related to a crossover regime [179], as the Kondo temperature in this system is about 17 K. Furthermore, the linewidth of the dynamical fluctuations does not vanish for $T \to 0$ as it should for a quantum critical point. This may indicate that $x = 0.075$ is a little off the QCP. Nevertheless, the linewidth is reduced by a factor of 5 compared to the pure compound while the correlation length for $T \to 0$ is only 1.5 times larger than for pure $CeRu_2Si_2$. The dynamic susceptibility $\chi(\mathbf{q},\omega)$ is interpreted in terms of the Random Phase Approximation (RPA),

$$\chi(\mathbf{q},\omega) = \frac{\chi_{ss}(\omega)}{1 - J(\mathbf{q})\chi_{ss}(\omega)} \quad (87)$$

where χ_{ss} is the single-site q-independent dynamical susceptibility, and $J(\mathbf{q})$ describes the magnetic interactions between $4f$ sites. Within this approximation an overall satisfactory agreement between the parameters of the LGW model as derived from inelastic neutron scattering and those derived from the specific heat is obtained [178]. As a final point we mention that in addition to the inelastic signal, [178] a very small elastic signal at $\mathbf{Q}$ = (0.69 1 0) was observed below 1.8 K for $x = 0.075$. This is surprising because this sample is believed to be just below the critical concentration x_c = 0.08 for the onset of magnetic order. The linewidth is only slightly larger than the instrumental resolution, and the ordered moment is $\sim 0.02\,\mu_B$ at the lowest temperature which may be due to concentration fluctuations. The evolution of spin dynamics was also studied for an ordered sample with $x = 0.13$ around the critical pressure p_c = 2.6 kbar [180].

To summarize, $Ce_{1-x}La_xRu_2Si_2$ appears to be a system with a QCP exhibiting 3D critical fluctuations. Hence the LGW model describes the thermodynamic transport and neutron-scattering data rather well.

Turning to $CeRu_2Ge_2$, this system exhibits magnetic order, due to its larger Ce–Ce separation as compared to $CeRu_2Si_2$. Unlike many other Ce compounds, it orders ferromagnetically below T_C = 8 K [181, 182]. Upon applying pressure, the ferromagnetism is quickly suppressed, and a rather complex magnetic phase diagram with several phases evolves. By *ac* susceptibility measurements under pressure, and in analogy with doping studies of $CeRu_2(Ge_{1-x}Si_x)_2$ where neutron scattering has been done, these phases are identified as antiferromagnetic. Magnetic order in $CeRu_2Ge_2$ disappears at $p_c \approx 6.5$ GPa; at p_c NFL behavior is observed in the electrical resistivity, $\rho = \rho_0 + A'T$ [183, 184]. The system provides a nice example of the equivalence between hydrostatic pressure and chemical pressure excerted by smaller Si atoms.

$CeRh_2Si_2$, which is isostructural to $CeRu_2Si_2$, is a local-moment antiferromagnet (T_N = 36 K) with a relatively large ordered moment of $\mu = 1.5\,\mu_B$/Ce-atom [185]. T_N can be driven to zero by hydrostatic pressure of p_c = 9 kbar [186]. $T_N(p)$ varies slowly first and then precipitously drops to zero upon approaching p_c [187]. This may be suggestive of a first-order transition near $T = 0$, perhaps even stronger than the one observed in MnSi [188], Sec. 4.2. Clearly, a first-order transition would cut off quantum fluctuations. Consequently, anomalies in the specific heat associated with NFL behavior were not observed in $CeRh_2Si_2$ around p_c [189]. In addition, the resistivity above p_c exhibits a FL-like T^2 behavior [187]. It is interesting to note that in the vicinity of p_c and beyond, $CeRh_2Si_2$ becomes superconducting with T_c between 0.2 and 0.4 K [187].

The magnetic phase diagram of $Ce(Ru_{1-x}Rh_x)_2Si_2$ has been studied by several groups [190, 191, 192] with a number of techniques, including neutron scattering. The data reveal a complex $T_N(x)$ phase diagram with two different antiferromagnetic phases, one between $x \approx 0.05$ and 0.27, with a maximum T_N of $\sim$ 5 K, and the other phase originating from $CeRh_2Si_2$ ($x = 1$), with T_N decreasing steeply to $\sim$ 13 K for $x = 0.95$ and then exhibiting a plateau at T_N = 12 K until $x = 0.6$, when it drops again and reaches $T_N = 0$ around $x_c \approx 0.5$. The magnetic structure in both concentration ranges has been determined by elastic neutron scattering. For the $CeRh_2Si_2$-derived phase, a wavevector $\mathbf{Q} = (\frac{1}{2}\,\frac{1}{2}\,0)$ corresponding to an antiferromagnetic coupling within (0 0 1) planes [185] was reported to exist down to $x = 0.6$ [190], with ordered moments 1.5 μ_B/Ce-atom for $x = 1$ and 0.65 μ_B/Ce-atom for $x = 0.65$. This structure was essentially confirmed earlier, although these authors observed for $x = 1$ additional reflections corresponding to $\mathbf{Q'}$ = $(\frac{1}{2}\,\frac{1}{2}\,\frac{1}{2})$ which already had been seen by [193]. It is not clear whether this behavior arises from a homogeneous single phase or a domain structure.

A large effort has been devoted to analyze the low-x magnetic phase centered around $x = 0.15$. A sinusoidal spin-density wave with the moments ordered along the propagation axis has been inferred [192, 194] which shifts from incommensurate to commensurate: $\mathbf{Q}$ = (0 0 0.42), (0 0 0.45) and (0 0 0.5) for x = 0.15, 0.2, and 0.25, respectively. This magnetic structure is quite different from that of La-diluted $CeRu_2Si_2$ discussed above. Recent inelastic neutron scattering experiments for $x = 0.03$ revealed a $T^{3/2}$ dependence of the linewidth at the antiferromagnetic wavevector Q = (0 0 0.35), compatible with the LGW scenario [195].

It should be mentioned that $\rho(T)$ increases strongly below T_N which is interpreted as arising from a spin-density wave [154, 196]. A detailed scenario for Fermi-surface nesting has been developed for $x = 0.15$ [194]. Deviations from FL behavior have been observed in the region where $T_N \approx 0$ for $x = 0.5$ [189] and $x = 0.4$ [197] in both specific heat and magnetic susceptibility. While initial data for $x = 0.4$ show $C/T \sim \ell n(T_0/T)$ between 10 K and 0.15 K, data taken by the same group with higher precision indicate a slight leveling off below 1 K [198] the data for $x = 0.5$ level off more strongly below $\sim$ 1 K. At least part of the data can be described by a Kondo disorder model with a distribution of T_K [189].

Let us point out that the $T_N(x)$ functional dependence of $Ce(Ru_{1-x}Rh_x)_2Si_2$ for $x > 0.5$ does not at all resemble the $T_N(p)$ dependence of $CeRh_2Si_2$. This might serve as a warning in taking "chemical pressure" effects too literally and reducing changes brought about by alloying to simple volume changes. Even for $CeCu_{6-x}Au_x$ where the functional dependence of $T_N(p)$ and $T_N(x)$ are the same (both with an exponent $\psi = 1$) as discussed above, and where isoelectronic substituents are used (Au vs. Cu) as opposed to the present case (Ru vs. Rh), the calculated change of T_N with volume from $T_N(p)$ via the compressibility does not correspond to the change calculated from $T_N(x)$ with the x-ray determined volume $V(x)$ [199, 143]. The comparison with the $CeRu_2(Si_{1-x}Ge_x)_2$ system above might suggest that the chemical pressure exerted by the polyvalent metal is more akin to hydrostatic pressure than that of transition metals with partly covalent bonding.

4.1.3 $CeCu_2Si_2$ and $CeNi_2Ge_2$

$CeCu_2Si_2$ was the first heavy-fermion superconductor, discovered nearly 30 years ago by Steglich et al. [126]. Like the $CeRu_2Si_2$ family, it exhibits the tetragonal I/4mmm structure. This is not the place to review the tremendous amount of work that has been done ever since on this compound, revealing a very complex ternary phase diagram in the vicinity of the 1:2:2 stoichiometric composition [200, 201]. In a small region around this point, three different types of crystals exist with different ground states: A (magnetically ordered ground state with superconducting minority phase), S (superconducting ground state), and AS (both states almost degenerate). Which type of crystal is prepared depends on the stoichiometry, on the preparation method (polycrystals versus single crystals), and heat treatment [200]. Unfortunately, the different types could up to now not be traced to different crystallographic and/or defect-induced properties. Recently, the type of magnetic order in A-type crystals was identified by neutron diffraction [202]. The observed incommensurate magnetic order is compatible with a spin-density wave, with the ordering wavevector determined by the band structure of heavy fermions as suggested by renormalized band-structure calculations. These findings support the early suggestion that the A/S transition can be viewed as a QCP [200, 201], where a variety of NFL features are observed.

The A phase in a polycrystalline sample can be suppressed by a moderate pressure $\sim$ 2 kbar [80]. At this pressure the resistivity exhibits $\Delta\rho \sim T^{3/2}$. At 6.7 kbar the specific heat can be consistently interpreted within the LGW theory for an antiferromagnet. (Specific-heat data at 2 kbar were not reported.) However, the zero-pressure starting point does not seem to correspond to a FL: while $\Delta\rho \sim T^2$ above the temperature T_A = 0.75 K of the A-phase transition, $\gamma = C/T$ is strongly T-dependent. Another puzzle results when an S-type single crystal lacking the A-phase signature is investigated (under zero pressure). In order to suppress superconductivity, a moderate field has to be applied. At $T > 0.2$ K and for $0.2\,\mathrm{T} \leq B < 6\,\mathrm{T}$, $\Delta\rho$ varies as $T^{3/2}$ and γ almost as $\gamma_0 - \beta\sqrt{T}$ suggesting the proximity of the QCP, and corroborating the high-pressure data of the A-phase crystal discussed above. However, when cooling below 0.2 K, γ shows a very steep upturn towards low T that cannot be ascribed to nuclear Zeeman contributions. In the same T range, $\Delta\rho \sim T^{3/2}$ is still observed. The latter is in keeping with spin-fluctuation theory for a rather disordered system ($\rho_0 \approx 30\,\mu\Omega$cm) as pointed out by Rosch [75, 76], cf. Sec. 3.5. For a disordered system at the AF QCP, a recovery of FL behavior $\Delta\rho \sim T^2$ towards $T \to 0$ is not expected [75]. On the other hand, Gegenwart et al. [80] suggested that the "disparate" behavior of transport and specific heat near the QCP in $CeCu_2Si_2$ is due to a "break up" of the heavy quasiparticles.

The compound $CeCu_2Ge_2$, isostructural to $CeCu_2Si_2$, orders antiferromagnetically with $T_N = 4.15$ K [203] which can be attributed to the larger volume and weaker c–f hybridization, in keeping with the Doniach model. Upon application of a rather large pressure, superconductivity occurs for $p \approx 70$ - 80 kbar with $T_c \approx 0.6$ K [204]. T_c remains independent of pressure up to 120 kbar, reaches a maximum $T_c = 2$ K at 160 kbar and drops again [205]. The magnetic transition temperature as determined from a kink of

$\rho(T)$ decreases only slightly with increasing pressure and appears to saturate around 2 K at 94 kbar; beyond this pressure a signature in $\rho(T)$ can no longer be observed [206]. The appearance of superconductivity near the pressure where magnetic order is suppressed is very interesting. As in $CeCu_2Si_2$ [207, 208], superconductivity and magnetic order (A phase) appear to be mutually exclusive. Because of the possible precipitous drop of $T_N(p)$, this transition might be of first order, thus suppressing quantum fluctuations and the concomitant NFL behavior expected only for a continuous magnetic–non-magnetic transition. It was already pointed out [206] that quantum critical behavior is not observed: the resistivity exponent α in $\Delta\rho \sim T^{\alpha}$ has almost exactly the FL value $\alpha = 2$ at the critical pressure.

$CeNi_2Ge_2$ is a stoichiometric HFS where pronounced NFL features are observed in thermodynamic properties [209, 83, 210]: The resistivity exponent α, $\Delta\rho \sim T^{\alpha}$, varies between 1.2 and 1.5 below 4 K dependent on the residual resistivity ρ_0. This variation of $\alpha(\rho_0)$ may be understood in terms of the competition between magnetic scattering and impurity scattering [76]. While early specific-heat data between 0.35 and 5 K were found to follow $C/T \propto -\ln T$ [209], later measurements by Küchler et al. could be described by $C/T = \gamma_0 - \beta\sqrt{T}$ as expected for the LGW scenario, when corrected for an (unexplained) steep low-T upturn C/T^2. (For a discussion of the sample dependence of the specific heat of this material, see Küchler et al. [210].) The divergence of the Grüneisen parameter $\Gamma\ T^{-1}$ for $T \to 0$ is also in line with this model [210]. It should be mentioned that in very pure $CeNi_2Ge_2$ samples traces of superconductivity [209], or even a complete resistive transition to superconductivity [83] were found. Thus, $CeNi_2Ge_2$ appears to be one of the few systems following the predictions of the LGW model for 3D antiferromagnets.

4.1.4 CeTIn$_5$ (T = Co, Rh, Ir)

A few years ago, members of the series $CeTIn_5$ (T = Ir, Rh, Co) were discovered as HFS with interesting ground states such as superconductivity at ambient pressure in $CeCoIn_5$ with T_c = 2.3 K [229] as well as antiferromagnetic order in $CeRhIn_5$ with T_N = 3.8 K that under pressure gives way to superconductivity [212]. The crystal structure of these compounds is derived from cubic $CeIn_3$ intercalated with TIn_2 layers along the (0 0 1) direction. This gives rise to a rather pronounced two-dimensionality in various physical properties. Alloys across this isoelectronic series display a rich variety of interplay between superconductivity and magnetic order, as can be seen in Fig. 27 below [213]. We will just mention a few findings about quantum criticality in these systems, focusing on the stoichiometric parent compounds. Space limitations force us to neglect the many interesting studies on alloy series, see, e.g., [214].

Although $CeCoIn_5$ does not order magnetically, intersite magnetic correlations seem to be rather strong: transport and thermodynamic data on La-diluted $CeCoIn_5$ were interpreted in terms of a single-ion T_K = 1.7 K and T_{coh} = 45 K $\gg T_K$ [215]. Pronounced NFL features are observed in $CeCoIn_5$ when superconductivity is suppressed above the upper critical field B_{c2} = 4.95 T [216]. Fig. 14(a) shows C/T vs. ℓnT for 5 T $\leq B \leq$ 9 T. The data can be reasonably well described by the LGW model where the parameter y_0 indicates the distance from the QCP. (y_0 is equivalent to r of Sec. 3.4)

At the same time, the data are shown to exhibit scaling in the form of $\gamma(B,T)$ –

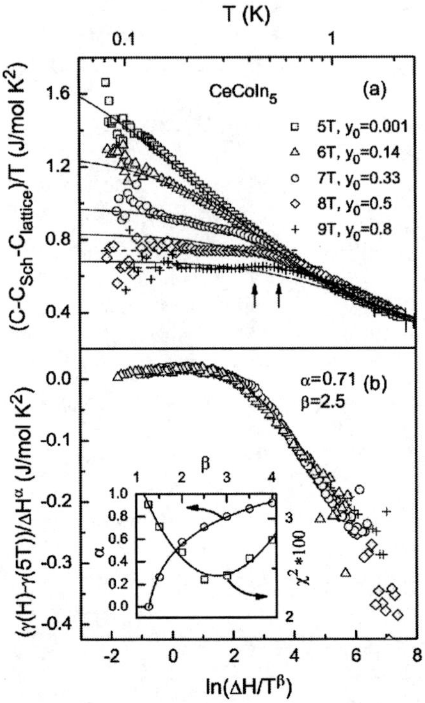

FIGURE 14. (a) Magnetic specific-heat coefficient $\gamma(T) = C_m(T)/T$ of $CeCoIn_5$ in magnetic fields $B = \mu_0 H \parallel [001]$. Dashed lines for 8 and 9 T emphasize the FL behavior with constant γ. Left(right) arrows indicate the crossover to FL behavior for 8 T (9 T). Solid lines are fits to the LGW spin-fluctuation model for each field, with the corresponding values of y_0 indicating the distance from the critical field B_c. (b) Scaling analysis of the data in (a) for $\alpha = 0.71$ and $\beta = 2.5$. Inset: Plots of α (β) which minimize χ^2 for a given β, and χ^2 for these α and β. From [280].

$\gamma(B_c,T) \approx (\Delta B)^{\alpha} f(\Delta B/T^{\beta})$ where $\Delta B = B - B_c$ and $B_c \approx B_{c2}$ is the (quantum-) critical field see Fig. 14b. Best scaling collapse is obtained for $\alpha = 0.71$ and $\beta = 2.5$. Note, however, that a value of $\alpha \neq 0$ is *not* consistent with the expected recovery of FL behavior for $T \to 0$ and $B > B_c$, which seems to be a more likely interpretation of the experiment. Scaling with $\alpha = 0$ has been previously obtained for $U_{0.2}Y_{0.8}Pd_3$ with $\beta = 1.3$ [217], for $YbRh_2Si_2$ with $\beta = 1.05$ [218], for $CeCu_{6-x}Ag_x$ with exponents $\beta = 0.85, 1.35, 1.6, 1.7$ for $x = 0.09, 0.48, 0.8, 1.2$ [82, 219]. At least in $YbRh_2Si_2$ and $CeCu_{5.8}Ag_{0.2}$ the scaling does not hold at the lowest T. In view of the different values of α and β, the physical significance of the scaling is not clear.

The electrical resistivity $\rho(T)$ was shown by [216] to follow the LGW model as discussed by [79], with a set of y_0 parameters consistent with the specific heat. The small value $y_0 \gtrsim 0.01$ suggests that $CeCoIn_5$ is indeed close to a QCP. The existence of a QCP in $CeCoIn_5$ very close to B_{c2} was also inferred independently from resistivity and magnetoresistivity measurements [220]. A surprising result is that when B_{c2} is suppressed upon alloying $CeCoIn_{5-x}Sn_x$, the field range where NFL behavior is observed tracks

B_{c2} [221]. However, recent resistivity measurements on $CeCoIn_5$ under pressure have shown that B_{c2} decreases much faster than the quantum-critical field, indicating that the two phenomena are not related [222].

While application of hydrostatic pressure drives $CeCoIn_5$ away from quantum criticality [223], the antiferromagnetic compound $CeRhIn_5$ is driven towards an instability and incipient superconductivity upon application of hydrostatic pressure of $p_c \approx 14.5$ kbar [212, 224]. The compounds $CeTIn_5$ consist of nearly two-dimensional $CeIn_3$ layers, separated by intercalated TIn_2 layers. However, the magnetic fluctuations of $CeRhIn_5$ as determined from neutron scattering are three-dimensional albeit with some anisotropy [225]. The pressure-induced transition to superconductivity appears to be of first order, thus "avoiding" a QCP. On the other hand, pronounced deviations from the Fermi-liquid T^2 dependence of $\rho(T)$ were reported by [226] to occur over a wide pressure range. Recent studies of the antiferromagnetic/superconducting phase boundary under pressure by [227, 228] revealed a quantum critical line between a phase of coexistence and a purely superconducting phase. Applying additionally a magnetic field, this line is suggested to end in a quantum tetracritical point.

Finally, turning to $CeIrIn_5$, this material exhibits bulk superconductivity below 0.4 K, with $\rho(T)$ vanishing already at 1.2 K [229]. The Ce-derived specific heat C_m, with lattice and nuclear contributions subtracted, exhibits when plotted as C_m/T vs. $\ell n T$ a broad hump for magnetic fields applied both along and perpendicular to the c axis [230]. For fields $B \parallel c$ larger than 12 T, a divergent C_m/T for $T \to 0$ is observed which is interpreted as NFL behavior. Below 0.5 K with field applied to suppress superconductivity, the resistivity follows a $T^{1.5}$ dependence, also signaling NFL behavior. These findings are interpreted as being due to a field-induced metamagnetic transition [230].

4.1.5 $YbRh_2(Si_{1-x}Ge_x)_2$

The compound $YbRh_2Si_2$ was the first Yb compound to show pronounced non-Fermi-liquid effects near a magnetic ordering temperature at very low temperature [218]. It crystallizes in the tetragonal I4/mmm structure (same structure as $CeCu_2Si_2$). At high temperatures $T > 200$ K the susceptibility follows a Curie-Weiss law with $\mu_{\text{eff}} = 4.5\,\mu_B$/Yb-atom but different Weiss temperatures $\theta_\perp = -9$ K and $\theta_\parallel = -180$ K, due to magnetocrystalline anisotropy, where $\parallel$ and $\perp$ refer to directions relative to the c axis. At 0.1 K, the large ratio $\chi_\perp/\chi_\parallel \approx 100$ classifies $YbRh_2Si_2$ as an easy-plane system, as opposed to the easy-axis system $CeCu_{6-x}Au_x$.

A sharp maximum in the *ac* susceptibility $\chi_\perp^{ac}(T)$ [218] and specific heat $C(T)$ [231] around 70 mK signal the onset of antiferromagnetic ordering, although to date no neutron scattering data are available to corroborate this assignment. Evidence for *ferromagnetic* quantum critical fluctuations was reported recently [232]. Well above the magnetic ordering temperature, i.e., between 0.3 and 10 K, the specific heat varies as $C/T = a\ell n(T_0/T)$ with $a = 0.17$ J/molK^2 and $T_0 = 24$ K (Fig. 15).

The electrical resistivity varies as $\rho = \rho_0 + A'T$ between 20 mK and 1 K, with no apparent effect of magnetic ordering on $\rho(T)$ (see Fig. 16) [218]. (In subsequent work [231], a kink in $\rho(T)$ was found, likely due to increased sample quality.) These NFL features look at first sight very similar to those in $CeCu_{6-x}Au_x$ and this prompted the

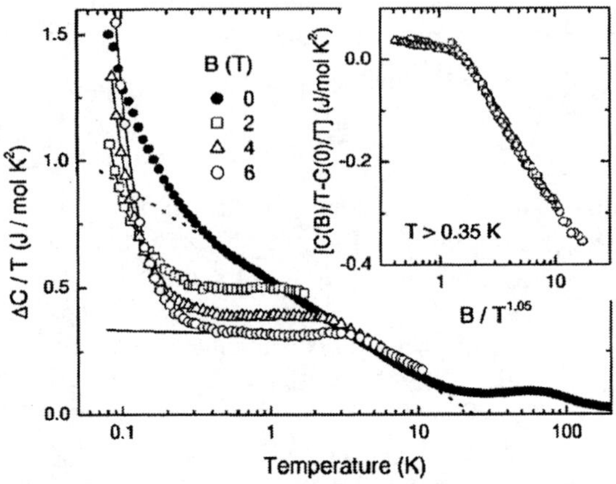

FIGURE 15. Yb contribution to the specific-heat coefficient, $\Delta C/T$, of Yb vs. T (logarithmic scale) for different magnetic fields B applied along the a axis. The dotted straight line represents $\Delta C \sim \ell n(T_0/T)$, the horizontal line the $B = 6\,\mathrm{T}$ data after subtraction of the hyperfine contribution. The inset demonstrates scaling of the data as $C(B)/T - C(0)/T = f(B/T^{\beta})$ with $\beta = 1.05$. From [218].

suggestion [218] of quasi-2D critical fluctuations by analogy with $CeCu_{6-x}Au_x$. Indeed, there is a surprising scaling when C/T is plotted as function of $\ell n(T_0/T)$: the data for $YbRh_2(Si_{0.95}Ge)_{0.05}$ and $CeCu_{5.8}Ag_{0.2}$ fall on top of each other in the range where $C/T \propto \ell n(T_0/T)$ [160]. The upturn of C/T below 0.4 K for $YbRh_2Si_2$ (Fig. 15) can be modelled as $C/T \sim T^{-\alpha}$ with $\alpha \sim 0.3$ [233]. This upturn suggests the presence of an additional low-energy scale below which the ℓnT behavior of C/T is cut off. It is an open question whether such crossover exists in $CeCu_{6-x}Au_x$ at correspondingly low reduced temperatures T/T_0 where $T_0 \approx 6\,\mathrm{K}$ is a factor of 4 smaller. Interestingly, the low-T upturn vanishes when $YbRh_2Si_2$ is diluted with La [234]. For $Yb_{0.95}La_{0.05}Rh_2Si_2$, $C/T \sim \ell n(T_0/T)$ is observed down to 0.35 K.

Measurements on $YbRh_2(Si_{1-x}Ge_x)$ with a nominal Ge concentration $x = 0.05$ showed that the magnetic ordering is suppressed down to 20 mK where a rather broad maximum in the specific heat is observed [233]. Above 100 mK, the C/T curves for $x = 0$ and $x = 0.05$ are nearly identical, indicating that the $T^{-0.3}$ upturn is not associated with classical short-range order effects in the proximity to magnetic ordering.

A further interesting discovery for $YbRh_2Si_2$, following the prediction by [65] discussed in Sec. 3.4, was the observation of a divergent Grüneisen parameter Γ near the QCP by [210]. As shown in Fig. 17, the Γ data for $YbRh_2(Si_{0.95}Ge_{0.05})_2$ follow a $T^{-0.7}$ dependence at lowest temperatures, i.e., between 50 mK and 0.3 K, with a slightly steeper slope approaching an exponent -1 at higher T. The $T^{-0.7}$ dependence of Γ at low T is certainly weaker that T^{-1} predicted for the 3D LGW scenario (and actually observed for $CeNi_2Ge_2$ by [210]), but also weaker than $T^{-1}(\ell n \ell n T)/(\ell n T)$ expected for a 2D LGW model. This observation may support the idea that degrees of freedom other than magnetism become critical at the QPT, see Sec. 3.8. However, the data would imply critical exponents with $1/(z\nu) = 0.7$ (at least if the transition is below its upper critical dimension), which has not been verified by other measurements.

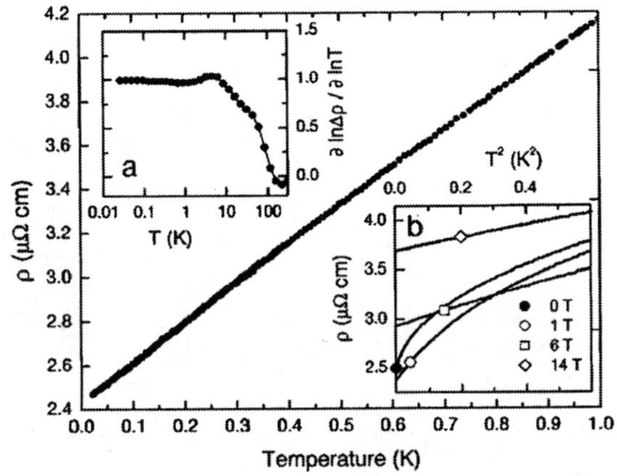

FIGURE 16. Low-temperature electrical resistivity ρ of $YbRh_2Si_2$ at $p = 0$ measured along the a axis as a function of temperature T, obeying $\rho(T) = \rho_0 + bT^{\alpha}$, with $\alpha \simeq 1$. (a) Temperature dependence of the effective exponent $\varepsilon = \partial \ell n \Delta\rho / \partial \ell n T$ with $\Delta\rho = \rho - \rho_0$. (b) $\rho(T)$ plotted as ρ vs. T^2, for $B \leq 14\,\mathrm{T}$ applied along the c axis. From [218].

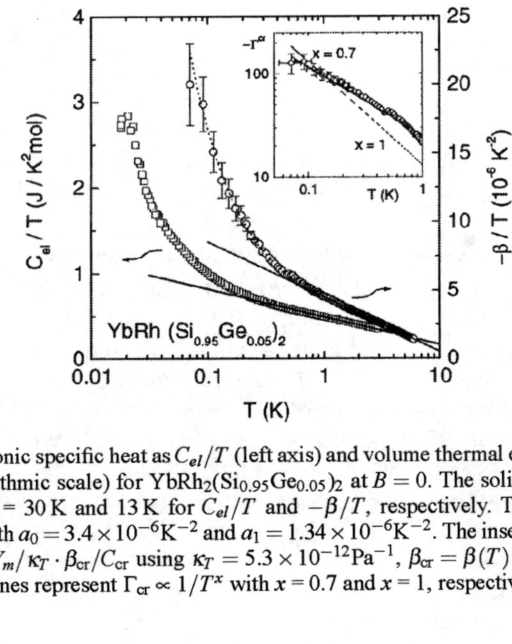

FIGURE 17. Electronic specific heat as C_{el}/T (left axis) and volume thermal expansion as $-\beta/T$ (right axis) vs T (on a logarithmic scale) for $YbRh_2(Si_{0.95}Ge_{0.05})_2$ at $B = 0$. The solid lines indicate $\ell n(T_0/T)$ dependencies with T_0 = 30 K and 13 K for C_{el}/T and $-\beta/T$, respectively. The dotted line represents $-\beta/T = a_0 + a_1/T$ with $a_0 = 3.4\times10^{-6}\mathrm{K}^{-2}$ and $a_1 = 1.34\times10^{-6}\mathrm{K}^{-2}$. The inset displays the log-log plot of $\Gamma_{cr}(T)$ with $\Gamma_{cr} = V_m/\kappa_T \cdot \beta_{cr}/C_{cr}$ using $\kappa_T = 5.3\times10^{-12}\mathrm{Pa}^{-1}$, $\beta_{cr} = \beta(T) + a_0 T$, and $C_{cr} = C_{el}(T)$. The solid and dotted lines represent $\Gamma_{cr} \propto 1/T^x$ with $x = 0.7$ and $x = 1$, respectively. From [210].

Rather detailed experiments were carried out to study $YbRh_2(Si_{1-x}Ge_x)_2$ in magnetic fields, exploring the field tuning of a quantum critical point, as previously done for the $CeCu_{6-x}Ag_x$ and $CeCu_{6-x}Au_x$ series. The magnetic $B-T$ phase diagram of $YbRh_2Si_2$ shown in Fig. 18 is universal with respect to the different field orientations when the field axes are scaled appropriately, similar to the behavior shown earlier for the anisotropic

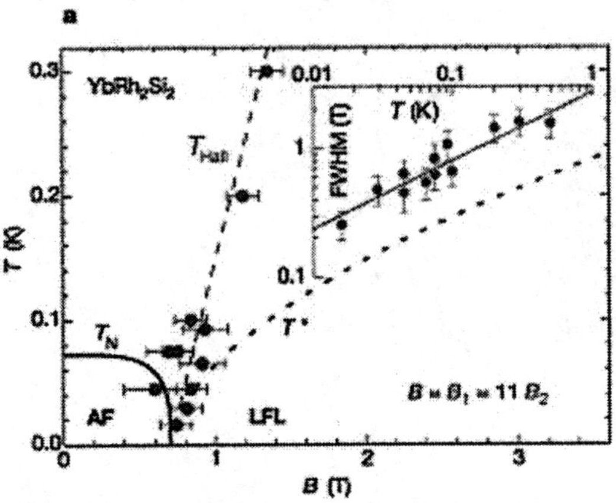

FIGURE 18. Temperature–field phase diagram of $YbRh_2Si_2$. Full and dotted black curves represent the field dependence of the Néel temperature T_N and the crossover temperature T^* below which the resistivity exhibits a Fermi-liquid T^2 dependence. Symbols present the crossover field where the Hall resistivity changes the slope. Light symbols: $B = B_1$ parallel to the c axis, dark symbols: crossed-field experiment with the tuning field $B = B_2$ perpendicular to the c axis. The relation $B = B_1 = 11B_2$ reflects the magnetic anisotropy of $YbRh_2Si_2$. The inset shows the width of the transition at B_c vs. T indicating that it vanishes as $T \to 0$. From [113].

easy-axis system $CeCu_{6-x}Au_x$ [138]. While for high magnetic fields constant Kadowaki-Woods ratios A/γ_0^2 and A/χ_0^2 typical for a Fermi liquid are observed, the ratio $A/\gamma_0^2 \sim (B-B_c)^{-0.33}$ appears to diverge close to B_c, as has been shown for $YbRh_2(Si_{0.95}Ge_{0.05})_2$ by [233]. In this alloy, $B_c^{\perp} = 0.027\,\mathrm{T}$ only and γ_0 diverges as $\gamma_0 \sim (B-B_c)^{-0.33}$. Yet, the coefficient A of the T^2 resistivity diverges more strongly. For a 2D LGW scenario, A/γ_0^2 diverges much faster, i.e., $A/\gamma_0^2 \sim 1/(B-B_c)$ with logarithmic corrections.

As discussed in detail in Sec. 3.8, one of the central questions for quantum-critical points in HFS is whether the Kondo effect breaks down at the critical point and whether the Fermi volume changes abruptly at the second-order transition [109, 105]. Indeed, Paschen et al. [113] found indications of such a behavior in measurements of the Hall effect across the field-driven transition in $YbRh_2Si_2$. In Fig. 19 the Hall constant R_H is shown for a geometry (see inset) where the field B_2 driving the QPT is parallel to the current I, and an additional small probing field B_1 perpendicular to I induces the Hall voltage. As can be seen in Fig. 19, R_H varies quite strongly, consistent with a change of the Fermi volume by one electron per unit cell (assuming a free-electron single-band picture). For lower T the crossover gets sharper, consistent with a scaling of the half-width with $\sqrt{T}$, see dark symbols in the inset of Fig. 18. The Hall effect was also measured in the usual geometry, where the field is identical to the field which induces the Hall voltage driving the QPT . This geometry is, however, more difficult to interpret, as the resulting differential Hall resistivity is expected to jump even for a standard SDW quantum-critical point. (Note that such a jump is expected to trace T_N, opposite to what is observed in $YbRh_2Si_2$.)

By using fits to the theory of the anomalous Hall effect arising from skew-scattering

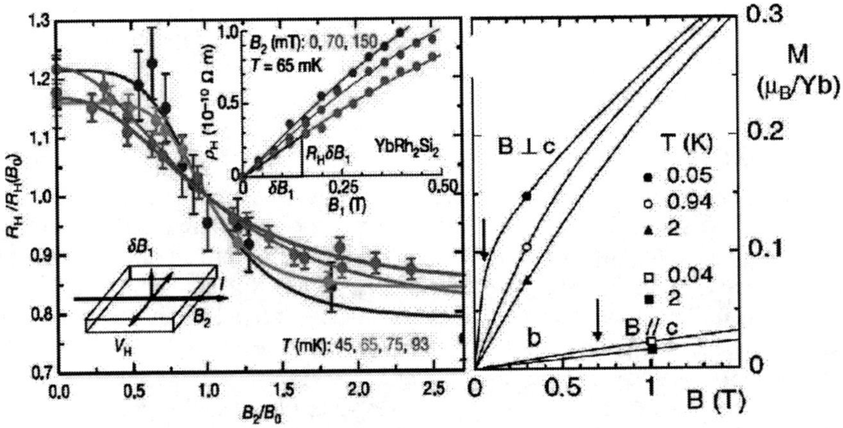

FIGURE 19. Left: Hall effect as a function of a magnetic field parallel to the current, from [113]. The crossover at the critical field $B_0 = 60\,\mathrm{mT}$ gets sharper towards lower T. Right: Magnetization as a function of B, arrows indicate the critical field. From [231].

[235], it was concluded that the former is sufficiently small to be either subtracted or neglected (depending on the geometry). In this case the extrapolated jump of the Hall coefficient at $T = 0$ would indeed prove rather unambiguously a jump of the Fermi volume at a second-order phase transition, certainly a very spectacular result. To establish this firmly, it would be important to track the sharpening of the crossover in Fig. 19 towards lower temperature, and check for similar effects in other systems as well. When interpreting the crossover at finite T one may also have to take into account that $YbRh_2Si_2$ is almost ferromagnetic (see below). Even the tiny critical field of 60 mT already induces a sizable magnetization of almost $0.1\mu_B$ per Yb which is strongly temperature dependent, see right-hand panel of Fig. 19. Such a magnetization could at least in principle lead to large changes of the Fermi surface and therefore of the Hall effect. The change in $\rho_{xy}(B)$ has also been interpreted in terms of quantum-critical valence fluctuations [236].

We now turn to the issue of critical fluctuations in $YbRh_2Si_2$ near the QPT. Unfortunately, neither elastic nor inelastic neutron scattering data exist to date which could identify the type of magnetic order and the nature of the critical fluctuations. Hence for a microscopic view one has to rely local probes such as NMR and ESR, and on macroscopic magnetization data. [237] reported ^{29}Si NMR data on aligned single crystals of $YbRh_2Si_2$ in fields parallel to the easy axis. While the nuclear spin-lattice relaxation rate divided by $T, 1/T_1T$, was found to vary as $\sim T^{-1/2}$ at small magnetic field of $B_\perp = 0.15\,\mathrm{T}$ down to the lowest temperature ($\sim$ 50 mK), the Knight shift K first also increases towards $T \to 0$ but turns over to a constant value at $T \approx 200\,\mathrm{mK}$ in the same field. While K probes the uniform static susceptibility $\chi'(\mathbf{q} = 0), 1/T_1T$ gives information over the imaginary part of the $\mathbf{q}$-averaged dynamical spin susceptibility $\chi''(\mathbf{q},\omega)$ at the (quasi-static) frequency corresponding to the nuclear Zeeman splitting. The difference between K and $1/T_1T$ has been attributed to the presence of finite-$\mathbf{q}$ critical fluctuations. For higher fields, K and $1/T_1T$ tend to a T-independent value and track each other, i.e., the Korringa relation $1/T_1T \sim SK^2$ is approximately fulfilled between 0.25 and 2.4 T. The strong deviation of S from the free-electron value $S_0 = \pi\hbar\gamma_n^2 k_B/\mu_B^2$, i.e., $S \approx 0.1 S_0$ as measured at 100 mK, suggests the presence of dominant $\mathbf{q}$ = 0, i.e., ferromagnetic fluctuations.

Similarly, the approximately constant ratio A/χ_0^2 discussed above and observed for $B \lesssim 2B_c$ is not easy to understand for dominant AFM fluctuations. In addition, the strong increase of the Wilson ratio χ_0/γ_0 by a factor of up to 30 compared to the free-electron value "highlights the importance of FM fluctuations in the approach to the QCP" [232]. However, by comparing tendencies to ferro- and antiferromagnetism in a Ge-doped sample with an undoped one, these authors also suggested that ferromagnetic tendencies are "not directly correlated to the AFM QCP".

There is a further intriguing observation in $YbRh_2Si_2$, possibly related to the ferromagnetic fluctuation component, i.e., the observation of an ESR signal by [238] that can be ascribed to a Yb^{3+} resonance. This observation is very surprising because the large Kondo temperature $T_K \approx 25$ K is associated with a very high spin-fluctuation rate that would according to current wisdom render any ESR signal below T_K unobservable. However, the magnitude of the ESR $Yb^{3+}4f^{13}$ signal well below T_K suggests that at least 60 % of the Yb^{3+} ions in $YbRh_2Si_2$ contribute.

It is tantalizing that the $T^{-0.7}$ dependence of Γ and the $T^{-0.3}$ dependence of C/T in $YbRh_2(Si_{1-x}Ge_x)_2$ at very low temperatures, but above the ordering temperature correspond to a 2D *ferromagnetic* (z = 3) scenario of a conventional LGW QPT. A speculative view could be that ferromagnetic planes are the source of the strong FM component of the fluctuations that have been discussed above, until the magnetic transition intervenes, perhaps induced by a weak antiferromagnetic interplane coupling.

4.2 Quantum-critical behavior of itinerant transition-metal magnets

Intermetallic transition-metal compounds have been traditionally viewed as systems close to a magnetic instability, and the LGW model has been formulated in the self-consistently renormalized (SCR) theory of spin fluctuations (Sec. 3.4) to describe these systems [66, 239]. While MnSi and $ZrZn_2$ have been outstanding examples under intense investigation recently – they will be discussed in this section – there have been a number of other systems, notably $Cr_{1-x}V_x$, that should be mentioned.

The nearly ferromagnetic metal Pd can be driven to static ferromagnetic order by introducing Ni, with a Ni critical concentration of $x = 0.026$. Here, the resistivity ρ follows a $T^{5/3}$ dependence [240]. This T dependence, as well as the $T_C \sim (x-x_c)^{3/4}$ dependence are in accord with the FM LGW model. Likewise, the specific heat ΔC (after subtraction of the phonon contribution) shows $\Delta C/T \sim \ell n(T_0/T)$ at x_c. The ferromagnet Ni_3Al has been investigated with resistivity measurements under pressure up to 10 GPa [241]. The critical pressure p_c where $T_c \to 0$ is estimated to be around 8 GPa. The T-dependent part $\Delta\rho$ of the resistivity does not reveal pronounced critical behavior, the exponent α depends on T and varies for 3.2 GPa between 1.9 and 1.5 at 1 K and 5 K, respectively, and smoothly shifts to lower values with increasing p. Below 1 K, the data can be approximated by $\Delta\rho = AT^2$, and A exhibits a rather sharp maximum at p_c. This behavior has been tentatively associated with a first-order transition near p_c [241]. YMn_2

is an itinerant antiferromagnet with a first-order transition around $T_C \approx 100$ K [242]. On the other hand, Mössbauer-effect measurements under pressure indicated a continuous reduction of the ordered magnetic moment [243] suggesting a second-order QCP.

More recently, the Hall effect near the QCP in $Cr_{1-x}V_x$ was investigated with concentration tuning ($x_c = 0.035$) [244], and by pressure tuning of a sample with $x = 0.032$ close to x_c [245]. This system derives from the itinerant *antiferromagnet* Cr, and is one of the few non-Kondo systems that can be driven through an antiferromagnetic QCP. For $x = 0.032$) $T_C = 52$ K compared to $T_C = 311$ K for pure Cr. While the transition upon concentration variation appears very sharp (i.e., almost of first order), using a finely spaced pressure tuning reveals it to be continuous [245]. The decrease of the inverse Hall constant, R_H^{-1}, upon opening of the spin-density wave gap, tracks the increase of ρ above the paramagnetic background resistivity. The Hall effect evolution was interpreted in terms of a Fermi-surface nesting effect [246].

We finally mention experiments on an, albeit complex, elemental metal, β-Mn [247], where the resistivity varies as $\Delta\rho \sim T^{3/2}$ between 5 K (the lowest temperature measured) and 25 K, suggesting the proximity to a AFM QCP. The spin fluctuations, however, defy a simple interpretation in terms of a unique scaling as observed in $CeCu_{5.9}Au_{0.1}$. As already mentioned at the beginning of this chapter, space limitations inhibit the discussion of QCP in transition metal oxides, such as high-T_c cuprates, manganites, or $Sr_3Ru_2O_7$.

4.2.1 MnSi

The itinerant-electron magnet MnSi orders below $T_C = 29.5$ K in a spiral magnetic structure with a long pitch of about 175. Early on it was realized [248, 249] that the helical order arises as the cubic B20 structure (space group $P2_13$) lacks inversion symmetry. Therefore the weak spin-orbit interactions assume a Dzyaloshinsky-Moriya form $\int \mathbf{S} \cdot (\nabla \times \mathbf{S}) d\mathbf{r}$ which twists the ferromagnetic alignment into a helix. The ordered moment of about 0.4 μ_B/Mn-atom is sizable for an itinerant magnet. With the possible exception of high-frequency optical-conductivity experiments [250], the ground state below T_C seems to be a standard three-dimensional weakly spin-polarized Fermi liquid [251, 252, 253]. MnSi, being almost a ferromagnet, played an important role in the development of spin-fluctuation theory for complex transition-metal compounds [239, 66].

When pressure is applied (see Fig. 20) the magnetic order is suppressed [254, 188] and vanishes above $p_c = 14.6$ kbar. Above this pressure, the resistivity shows an anomalous power-law, $\Delta\rho \sim T^{3/2}$ [255, 256]. It has to be emphasized that this power-law holds over almost three decades in temperature from about 6 K down to a few mK and over a large pressure range.

Therefore the question arises whether the anomalous resistivity can be associated with critical fluctuations. For example, due to the long pitch of the spiral, one might expect a quantum-critical behavior similar to that of a ferromagnet with $\Delta\rho \sim T^{5/3}$, not too far from the observed $\Delta\rho \sim T^{3/2}$. There are two main arguments against such an interpretation. First, the anomalous transport is observed far away from the putative quantum critical point. While a pressure of 14.6 kbar is sufficient to suppress T_C from

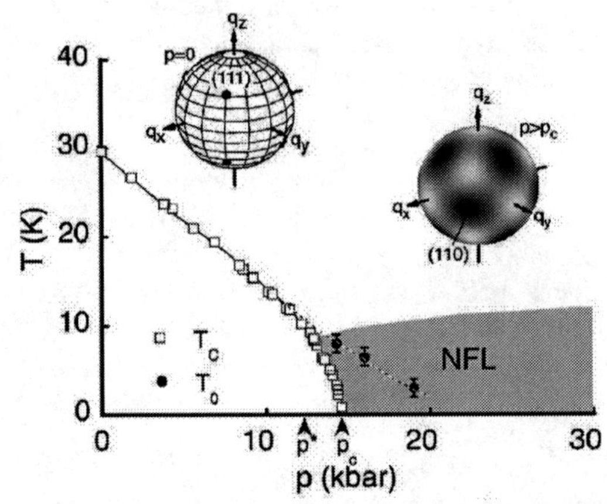

FIGURE 20. Temperature–pressure phase diagram of MnSi. Above p^*, the transition is first order, see Fig. 21 and vanishes at p_c. In the shaded region, the resistivity shows an anomalous temperature dependence, $\Delta\rho(T) \sim T^{3/2}$. The insets qualitatively show main features of the neutron scattering intensity. From [258]

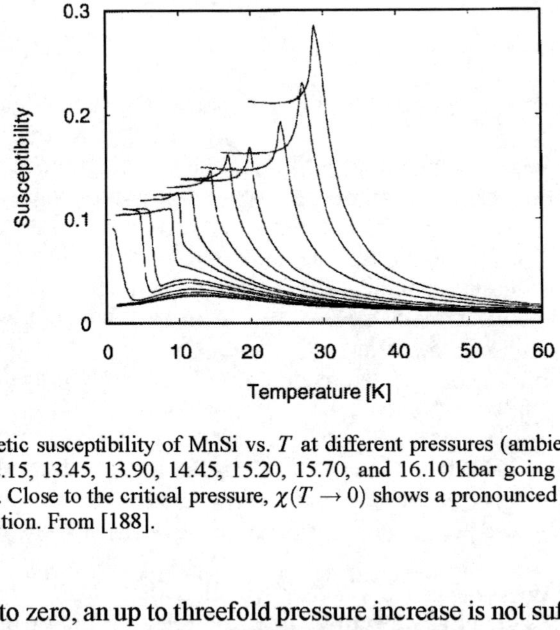

FIGURE 21. Magnetic susceptibility of MnSi vs. T at different pressures (ambient, 1.80, 3.80, 6.90, 8.60, 10.15, 11.25, 12.15, 13.45, 13.90, 14.45, 15.20, 15.70, and 16.10 kbar going down, starting from the top curve at 30 K). Close to the critical pressure, $\chi(T \to 0)$ shows a pronounced jump indicative of a first-order phase transition. From [188].

almost 30 K down to zero, an up to threefold pressure increase is not sufficient to recover Fermi-liquid behavior at temperatures of the order of 50 mK. Recent measurements by [257] indicate, however, that for $p > 3p_c$ the exponent changes and Fermi-liquid behavior might be recovered at the lowest temperatures.

More relevant is a second argument: Susceptibility measurements, see Fig. 21, show that close to p_c the phase transition is of first order [188] and critical fluctuations should

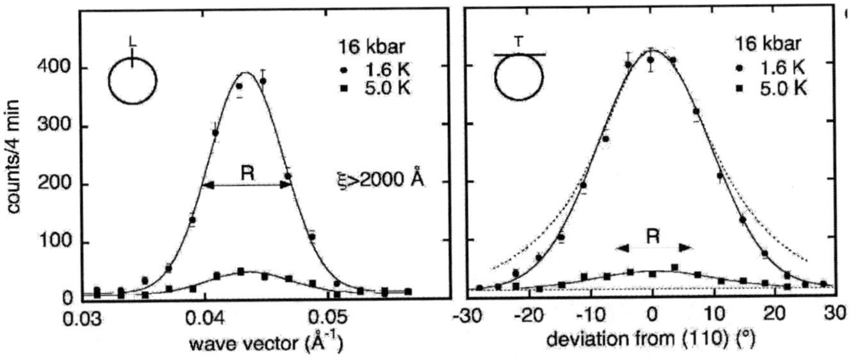

FIGURE 22. Scans of the elastic neutron scattering intensity in MnSi at 16 kbar perpendicular (left) and parallel (right) to the surface of the sphere shown in Fig. 20 [in (110) direction] [258].

be absent. An estimate suggested the consequence of such a first-order transition quantitatively and concluded that within a standard spin-fluctuation scenario a T^2 resistivity should be observable below 1 or 2 K, in contradiction to experiment [256, 257]. The situation is, however, complicated by the fact that not all experimental probes show indications of a strong first-order transition. For example, the hysteretic part of the signal in elastic neutron scattering [258] is rather small close to p_c and on the ordered side of the phase diagram the A coefficient of the $T \to 0$ resistivity $\rho = \rho_0 + AT^2$ seems to diverge upon approaching p_c indicative of a second-order transition (possibly of percolative type).

A consistent explanation of the precise nature of the quantum phase transition is presently lacking. Theoretically, it has been argued by [259] that strong fluctuations of the direction of the ordering vector generically drive the QPT towards first order. If quantum-critical fluctuations can be ruled out as the origin of the NFL behavior, the alternative scenario is the existence of an *extended* new NFL phase which is stable against changes of pressure and temperature in a sizable region. This would be remarkable for a three-dimensional cubic system where disorder effects should be negligible due to high-quality single crystals with a mean free path of several thousand Å.

Elastic neutron scattering experiments by [258] have surprisingly shown that even above p_c the magnetic order has not vanished completely. While for $p < p_c$ one observes Bragg reflections in the $\langle 111 \rangle$ directions with an ordering vector of $|\mathbf{Q}| \approx 0.037\,\text{Å}^{-1}$ (increasing under pressure), one finds for $p \gtrsim p_c$ and below a crossover temperature T_0 (shown in Fig. 20) a strange type of "partial order": the neutron scattering intensity is concentrated on the *surface* of a sphere in reciprocal space with a radius $Q \approx 0.0437\,\text{Å}^{-1}$. Resolution-limited scans along the radial direction of this sphere, see Fig. 22, reveal that the helical order survives above p_c on length scales of at least ~ 2000 Å. The broad distribution of the $\mathbf{Q}$ vector in tangential direction on the surface of the sphere (Fig. 22 and inset of Fig. 20) implies the absence of long-range order: the helices have lost their directional order. Interestingly, the total intensity of the elastic neutron signal is comparable to that at ambient pressure implying that a sizable fraction

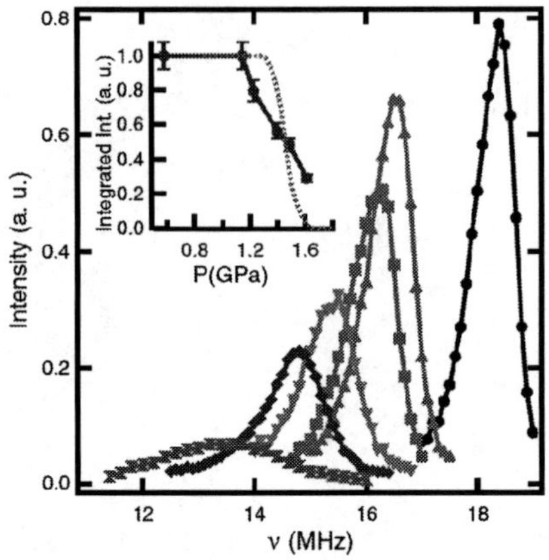

FIGURE 23. Distribution of ^{29}Si NMR frequencies in MnSi for pressures $p = 0.58, 1.15, 1.23, 1.40, 1.49$, and 1.62 GPa (right to left). The inset shows the corresponding total intensities. From [260].

of the local order survives above p_c.

Likewise, zero-field ^{29}Si NMR experiments by [260] show this local order below T_0. The broadening of the distribution of NMR frequencies shown in Fig. 23 is consistent with a picture of intrinsic phase inhomogeneities above p^* (note that the powdered samples used in this experiment may have extra pinning centers). Several theoretical attempts have been made to identify possible unconventional order-parameter structures and lattices of defects [261, 262, 263] which may serve as a starting point to understand the observed partial order. The role of disorder and phase inhomogeneities in this extremely clean system remains, however, presently unclear.

The question arises whether the anomalous partial order or a new QCP, associated with the vanishing of the crossover scale T_0 (see Fig. 20), can be the origin of the anomalous transport properties in MnSi. For example, anomalously soft Goldstone modes of a helimagnet give rise to singular behavior of thermodynamic and transport quantities [264]. However, for clean systems $\Delta\rho \sim T^{5/2}$ rather than the observed $T^{3/2}$ has been predicted [265] (neglecting, however, the role of spin-orbit coupling [266]). A challenge is to understand the role of the scale T_0 above which the partial order seems to vanish. Surprisingly, no signatures of T_0 can be seen in transport, $\Delta\rho \sim T^{3/2}$ both above and below T_0. This is unexpected as at T_C, when long-range order sets in, the resistivity shows a pronounced kink and a strong reduction. It has therefore been speculated [258] that the partial order survives on intermediate length and time scales and gives rise to an anomalous scattering of electrons (e.g, from fluctuating topological defects) even above T_0 everywhere in the shaded area in Fig. 20. Within this interpretation T_0 is just a freezing temperature below which the partial order gets static and is observable by elastic neutron scattering or NMR. As both inhomogeneities and fluctuating order on intermediate time and length scales are expected to be of importance in other quantum-critical systems as well, a further understanding of the NFL behavior in MnSi is highly desirable. Recent μSR experiments have put severe constrictions on the time scale of the partial order [267].

4.2.2 $ZrZn_2$

$ZrZn_2$ is an itinerant electron ferromagnet. It has a cubic structure (C15, Fd$\bar{3}$m). The small ordered magnetic moment $\mu_0 = 0.17\,\mu_B$/Zr-atom contrasts with the large Curie-Weiss moment $\mu_{eff} = 1.9\,\mu_B$/Zr-atom. This difference and the low Curie temperature T_C = 28.5 K classify the system as weak itinerant magnet. It has long been known that T_C and μ_0 can be suppressed by hydrostatic pressure while μ_{eff} stays practically constant [268]. However, the detailed $T_C(p)$ dependence depends strongly on the purity of the sample. $ZrZn_2$ has long been considered a candidate for p-wave superconductivity [269]. Recent experiments reported superconductivity in very pure $ZrZn_2$ [270] which was, however, subsequently shown to arise from a surface layer treated by spark erosion [271]. Magnetization experiments on very pure crystals (RRR $\approx$ 100) suggest that μ_0 and T_C decrease linearly and drop discontinuously at a pressure p_c = 16.5 kbar as shown in Fig. 23 [272]. This is in contrast with the continuous transition observed as a function of temperature at $p = 0$. de Haas-van Alphen experiments under pressure show that the exchange splitting between two nearly spherical Fermi-surface sheets that can be identified with minority and majority spin sheets, respectively, decreases with increasing pressure [273].

The magnetization data as function of p and B suggest the existence of two distinct ferromagnetic phases FM1 and FM2. This rather complicated phase diagram may be a consequence of a double-peak structure in the electronic density of states close to E_F [274]. These phases are shown in Fig. 24 have indeed been identified in magnetization measurements. This scenario possibly causes the first-order transition near p_c. On more general grounds, the transition close to the QPT may be generically of first order due to a coupling of long-wavelength magnetic modes to particle-hole excitations, see Sec. 3.6. In this scenario [94] the second-order transition at low T and zero field becomes first order at a tricritical point, while second-order phase transition lines seam a surface of first-order transitions in the three-dimensional (p,B,T) space.

4.3 Superconductivity near the magnetic–non-magnetic quantum phase transition

For many years, $CeCu_2Si_2$ stood out as the single example of superconductivity in HFS, followed by the discovery of superconductivity in UPt_3, UBe_{13} and URu_2Si_2. $CeCu_2Ge_2$ becomes superconducting above a pressure of 8 GPa [204, 206]. A systematic search by several groups, notably Lonzarich and coworkers, for superconductivity in Ce compounds at the brink of magnetic order led to the discovery of several new heavy-fermion superconductors, $CePd_2Si_2$ [84, 278, 279, 85], $CeRh_2Si_2$ [187], $CeIn_3$ [84, 279, 85], and possibly $CeCu_2$ [275]. In order to observe superconductivity in these systems at the magnetic instability where $T_N \to 0$, pure samples are often essential. This, together with the very fact that superconductivity appears at a point where low-lying magnetic fluctuations abound, suggest that the superconductivity may be magnetically mediated [85].

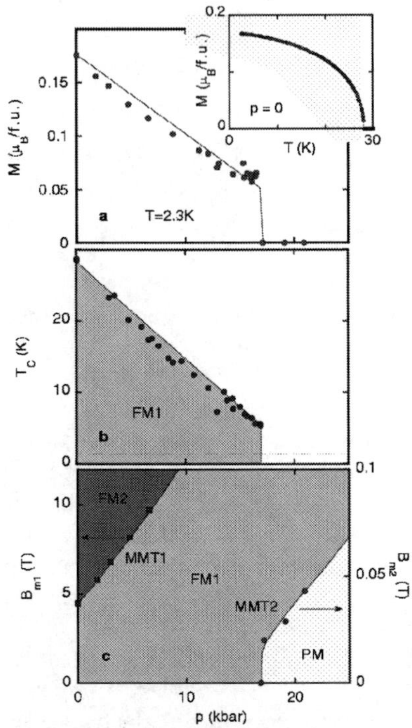

FIGURE 24. Pressure dependence (a) of the spontaneously ordered magnetic moment M in units $\mu_B/f.u.$ and (b) of the curic temperature T_c of $ZrZn_2$. Dashed line in (b) indicates the lowest temperature measured. (c) (B,p) phase diagram at low T with two ferromagnetic phases FM1 and FM2, separated by phase lines MMT1 and MMT2. Inset: Continuous decreas of $M(T)$ at $p = 0$. From [272].

Fig. 25 shows the phase diagram for $CePd_2Si_2$ where T_N drops from 10.5 K at ambient pressure to below 1.6 K around 25 kbar. T_N extrapolates to zero by continuing the linear $T_N(p)$ dependence at p_c = 27 kbar. Around this critical pressure the superconducting transition temperature T_c has its maximum of $\approx$ 0.6 K and extends almost symmetrically to $\pm$ 5 kbar around p_c. However, it is not clear whether antiferromagnetism gives way to superconductivity just below p_c or both types of order coexist. It should be mentioned that besides systems exhibiting a narrow dome of superconductivity near a magnetic QCP, exemplified by $CePd_2Si_2$, wide pressure ranges of superconductivity are observed, e.g., in $CeCu_2Ge_2$. Indeed, it has been suggested that superconductivity in $CeCu_2Ge_2$ is mediated by valence fluctuations rather than spin fluctuations which may cause a rather wide range of superconductivity. This idea is corroborated by recent experiments on $CeCu_2Si_2$ doped with Ge where indeed two disconnected superconducting regions are observed as a function of pressure as shown in Fig. 26 [276, 277]. The relation to quantum criticality is manifest through the pronounced pressure dependence of the residual resistivity ρ_0 and of the resistivity exponent α.

At p_c the electrical resistivity ρ varies with $T^{1.2}$ over almost two orders of magnitude up to $T > 30$ K (see inset of Fig. 25). This quasi-linear T dependence of ρ, together

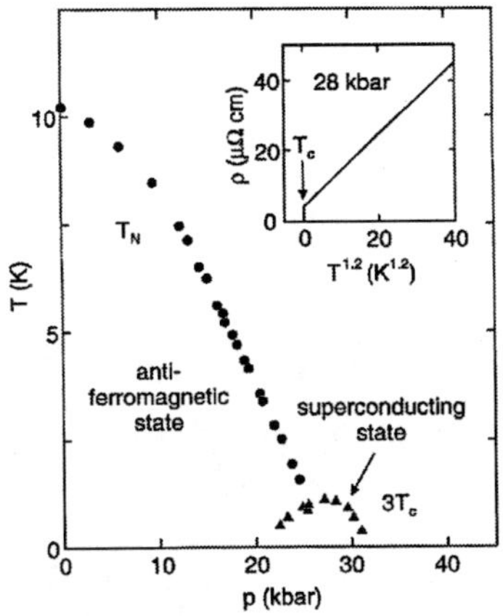

FIGURE 25. Temperature–pressure phase diagram of a high-purity $CePd_2Si_2$ single crystal. Superconductivity appears below T_c in a narrow window where the Néel temperature T_N tends to absolute zero. For clarity, the values of T_c have been scaled by a factor of three, and the origin of the inset has been set below absolute zero. The inset shows that the resistivity ρ exhibits a $T^{1.2}$ dependence on temperature over a wide T range. From [85].

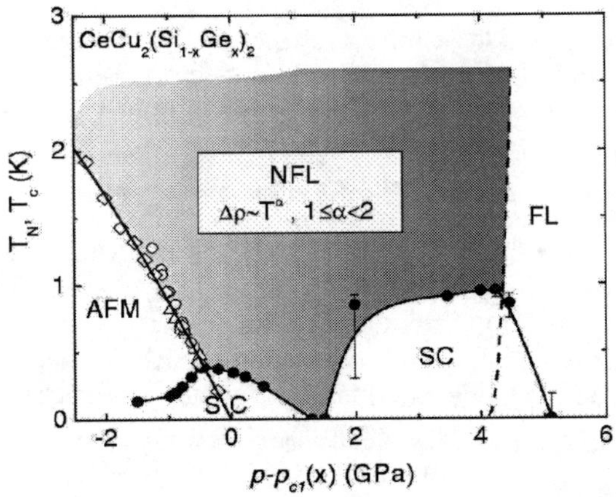

FIGURE 26. Combined (p,x,T) diagram of $CeCu_2(Si_{1-x}Ge_x)_2$. $p_{c1}(x)$ denotes the pressure-tuned QCP where the Néel temperature $T_N \to 0$; $p_{c1}(0) = 0$, $p_{c1}(2.5) = 2.4$ GPa. Open symbols denote T_N, closed symbols the superconducting transition temperature T_c. In the wide shaded area, non-Fermi-liquid behavior is observed in the resistivity $\rho(T)$. From [276]

with the linear relationship $T_N \sim |p - p_c|^{\psi}$, $\psi = 1$, have led to the suggestion of 2D fluctuations [85], as in $CeCu_{6-x}Au_x$. On the other hand, the anomalous $T^{1.2}$ dependence may arise from an effective crossover of $\rho(T)$ in the 3D spin-fluctuation scenario [75, 76] as discussed in Sec. 3.4. $CeNi_2Ge_2$ exhibits a similar anomalous power law of $\rho(T)$ and a $C/T = \gamma_0 - \beta\sqrt{T}$ dependence at ambient pressure [278, 279, 210]. Some very pure samples even exhibit traces of superconductivity at $p = 0$ [209]. These findings

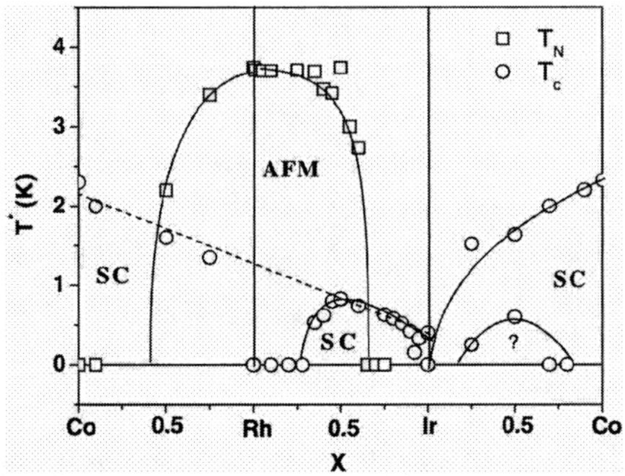

FIGURE 27. Phase diagram of $Ce(Rh,Ir,Co)In_5$ displaying the interplay of antiferromagnetic order (AFM) and superconductivity (SC). From [213].

suggest that $CeNi_2Ge_2$ at $p = 0$ is right at the magnetic instability, although other samples become superconducting at $p > 15$ kbar only [278].

$CeIn_3$ has been suggested to be an antiferromagnet with 3D critical fluctuations, giving way to superconductivity again below 1 K under pressure, on account of the $T^{1.5}$ dependence of $\Delta\rho(T)$ close to $p_c \approx 25$ kbar [279, 85]. In line with this interpretation, the dependence of T_N on $|p - p_c|$ appears to be sublinear; an exponent $\psi = 2/3$ is predicted for a 3d antiferromagnet. The $CeIn_3$-derived $CeTIn_5$ samples exhibit superconductivity over a wide range in alloying among each other (T = Co, Ir, Rh), as shown in Fig. 27. Particularly interesting is $CeCoIn_5$ (see also Sec. 4.1), where a magnetic field induces a second superconducting phase below B_{c2} [280]. This phase is a prime candidate for a modulated superconducting state, i.e., The Fulde-Ferrel-Larkin-Ovchinikov (FFLO) state [281, 282]. Last not least, we wish to mention $CeRh_2Si_2$ which becomes superconducting above p_c = 6 kbar [187], however, here antiferromagnetism disappears through a first-order transition.

A few general remarks are in order. Although the various Ce compounds have vastly different magnetic ordering temperatures, the superconducting T_c are quite similar of the order of 0.5 K with the exception of the "high" T_c = 2.3 K for $CeCoIn_5$. Further, the diverse role of impurities in HFS superconductors is not understood: while some of them are extremely sensitive to impurities, e.g. $CePd_2Si_2$, others are not. In most of the HFS, the question of whether magnetism and superconductivity coexist cooperatively has not been investigated in detail. A prominent counterexample is $CeCu_2Si_2$ where superconductivity competes with A-phase magnetism. We finally note that it has been speculated by [283] that conventional LGW critical points favor unconventional superconductivity, while non-LGW QCP disfavor superconductivity in their vicinity.

While all the above systems exhibit superconductivity in the vicinity of antiferromagnetic instabilities, there has been a long-standing quest for superconductivity of an incipient ferromagnet [269]. Initially such type of superconductivity, presumably of spin-parallel pairing, was sought among weak itinerant ferromagnets, e.g., $ZrZn_2$. However, the first unambiguous observation of a ferromagnetic superconductor was found in UGe_2 under pressure close to the FM instability, with a maximum $T_c \simeq 0.4$ K [284], while the related system URhGe becomes superconducting even at ambient pressure ($T_c \simeq 0.3$ K) [285]. The pressure-temperature phase diagram of UGe_2 reveals two different ferromagnetic phases, overall similar to $ZrZn_2$ [286] The different phases are clearly identified by their different ordered moments. As in $ZrZn_2$, the pressure-driven transitions between the FM phases and from ferromagnet to paramagnet are clearly first order. Consequently, the electricall of UGe_2 resistivity shows FL-like behavior across the transition, with $\Delta\rho = AT^2$ and A peaking at the critical pressure p_c [284]. The superconductive phase resides completely inside the FM domain, giving strong evidence for spin-parallel pairing in coexistence with ferromagnetism. These observations have instigated numerous theoretical studies, since the early weak-coupling calculation [269] suggested two superconductive "domes" just below and above p_c.

Of particular interest is the observation of two distinct superconductive phases in URhGe as a function of magnetic field [287]. The low-field phase subsides at a critical field of $B_{c2} = 2$ T applied along the b axis of the orthorhombic crystal structure. In a field of 11.7 T the spin direction changes abruptly. A second superconductive phase appears in the vicinity of this transition, extending between 8 and 13 T. Hence, viewing the spin reorientation as a quantum phase transition, the superconductivity is clearly linked to this QPT.

It has been suggested that the absence of superconductivity in MnSi, even in very pure samples, is due to the lack of inversion symmetry in this material [284]. However, superconductivity below $T_c = 0.75$ K has been observed recently in a HFS lacking inversion symmetry, $CePt_3Si$, albeit in coexistence with *anti*ferromagnetism, with $T_c \approx 2.2$ K [288].

5. METAL-INSULATOR TRANSITION IN HEAVILY DOPED SEMICONDUCTORS

The problem of the metal-insulator (MI) transition in heavily doped semiconductors has come of age. Although these materials have become prototype systems for the study of a MI transition driven by the combined effects of disorder (Anderson transition) and electron-electron interactions (Mott-Hubbard transition) [289], these effects are only beginning to be disentangled experimentally. In a strict sense, the MI transition viewed as a transition from extended to localized states at the Fermi level, occurs only at $T = 0$ because at finite temperature T thermally activated carrier transport may occur. This is distinctly different from the phase diagrams near a magnetic QPT, see Figs. 1 and 3. Furthermore, it is not clear how to define an order parameter for the MI transition. Hence a metal is defined by a finite *dc* conductivity $\sigma(T)$ for $T \to 0$, while for an insulator $\sigma(0) = 0$. The determination of the critical exponent μ with which $\sigma(0)$ vanishes at $T = 0$ as a function of a control parameter δ at a critical value δ_c, i.e., $\sigma(0) \sim |\delta - \delta_c|^\mu$, has been a subject of considerable controversy, as will be discussed below. However, using both carrier concentration N and uniaxial stress S to tune the

MI transition, identifying the critical region of δ where critical behavior is expected to occur, and employing a dynamic scaling analysis, a consistent picture is beginning to emerge for Si:P which is the prototype material for a localization transition in three dimensions.

The long-range electron interactions around the MI transition have been analyzed with a lot of scrutiny. On the metallic side, the progressive loss of screening upon approaching the MI transition, due to the slow diffusive electron motion arising from the disorder, leads to the well-known Altshuler-Aronov reduction of the density of states $N(E)$ around the Fermi energy $E_{\rm F}$ and corresponding corrections to the Drude conductivity [52]. On the insulating side, the dip at $E_{\rm F}$ develops into a soft Coulomb gap, $N(E) \sim\mid E - E_{\rm F} \mid^2$, giving rise to Efros-Shklovskii variable range hopping [290]. Another consequence of the long-range Coulomb interaction, together with the Pauli principle, is the formation of antiferromagnetically coupled pairs of localized moments on the insulating side of the MI transition [291]. A direct comparison of thermodynamic quantities within the Bhatt-Lee model has been possible [292, 293].

Here we focus on the short-range, i.e. on-site, Coulomb interaction leading on the metallic side, together with disorder, to the formation of localized magnetic moments and to the Kondo effect. Deep in the insulating side, double occupancy of the donor impurity level – more exactly: of the $1s(\mathrm{A}_1)$ groundstate level of the donor state whose sixfold degeneracy is split by the crystalline electric field of the Si host (valley-orbit splitting) – is energetically unfavorable, with an on-site repulsion that is close to the ionization energy of 45 meV. Upon increasing the concentration the singly and doubly occupied states develop into Hubbard bands where corresponding transport properties suggest a scenario of the MI transition involving states at the edges of the lower and upper Hubbard band. Anomalies occurring in the transport data for Si:P below a concentration $N_0 \approx 2.7 \cdot 10^{18}\mathrm{cm}^{-3}$, i.e., well below the critical concentration $N_c \approx 3.52 \cdot 10^{18}\mathrm{cm}^{-3}$ at ambient pressure, can be consistently attributed to the opening of a Hubbard gap below N_c, as supported by the absence of such features in compensated Si:(P,B) which of course is away from half-filling of the $1s(\mathrm{A}_1)$ band.

In this chapter, we will discuss a few aspects of disorder, electron-electron interactions and the critical behavior at the MI transition focusing on three-dimensional P-doped Si and illustrating the advance that has been reached over the last decade [294]. For a discussion of the MI transition in two-dimensional systems (at zero magnetic field) which has received a lot of attention recently, the reader is referred to [295]. In Sec. 5.1 we address the question of disorder by identifying single P donors at freshly cleaved Si(111) surfaces with scanning tunneling microscopy and spectroscopy and determining the distribution of nearest neighbor distances. In Sec. 5.2 we treat the on-site electron-electron interactions. The theoretically derived density of local magnetic moments on the metallic side is compared with experiments and evidence for a Kondo effect already mentioned will be presented. The Hubbard scenario briefly touched upon above will be discussed in more detail. In Sec. 5.3 we present new results we present new results on the critical behavior of the conductivity at the MI transition. In particular, we will discuss dynamic scaling properties and compare the critical behavior with stress-tuning and concentration-tuning the MI transition.

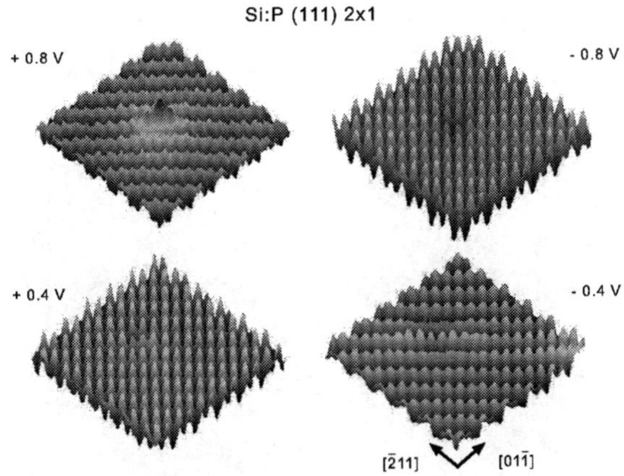

FIGURE 28. STM images of a Si:P (111) 2x1 surface (50x50 Å^2), bulk P concentration $N = 6.7 \cdot 10^{18}$cm^{-3}), taken at different sample voltages with 0.3 nA tunneling current. The feature at the center is attributed to the change of the surface density of states around an ionized P atom. From [296].

5.1 Observation of P donors and their spatial distribution in Si by scanning tunneling microscopy and spectroscopy

Considering the MI transition in doped semiconductors, the question of whether or not dopant atoms are distributed at random is at the heart of the issue. Clustering or short-range ordering could drastically affect the behavior of the MI transition. For instance, extreme clustering, viz. phase separation, could drive the transition towards a classical percolation transition. The distribution of dopants can be investigated by scanning tunneling microscopy (STM) of freshly cleaved surfaces, provided that the observed features can be clearly attributed to dopant atoms. This has been achieved by using scanning tunneling spectroscopy (STS), as demonstrated for Si:P [296, 297], Si:B [298] and GaAs:Si_{Ga} [299] and n-doped InAs [300]. Here we confine the discussion to Si:P. Fig. 28 shows scans of a small section of a Si:P(111) surface after (2x1) reconstruction for different bias taken at a constant current of 0.3 μA at room temperature [296, 297]. The sample with a P concentration of $N = 6.7 \cdot 10^{18}$cm^{-3}, i.e., roughly twice the critical concentration N_c, was previously cleaved *in situ* in UHV along the [211] direction. Positive and negative sample voltages image empty and occupied surface states, respectively. Simultaneous data acquisition was performed by recording the tip height during forward and backward scan of each line with reversed polarity. At +0.8 V, the two Si atoms belonging to the elevated (reconstructed) chain are imaged as zig-zag pattern whereas the image taken at -0.8 V shows only one maximum per surface 2x1 unit cell. The protrusion of apparently 0.5 Å height at positive voltage corresponds to an indentation of 0.2 Å depth at negative bias, located at exactly the same position. For a smaller absolute voltage, we observe an only barely visible depression at +0.4 V, while for -0.4 V the regular atomic arrangement is clearly disturbed: An additional maximum appears at the regular position of a minimum. This can be unmistakenly identified by counting the number of maxima along the horizontal rows of Fig. 28. Locally, an empty

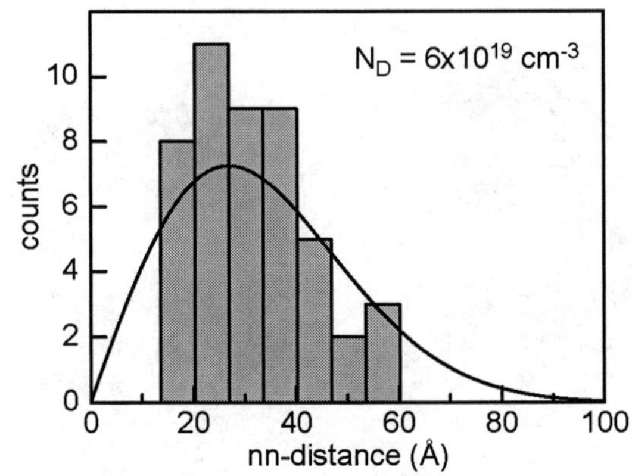

FIGURE 29. Histogram of measured nearest-neigbor distances compared to a distribution function assuming a random spatial arrangement of P donors. From [296].

orbital becomes sufficiently occupied to be imaged at negative sample voltage. This occurs because for the heavily n-doped samples investigated, the Fermi level in the bulk is located near the conduction-band minimum E_c and the resulting upward band-bending due to the negatively charged surface states leads to an ionization of the P atoms. Locally then, the Coulomb potential of the positively charged P atom will slightly reduce the upward band-bending, lowering the energetic position of the surface states and of the valence and conduction bands near the surface. Keeping in mind that the dangling bonds along the (2x1)-reconstructed Si chains alternate between being predominantly occupied or empty, hence being visible with the STM at positive or negative bias, respectively, the additional maximum observed at negative bias at the P site is readily explained as arising from locally shifting the lower edge of the usually unoccupied upper surface band below E_{F}. Even the slight lateral shift of the maximum towards the [211] direction [296] (not visible in Fig. 28) is in line with this assignment.

Having identified the P atoms, we are able to investigate their spatial distribution at the [111] surface. For a random arrangement according to a Poisson distribution, the nearest-neighbor (nn) distances r should be distributed as $f(r) = 2\pi\rho r \cdot e^{-\pi\rho r^2}$, with ρ denoting the surface dopant density. This function is plotted in Fig. 29, together with a histogram of the measured nn distances for a heavily doped sample with $N = 6 \cdot 10^{19}\text{cm}^{-3} \approx 17 N_c$. The general agreement with the above Poisson-derived distribution function does suggest a random spatial arrangement of the P atoms, although a cut-off at $r_{min} \approx 13$ is seen. Interestingly, r_{min} corresponds to the average spacing of 16.3 of P atoms at the solubility limit (assuming a simple cubic lattice) of 2.4 at% of P in Si. This indicates the length scale at which dopant-dopant interactions will come into play. The rather wide distribution of nn distances observed experimentally excludes significant clustering even at this high P concentration far above the MI transition. Of course, at a macroscopic level, segregation and striations during crystal growth will lead to concentration fluctuations at the 1 % level in Si:P.

5.2 On-site electron-electron interactions

5.2.1 Metallic Si:P: formation of local magnetic moments and Kondo effect

It has been known for a long time that a small fraction of localized moments derived from the spin-degenerate groundstate of the hydrogen-like donor or acceptor wave function on the insulating side, survive on the metallic side of the MI transition. These localized magnetic moments have been mostly detected with magnetization [301, 292], magnetic resonance [53, 302] and specific-heat measurements [303, 304]. The dependence of the concentration of localized moments on the P concentration N has been mapped out systematically from specific-heat measurements for Si:P [305, 55] and for Si:(P,B) [306]. The data for Si:P are shown in Fig. 30 with the overall behavior of Si:(P,B) being quite similar. The concentration of local moments has been determined (i) from zero-field excess specific heat $\Delta C = C - \gamma T - \beta T^3$, i.e. conduction electron and phonon contributions subtracted from the measured specific heat C, via the entropy $S = \int(\Delta C/T)\mathrm{d}T = N_S ln2$, and (ii) from the Schottky anomaly observed in applied magnetic fields $B = 1.5$ and 5.7 T, yielding N_{Sch}. In both cases, $S = 1/2$ is assumed for the spin. N_S and N_{Sch} vary smoothly through N_c and localized moments survive in the metallic state due to the disorder and on-site Coulomb repulsion as will be discussed in detail below. The fact that $N_S < N_{Sch}$ on the insulating side can be readily explained because truly isolated spins do not contribute to N_S but, of course, undergo a Zeeman splitting and therefore contribute to C in a magnetic field. On the other hand, due to interactions between moments the Schottky anomalies become broadened and hence N_{Sch} as obtained from the height of the Schottky anomaly underestimates the number of local moments for heavily doped Si:P. Likewise, the (predominantly antiferromagnetic) interactions can be overcome by large applied magnetic fields B and this leads to $N_{Sch}(5.7\,\mathrm{T}) > N_{Sch}(1.5\,\mathrm{T})$, as seen in Fig. 30. The increasing importance of antiferromagnetic interactions can also be inferred from the concentration dependence of the effective field B_{eff} as determined from the temperature of the Schottky maximum: while $B_{eff}/B \approx 1$ for $N < 10^{18}\mathrm{cm}^{-3}$, this ratio is ≈ 0.7 at N_c and decreases to 0.5 at $2N_c$ [55].

The random distribution of dopant atoms, as discussed in Section 2, leads to a wide distribution of nn distances and, hence, nn exchange couplings. This wide distribution of exchange couplings led Bhatt and Lee [291] to a description of the magnetic properties of doped semiconductors in terms of a hierarchical coupling of antiferromagnetic pairs of localized moments. In this model, only free spins that are not coupled to pairs at a given T (with density $N_{free}(T)$) contribute to the specific heat at that temperature with

$$C_{loc} = k_B T \cdot \ell n2 \cdot \mathrm{d}N_{free}(T)/\mathrm{d}T \tag{88}$$

while the magnetic susceptibility is given by

$$\chi_{loc} = \mu_0 \mu_B^2 N_{free}(T)/k_B T. \tag{89}$$

In order to test the validity of the Bhatt-Lee model one can calculate C_{loc} from χ_{loc} with the above equations and compare it to ΔC as measured directly [305, 55]. The overall good agreement found on the insulating side [292, 293] indicates the validity of the

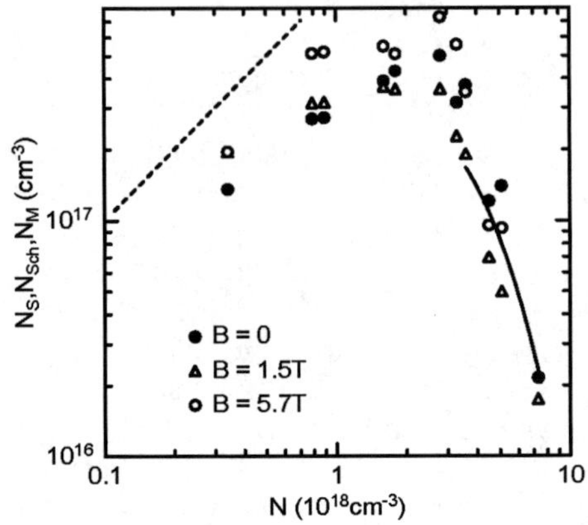

FIGURE 30. Concentration of magnetic moments N_S determined experimentally in zero field and N_{Sch} in finite magnetic fields 1.5 and 5.7 T versus doping concentration. Dashed line indicates $N_{Sch} = N$ expected in the dilute insulating limit. Solid line gives the concentration of local moments N_M in the metallic phase as calculated theoretically. From [55].

Bhatt-Lee model. However, with increasing $N \geq 3.3 \cdot 10^{18}\mathrm{cm}^{-3}$, i.e, on the metallic side of the MI transition, the agreement quickly deteriorates [293]. This clearly indicates that the Bhatt-Lee model is unsuited for the metallic state although its validity has often been tacitly assumed in this range.

The origin of localized moments in the metallic state has been investigated in a number of approaches [307, 308, 309, 310, 55], most of them starting with an Anderson-Hubbard model which is the simplest model Hamiltonian featuring the essential elements of disorder, through random hopping integrals t_{ij}, and on-site Coulomb interaction U:

$$H = \sum_{i\sigma}(\varepsilon_i - \mu)n_{i\sigma} + \sum_{ij\sigma} t_{ij} c^+_{i\sigma} c_{j\sigma} + U \sum_i n_{i\uparrow} n_{i\downarrow} \tag{90}$$

$c_{i\sigma}(c^+_{i\sigma})$ is the annihilation (creation) operator for an electron with spin projection σ in the $1s(A_1)$ groundstate of the dopant atom at site i, and $n_{i\sigma} = c^+_{i\sigma} c_{i\sigma}$. It is usually assumed that the five excited states generated by valley-orbit splitting are sufficiently far away (experimentally $\sim$ 10 meV, see below) that a single-band model is applicable. The host semiconductor properties enter only through the fact that the positions i are random sites of the Si lattice and that the effective mass and interaction are renormalized. The various approaches differ in the type of mean-field model employed. In the approach taken by Langenfeld and Wölfle [310] the starting point is the observation that the fraction of dopant atoms carrying a localized moment in the metallic state is relatively small, of the order of several percent at N_c (see Fig. 30). The sites i are assumed to be on a regular simple-cubic lattice and the disorder is introduced through a variation of t_{ij}. Therefore an isolated moment ("impurity") forming in an effective homogeneous medium is considered. The effective Hamiltonian is that of the Anderson magnetic-impurity model with the local spin occupation number of an impurity atom, which

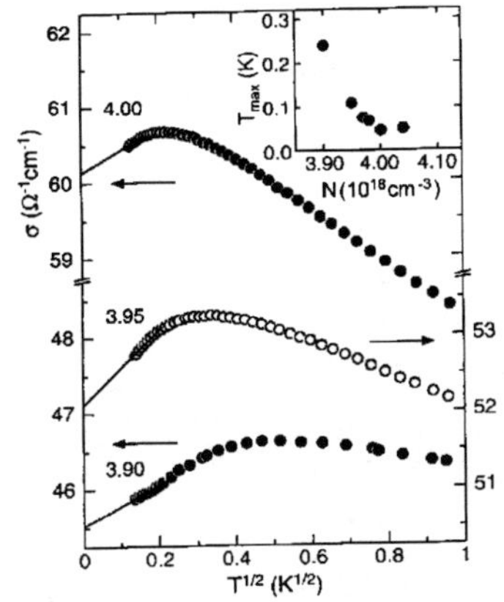

FIGURE 31. σ vs. $\sqrt{T}$ for some Si:P samples. P concentrations N (in $10^{18}\mathrm{cm}^{-3}$) from top to bottom: 4.00, 3.95, and 3.90. The inset shows the temperature T_{max} of the conductivity maximum vs. N. From [313].

is proportional to the magnetization, to be determined self-consistently. The resulting concentration N_M of local moments (obtained applying the usual Anderson criterion for the stability of a local moment) is shown in Fig. 30 where good agreement between N_M and the experimentally determined values N_S and N_{Sch} on the metallic side is seen.

The coupling of the localized moments to the itinerant electrons by an effective exchange interaction J_{eff} gives rise to the Kondo effect. Because of disorder, both the density of states at the Fermi level $N(E_F)$ and the effective exchange interaction $J_{eff} \approx \bar{t}^2/U$ where $\bar{t}$ is the average hopping amplitude from the local-moment site to the neighboring sites, will fluctuate. This leads to a distribution $P(T_K)$ of Kondo temperatures $T_K \sim \exp(-(1/N(E_F)J_{eff})$. An approximate power-law behavior $P(T_K) \sim T_K^{-\alpha_K}$ is found, with $\alpha_K \approx 0.9$ [55] which translates to a specific-heat contribution $C_K \sim T^{1-\alpha_K}$ because the Kondo specific heat is a universal function of T/T_K (see Sec. 2.6). Attributing ΔC in the metallic state of Si:P to the Kondo effect, $\alpha_K = 1 - \alpha_c \approx 0.8$ is inferred experimentally, in rough agreement with the theoretical calculation. However, the Kondo susceptibility for the same distribution $P(T_K)$ should vary roughly as $\chi_K \sim T^{-\alpha_K}$ which is not observed. Rather, the experimental T dependence of χ can be approximated by an exponent of 0.5 [293]. Further work is necessary to resolve this discrepancy. In the theory, interactions among localized moments (RKKY interaction) or a possible quenching of the Kondo effect by disorder have not been considered. The effect of localized magnetic moments on the electrical resistivity via the process of spin-flip scattering has not been positively identified because $\rho(T,B)$ is dominated by weak localization, long-range electron-electron interactions, and wave-function shrinkage induced by a magnetic field B. The strong influence of the two latter mechanisms, for instance, is directly borne out by the negative magnetoconductance observed in metallic Si:P [311, 56, 312] while for spin-flip scattering (and also for weak localization) a positive magnetoconductance, i.e., negative magnetoresistance, is expected. Hence, while a direct effect of spin-flip scattering in $\rho(T,B)$ appears to be negligible, the discovery [313] of a shallow maximum in $\sigma(T)$ in Si:P for concentrations around $N \approx 1.1N_c$ (see Fig. 31) does suggest an indirect influence of local moments on the conductivity by virtue of their influence on electron-electron interactions. As pointed out by Paalanen et al. [53], the contribution from the triplet particle-hole channel is suppressed if the electron spin-flip rate τ_s^{-1} becomes sufficiently large. The situation is analogous to the case of Si:P at higher concentrations in a magnetic field where it is well known that the sign change of the temperature coefficient m of the conductivity $\sigma(T) = \sigma(0) + m\sqrt{T}$ from $m < 0$ for $g\mu_B B \lesssim k_B T$ to $m > 0$ for $g\mu_B B \gtrsim k_B T$ is due to spin-splitting of the triplet term where long-range electron-electron interactions are cut off for $g\mu_B B > k_B T$ [52, 311, 56, 312]. In the concentration region of the $\sigma(T)$ maximum, an anomalously strong shift of T_{max} with field is observed [313].

The thermoelectric power S, on the other hand, clearly gives evidence for scattering by localized magnetic moments [314] because it is particularly sensitive to the Kondo effect. Fig. 32 shows $S(T)$ for a Si:P sample with $N = 4.1 \cdot 10^{18}\text{cm}^{-1}$. The maximum of $S(T)$ is attributed to magnetic scattering since it is observed only for N slightly above N_c where an appreciable density of localized moments exists. (It is the same N range where the maximum in $\sigma(T)$ has been found.) Even more convincing is the suppression of the $S(T)$ maximum in large magnetic fields, also shown in Fig. 32. In B = 6 T we recover the negative diffusion thermoelectric power $S \sim -T$ observed for $N \gg N_c$ in zero field [314]. A very similar behavior of $S(T)$ has been found for compensated Si:(P,B) [315]. Assuming a Kondo-derived thermoelectric power one can compare the $S(T)$ maximum to a corresponding single-ion expression derived by Maki [316], cf. Fig. 32. The deviations between data and best fit might be due to the neglect of a T_K distribution in the fit where a single-valued $T_K = 0.8$ K is assumed. Moreover, the low-T behavior is only approximately correct in this model. It is reassuring that the change of the conductivity due to Kondo scattering as calculated from the fit [314] amounts to only a few percent of the total variation of $\sigma(T)$, in line with the lack of a sizable Kondo effect on $\sigma(T,B)$.

Despite a number of important differences, transport processes of compensated vs. uncompensated Si:P on the metallic side not too close to the MI transition are overall similar. The critical behavior at the MI transition is discussed in Section 5.3. For instance, we have just seen that the scattering by magnetic moments leads to a similar behavior of the thermoelectric power in both types of materials. It has been demonstrated with infrared reflection measurements that the MI transition in Si:P occurs for electronic states in the impurity band which even for P concentrations $N \approx 2N_c$ is still energetically separated from the conduction band [317, 318]. This was inferred from features at 10 and 45 meV in the reflection spectra that could be attributed to optical transitions within the valley-orbit split impurity multiplet, i.e., from the $1s(A_1)$ state to the (nearly degenerate) $1s(E)$ or $1s(T_2)$ states, and to the conduction band, respectively. That the impurity band is energetically separated from the conduction band at the MI transition was actually suggested long time ago by Mott, based on some more indirect evidence [319]. Again, overall similar behavior is found for the infrared reflection of compensated Si:(P,B) [317, 318].

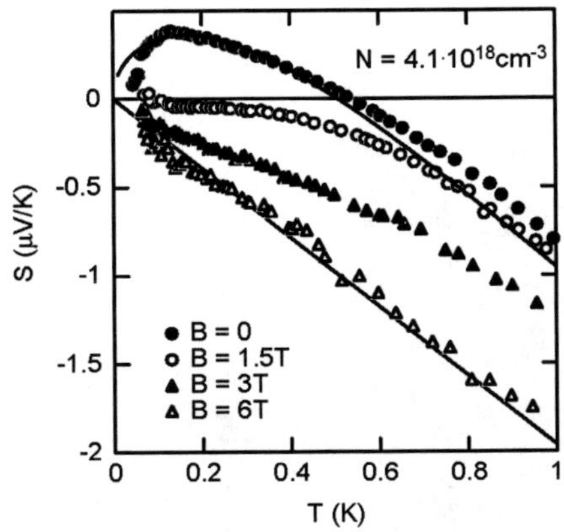

FIGURE 32. Thermoelectric power S vs. temperature T for different magnetic fields B for a metallic Si:P. Solid line is the best fit of a Kondo thermoelectric power with $T_K = 0.8$ K to the zero-field data. From [314].

5.2.2 Insulating Si:P: evidence for Hubbard splitting from transport properties

Marked differences in transport properties between Si:P and Si:(P,B) occur well on the insulating side [320]. Fig. 33 shows the thermoelectric power plotted as $-S \cdot N$ vs. T on a log-log plot. (Here the T range is above 1 K so that the Kondo anomaly in metallic samples, i.e., the positive $S(T)$ maximum superimposed on an overall negative diffusion thermopower with a concomitant sign change of $S(T)$, is not seen and should not be confused with the behavior above 1 K.) For sufficiently high T ($\gtrsim$ 15 K) S is always negative. While the $-S \cdot N$ curves for compensated Si:(P,B) fall on an almost universal curve for carrier concentration N between 1.5 and $2.95 \cdot 10^{18}\text{cm}^{-3}$, such a scaling is seen for uncompensated Si:P only in the concentration range above $N_0 = 2.78 \cdot 10^{18}\text{cm}^{-3}$. Below that concentration $S(T)$ exhibits a sign change at a temperature $T_{S=0}$ which rapidly shifts to higher values with decreasing N. This sign change from $S < 0$ to $S > 0$ with decreasing T is visible in Fig. 33 as a precipitous drop of $\log(-S \cdot N)$ with decreasing T. A further strong difference is seen in the electrical resistivity $\rho(T)$ (Fig. 34). For low carrier concentration, $N < N_0$, ρ rises much faster with decreasing T for Si:P than for Si:(P,B), while the behavior is similar for both types of material $N > N_0$. The strong qualitative difference in $S(T)$ and $\rho(T)$ in Si:P upon crossing N_0 points to different dominant transport processes above and below N_0. Interpreting the steep $\rho(T)$ increase of Si:P below N_0 as an activated process, we can extract an activation energy E_2. Fig. 35 shows a comparison of E_2 and $T_{S=0}$. Although there is an order-of-magnitude difference in absolute values of E_2/k_B and $T_{S=0}$, the similarity of the concentration dependence is striking. Hence we interpret the sign change of $S(T)$ as the onset of an activated process. If one were forced to assume an E_2 process also for Si:(P,B) – although the data actually suggest Efros-Shklovskii VRH with a hopping exponent $p = 1/2$, see inset of Fig. 34 – one would obtain the open circles in Fig. 35,

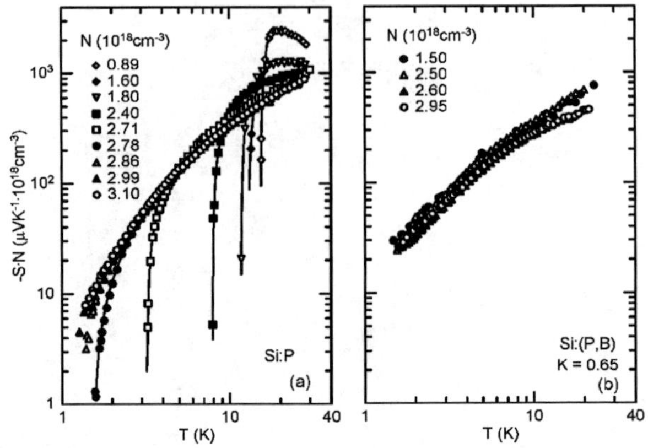

FIGURE 33. Negative thermoelectric power times carrier concentration $-SN$ vs. temperature T on a log-log plot for insulating samples of (a) Si:P and (b) Si:(P,B). From [320].

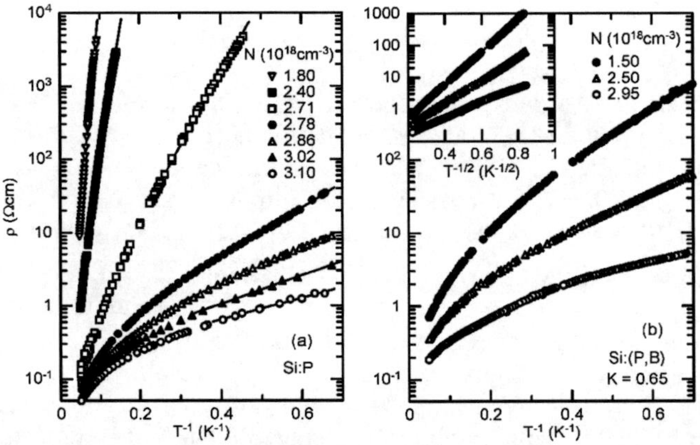

FIGURE 34. Electrical resistivity ρ plotted vs. inverse temperature for (a) Si:P and (b) Si:(P,B). Straight lines in (a) indicate fits to obtain the activation energy E_2. Inset in (b) shows the Si:(P,B) data plotted as $\log \rho$ vs $T^{-1/2}$. Straight lines indicate Efros-Shklovskii hopping. From [320].

i.e., no feature appears at N_0.

The sudden appearence of a hard gap only in Si:P at N_0 well below N_c suggests that we are observing the Hubbard gap due to the on-site Coulomb repulsion. It has been speculated already many years ago that the E_2 process is indeed due to the Hubbard gap [321]. Some time ago the Hubbard gap was inferred for very dilute samples ($N < 10^{17}\text{cm}^{-3}$) from optical measurements [322] while more recent infrared transmission [323] and reflection [317, 318] did not see any signature of a Hubbard gap closer to the transition. In fact, from a calculation of the electronic properties of P clusters in Si:P a stability of a negatively charged cluster of four P sites was found, indicating a

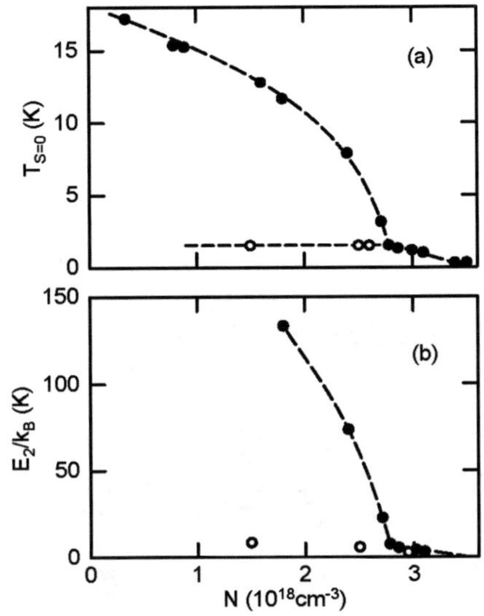

FIGURE 35. (a) Temperature $T_{S=0}$ of the thermoelectric power zero and (b) activation energy E_2 vs. carrier concentration N for uncompensated Si:P (closed circles) and Si:(P,B) (open circles). From [320].

lack of electronic repulsion [324]. On the other hand, the negatively charged isolated P donor in Si is barely stable (binding energy 1.7 meV), i.e., U is of the order of the ionization energy of 45 meV. Therefore it is very likely that the on-site electron repulsion weakens progressively as the P concentration (and carrier concentration) increases, until at $N_0 = 2.78 \cdot 10^{18}\text{cm}^{-3} \approx 0.8N_c$ the two Hubbard bands start to overlap. Because of disorder, however, metallic behavior does not incur immediately since the tail states of the Hubbard bands are localized. It is only at the critical concentration N_c that the Hubbard bands are so close that the chemical potential is within the range of extended states. This scenario is schematically depicted in Fig. 36. It gives a physical picture of how disorder and electron-electron interactions drive the MI transition in Si:P. Of course, the Hubbard features are absent in Si:(P,B) as they should because compensated semiconductors are away from half-filling.

We end this Section by mentioning two observations that may be related to the Hubbard scenario although a direct connection has yet to be worked out. First, it is surprising to see that the density of localized magnetic moments in Si:P is not maximal at N_c but peaks rather precisely at N_0 where the two Hubbard bands in our scenario are suggested to merge, see Fig. 30. Second, in the dynamical mean field theory (DMFT) treatment of the correlation-induced MI transition featured by the Hubbard model which is mapped onto a generalized Anderson impurity model, a peak in the density of states occurs at the chemical potential right at the transition [324]. However, a DMFT extension including non-local dynamics shows that such a peak may in fact be absent [325]. Experimentally, from our earlier data [55] there is no tendency that C/T for $T \to 0$, i.e. below 100 mK, exhibits a pronounced maximum, neither at N_c nor at N_0.

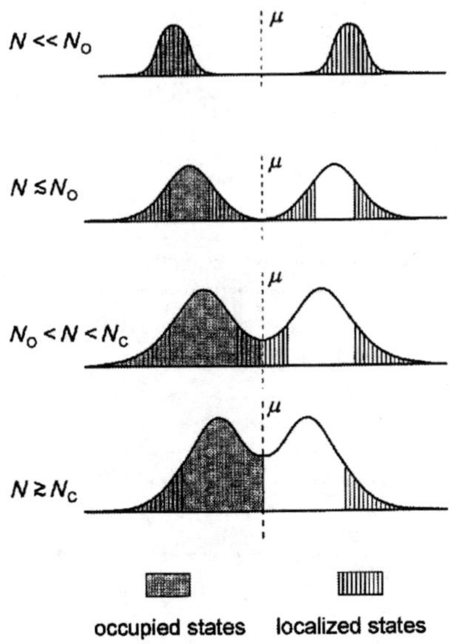

FIGURE 36. Qualitative sketch of the density of states of the impurity band of uncompensated Si:P for several P concentrations N, indicating the splitting into lower and upper Hubbard band around the chemical potential. See text for details.

Between N = 3.6 and $1.6 \cdot 10^{18}\text{cm}^{-3}$, C/T at 100 mK varies by roughly 30% only. Likewise, the linear specific-heat coefficient $\gamma = (C - \Delta C)/T - \beta T^3$, after subtraction of ΔC, – the phonon contribution C_{ph} can be neglected at these low temperatures – decreases smoothly when passing through N_c and N_0 from above. Moreover, there is no strong difference in the behavior between Si:P and Si:(P,B) although the density of localized moments is reduced by a factor of two in the latter [306].

5.3 Critical behavior of the conductivity at the metal-insulator transition

While early suggestions considered the disorder-driven Anderson MI transition to be discontinuous [319], scaling approaches for noninteracting electrons suggested the existence of a continuous second-order phase transition for three-dimensional systems [326] although this is still debated. For a continuous MI transition at a critical value t_c of the parameter t driving the transition, occurring in the strict sense only at zero temperature, the dc conductivity at $T = 0$ is expected to vary as $\sigma(0) \sim \mid t - t_c \mid^{\mu}$. Theoretically, μ is usually inferred from the correlation-length critical exponent ν via Wegner scaling $\mu = \nu(d-2)$ where d is the spatial dimension of the MI systems. In a self-consistent theory of Anderson localization (neglecting interactions), an exponent $\mu = 1$ for $d = 3$ is suggested [327]. Values of ν derived from numerical studies of noninteracting systems range between 1.3 and 1.6 [328, 329, 330]. A review on the theory was given by Belitz and Kirkpatrick [289].

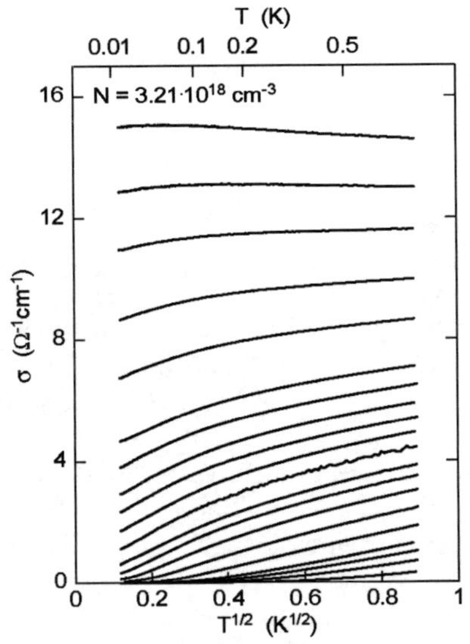

FIGURE 37. Conductivity σ of a Si:P sample with P concentration $N = 3.21 \cdot 10^{18}\mathrm{cm}^{-3}$ versus $\sqrt{T}$ for several values of uniaxial stress applied along the [100] direction. From top to bottom: $S = 3.05$, 2.78, 2.57, 2.34, 2.17, 2.00, 1.94, 1.87, 1.82, 1.77, 1.72, 1.66, 1.61, 1.56, 1.50, 1.41, 1.33, 1.26, 1.18, 1.00 kbar. Solid lines are connecting the very finely spaced individual data points. From [335].

Several investigations, notably on uncompensated semiconductors, have reported values of the critical exponent $\mu = 0.5$, in contrast to $\mu = 1$ generally found for compensated semiconductors and most amorphous metals [331]. However, there seems to be no clear-cut physical distinction between these materials that would justify grouping them in different universality classes. The exponent $\mu = 0.5$ was largely based on the stress-tuning experiments by Paalanen and coworkers [333, 332, 311] where uniaxial stress was used to drive an initially insulating Si:P sample metallic. We suggested several years ago, on the basis of an observed crossover in concentration-tuned uncompensated Si:P from an exponent $\mu \approx 1.3$ close to the critical concentration N_c where $\mathrm{d}\sigma/\mathrm{d}T > 0$, to $\mu = 0.64$ above a crossover concentration $N_{cr} \approx 1.1 N_c$ where $\mathrm{d}\sigma/\mathrm{d}T$ changes sign, to limit the critical region of the MIT on the metallic side to concentrations where the conductivity decreases with decreasing T, i.e., to the range $N_c < N < N_{cr}$ [334]. The value of the conductivity at this crossover is $\sigma_{cr} \approx 40\,\Omega^{-1}\mathrm{cm}^{-1}$. The situation concerning the $T \to 0$ extrapolation to obtain $\sigma(0)$ as a function of N or S being controversial, one may ressort to exploiting the scaling properties of a quantum phase transition in order to shed more light on the critical behavior of the MIT. However, the earlier stress-tuning data were shown to fail scaling behavior at finite T (often called dynamic scaling) [289]. Therefore, we recently reinvestigated the conductivity of Si:P under uniaxial stress with the aim at (i) looking for dynamic scaling properties and (ii) comparing the concentration-tuned and stress-tuned critical behavior of uncompensated Si:P with the same set of samples. For the experiments under stress, the sample was mounted in a ^{4}He-activated uniaxial pressure cell equipped with a piezoelectric force sensor. The stress S was applied along the [100] direction which was the most elongated dimension of the sample. The device allows a continuous change of S while the sample remains at low T.

Fig. 37 shows the electrical conductivity $\sigma(T)$ of a barely insulating sample ($N = 3.21 \cdot 10^{18}\text{cm}^{-3}$) for uniaxial pressures between 1 and 3.05 kbar. The data are plotted vs. $\sqrt{T}$ which is the T dependence expected due to e-e interactions and indeed observed well above the MI transition, as mentioned above [311]. The curves are in fact polygons connecting adjacent data points. Under uniaxial stress between 1 and 2.57 kbar the $\sigma(T)$ curves evolve smoothly from insulating to metallic behavior with $m > 0$, and $\sigma(T)$ becomes nearly independent of T with a value $\sigma_{cr} \approx 12\,\Omega^{-1}\text{cm}^{-1}$ at ≈ 2.7 kbar. For larger stress $\sigma(T)$ passes over a shallow maximum signaling the crossover to $m < 0$, as observed with concentration tuning. It is interesting to note that $\sigma_{cr}(S) \approx 0.3\,\sigma_{cr}(N)$, thus severely limiting the critical region. Our data do not exhibit the precipitous drop of $\sigma(T)$ below ~ 40 mK for pressures closest to the MIT, in distinction to the earlier stress-tuning data on Si:P extending to 3 mK [333, 332].

Closer inspection shows that the data near the MIT are actually better described by a $T^{1/3}$ dependence for low T, see Fig. 38(a). $\sigma(0)$ obtained from the $T^{1/3}$ extrapolation to $T = 0$ is shown in Fig. 38(b), together with data for a sample closer to the critical concentration ($N = 3.43 \cdot 10^{18}\text{cm}^{-3}$). Here $\sigma(0)$ is plotted linearly vs. S, yielding S_c = 1.75 kbar and 1.54 kbar for two samples with 3.21 and $3.43 \cdot 10^{18}\text{cm}^{-3}$, respectively. Note that the critical stress S_c is quite well defined, as $\sigma(0)$ breaks away roughly linearly from zero within less than 0.1 kbar. Applying our criterion for the critical region, the analysis should be limited to data with $\sigma < \sigma_{cr} \approx 12\,\Omega^{-1}\text{cm}^{-1}$. In this range the critical exponent μ is 0.96 and 1.09 for the two samples, respectively. $\mu \approx 1$ is found also when the more conventional $\sqrt{T}$ extrapolation is employed. This behavior contrasts with the earlier stress-tuning data [333] reproduced in the inset of Fig. 38b, where appreciable rounding close to N_c is visible when plotted against S as compared to our samples. However, those $\sigma(0)$ data between 4 and $16\,\Omega^{-1}\text{cm}^{-1}$ are compatible with a linear dependence on uniaxial stress. Stress and concentration tuning lead to strikingly different T dependences of the conductivity in the vicinity of the MI transition. Concentration increase through the MI transition yields – for a given magnitude of σ – a much stronger T dependence $m = \mathrm{d}\sigma/\mathrm{d}\sqrt{T}$ than stress increase, as shown in Fig. 39. This is in line with the fact that $\sigma_{cr}(S)$ is much lower than $\sigma_{cr}(N)$. As the exact origin of the $\sigma(T)$ behavior close to the MIT is unknown, we cannot offer an explanation for the different behavior which, of course, must arise from the change of donor wave functions under uniaxial stress. This nonuniversal behavior does not affect the notion of universality close to the MI transition, In fact, it has been suggested that tuning with S or N should yield the same critical exponents [333, 332, 311, 291]. The decrease of N_c with uniaxial stress is attributed to the admixture of the more extended 1s(E) and 1s(T_2) excited states to the 1s(A_1) groundstate of the valley-orbit split sixfold donor 1s multiplet [336]. The fact that stress was applied to different directions in the previous and present studies, i.e., [123] and [100], respectively, may well be one reason for the different behavior of $\sigma(T)$. Figs. 37 and 38 might suggest that concentration inhomogeneities in a given sample exposed to stress might also play a role.

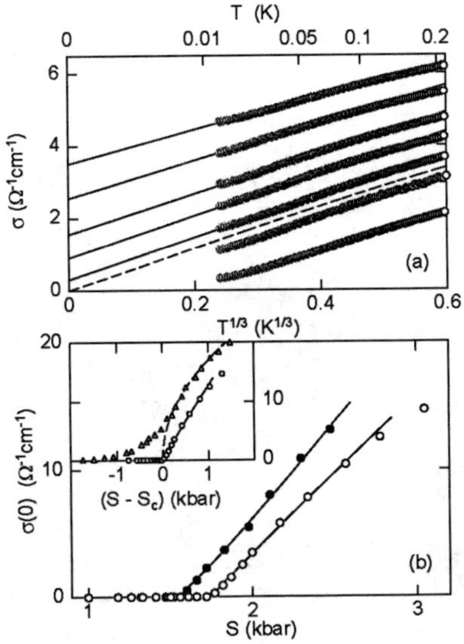

FIGURE 38. (a) Low-temperature data of σ of Fig. 6 for stress in the immediate vicinity of the metal-insulator transition plotted against $T^{1/3}$. Dashed line indicates the conductivity at the critical stress (see text). (b) Extrapolated conductivity $\sigma(0)$ for $T \to 0$ versus uniaxial stress S for two P concentrations $N = 3.21$ and $3.43 \cdot 10^{18}\text{cm}^{-3}$ (open and closed circles, respectively). The inset shows earlier $\sigma(0)$ versus $S - S_c$ data (triangles) from [333] in comparison to our data for sample 1 (circles)[335].

In order to analyze the scaling behavior of σ at finite temperatures using the data of the sample with $N = 3.21 \cdot 10^{18}\text{cm}^{-3}$, we employ the scaling relation [337]

$$\sigma(t,T) = \sigma_c(T)\mathscr{F}'((t-t_c)/T^y) \tag{91}$$

where $\sigma_c(T) = \sigma(t_c,T)$ is the conductivity at the critical value t_c of the parameter t driving the MI tranistion. This scaling relation is derived from the general scaling relation

$$\sigma(t,T) = b^{-(d-2)}\mathscr{F}''((t-t_c)b^{1/\nu}, b^z T) \tag{92}$$

where b is a scaling parameter. If the leading term to $\sigma_c(T)$ is proportional to T^x, one obtains $x = \mu/\nu z$ and $y = 1/\nu z$ from a scaling plot. Fig. 37 shows that σ for S close to S_c does not exhibit a simple power-law T dependence over the whole T range investigated. We therefore describe $\sigma_c(T)$ by the function $\sigma_c(T) = aT^x(1 + dT^w)$ with $a = 6.01\Omega^{-1}\text{cm}^{-1}, x = 0.34, d = -0.202, w = 0.863$, and T is expressed in K. Here the dT^w term presents a correction to the critical dynamics. All $\sigma(S,T)$ curves with $1.00\,\text{kbar} < S < 2.34\,\text{kbar}$ up to 800 mK are then used for the scaling analysis according to Eq. (91). The same procedure was repeated for other choices of $\sigma_c(T)$ between the two measured $\sigma(T)$ curves embracing the critical stress with clearly less satisfactory results.

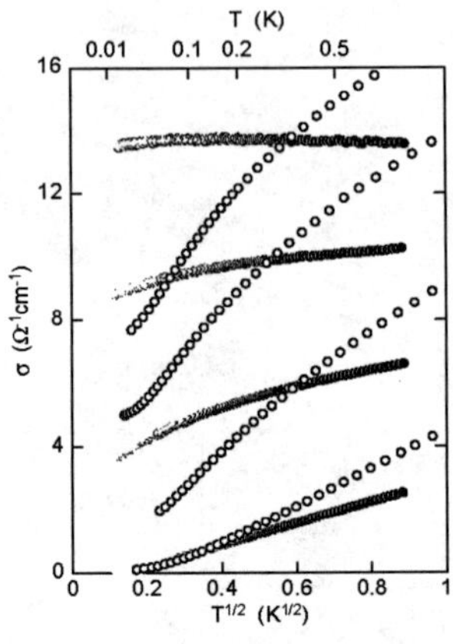

FIGURE 39. Comparison of the concentration dependence of $\sigma(T)$ of Si:P (open symbols, from top to bottom: $N = 3.60, 3.56, 3.50, 3.38 \cdot 10^{18}\text{cm}^{-3}$, from Ref. [14]) and stress dependence of $\sigma(T)$ (closed symbols, $N = 3.21 \cdot 10^{18}\text{cm}^{-3}$, from top to bottom: $S = 2.78$, 2.34, 1.94, 1.50 kbar). From [335].

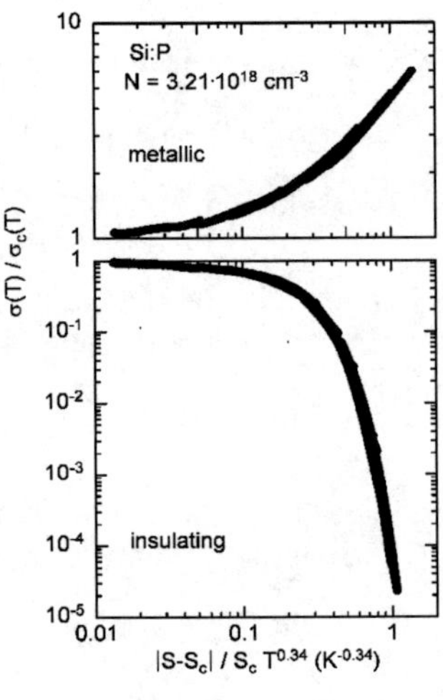

FIGURE 40. Scaling plot of σ/σ_c vs. $\mid S - S_c \mid /S_c T^y$ for Si:P with $N = 3.21 \cdot 10^{18}\text{cm}^{-3}$ at different uniaxial pressures S, with $S_c = 1.75$ kbar and $y = 0.34$. From [335].

Fig. 40 shows the scaling plot of $\sigma(S,T)/\sigma_c(T)$ vs. $\mid S - S_c \mid /S_c T^y$. The data are seen to collapse on a single branch each for the metallic and insulating side, respectively. The best scaling, as shown, is achieved for $y = 1/z\nu = 0.34$. Together with $\mu = 1.0$ as obtained from Fig. 38b and assuming Wegner scaling $\nu = \mu$ for $d = 3$, we find $z = 2.94$, consistent with $\sigma_c \sim T^{1/z} \sim T^{1/3}$ for $T \to 0$ used to obtain $\sigma(0)$, see Fig. 38a. An alternative way to express dynamic scaling is to plot isotherms $\sigma(T)/\sigma_c(T)$ vs. S, as shown in Fig. 41. This serves as a consistency check for the correct determination

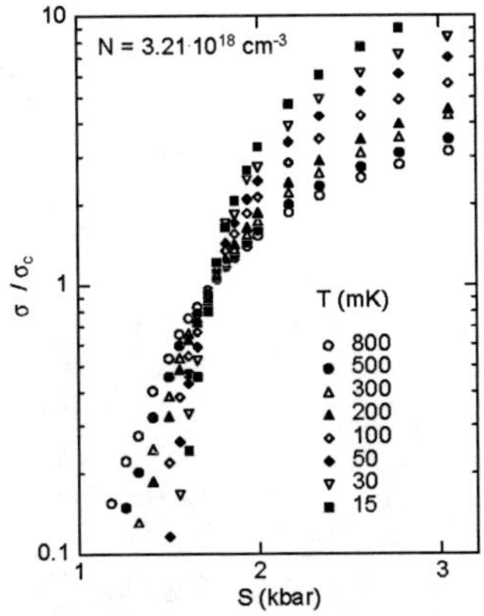

FIGURE 41. Conductivity isotherms $\sigma(T = \text{const}, S)$ normalized to the critical conductance $\sigma_c = \sigma(T, S_c)$ vs. uniaxial stress S for the Si:P sample with $N = 3.21 \cdot 10^{18}\text{cm}^{-3}$.

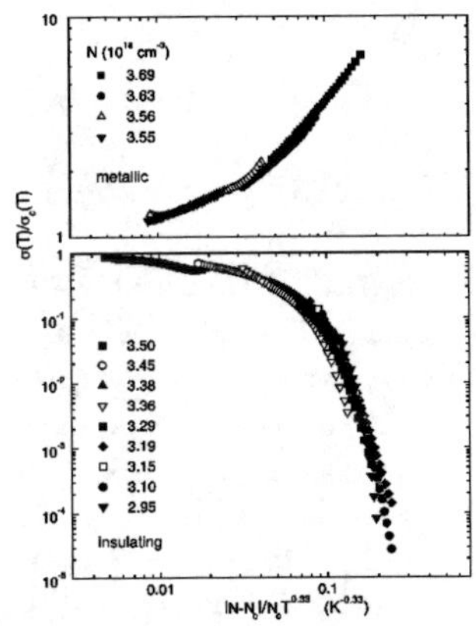

FIGURE 42. Scaling plot of σ/σ_c vs. $|N - N_c|/N_c T^y$ for Si:P with different P concentrations N, with $N_c = 3.52 \cdot 10^{18}\text{cm}^{-3}$ and $y = 0.33$ [339], data after Ref. [334].

of the critical stress S_c where all isotherms should collapse onto one point, as indeed observed. The critical stress is determined with a relative accuracy to better than 0.1 kbar. It is important to note that either $\sigma(0)$ scaling (Fig. 38) or dynamic scaling (Fig. 40) when taken by itself, may lead to a rather large error in μ and/or z, just because of the ambiguity of determining the critical region. It is the consistent combination of both approaches that lends confidence to the results.

From an analysis of $\sigma(T)$ of Si:P for different P concentrations N we had previously inferred $\mu = 1.3$ from $\sigma(0)$ vs. N [334] and $z = 2.4$ from a dynamic scaling analysis of metallic samples only [338]. One can of course check whether a dynamic scaling of the form of Eq.(91) is possible using N instead of S as well. The result is shown in Fig. 42, with the following parameters of $\sigma_c(T)$: $a = 12.15\Omega^{-1}\mathrm{cm}^{-1}, x = 0.43, d = -0.093, w = 1.16$. The best scaling is achieved with $y = 0.33$, or $z\nu = 3.0$. With $\nu = 1.1$ as obtained from a variable-range hopping analysis of the conductivity in the insulating region [339] one obtains $z = 2.75$. The value $\mu = x/y = 1.3$ agrees fully with that derived from the $\sigma(0)$ extrapolation [334]. In view of the scatter of the data in the scaling plot, these values are consistent with the previous dynamic scaling analysis. Moreover, given the uncertainty in the concentration determination, these values are broadly consistent with those obtained from the present stress-tuning study for which we estimate an error of $\pm 10\%$ for μ and z. Bogdanovich et al. [337] demonstrated that conductivity data for Si:B under uniaxial stress obey very nicely the dynamic scaling on both metallic and insulating sides, yielding $\mu = 1.6$ and $z = 2$, while concentration tuning of $\sigma(0)$ on the same system had suggested $\mu = 0.63$ [340]. This large difference is not understood at present. For Si:P, the exponents μ and z are in broad agreement with the expectation for a noninteracting system. This is in line with our above schematic scenario for the MI transition (cf. Fig. 36) where the transition proper is driven by disorder although interactions are vital in inducing the Hubbard splitting.

CONCLUSIONS

In general, metallic systems of interacting fermions at low temperature are well described by the Landau Fermi-liquid theory. This holds for heavy-fermion systems and for heavily-doped semiconductors sufficiently above the critical dopant concentration. Non-Fermi-liquid behavior over an extended range of temperatures down to absolute zero may occur near the borderline between two qualitatively different ground states. The order-parameter fluctuations in the neighborhood of such a quantum critical point can induce singular scattering between the fermions, leading to a breakdown of the usual phase space arguments on which Fermi-liquid theory is based.

In this review we aimed to give a systematic, balanced and critical account of the present knowledge in the area of Fermi-liquid instabilities near quantum critical points. We focused on magnetic QCP, which are perhaps the best studied cases at present. While a large body of experimental and theoretical work on such systems has been accumulated over the past 20 years, several principal questions appear to be far from answered. The situation for the metal-insulator transition driven by the combined effects of disorder and electron-electron interactions is likewise not well understood. Here, we do not even have a proper notion of an order parameter. Hence, more work, both theoretically and experimentally needs to be done, with the prospect of finding novel behavior and novel phases near quantum phase transitions.

ACKNOWLEDGMENTS

We thank our many colleagues for stimulating discussions. We are particularly grateful to A. Rosch and M. Vojta for sharing their theoretical insights in quantum phase transitions with us. This work was partly supported by the Deutsche Forschungsgemeinschaft via the Research Unit “Quantum Phase Transitions”.

REFERENCES

1. G. R. Stewart, *Rev. Mod. Phys.* **73**, 797 (2001); G. R. Stewart, Rev. Mod. Phys. **78**, 743 (2006).
2. J. A. Hertz, *Phys. Rev. B* **14**, 1165 (1976).
3. A. J. Millis, *Phys. Rev. B* **48**, 7183 (1993).
4. H. v. Löhneysen et al., *Rev. Mod. Phys.* **79**, 1015 (2007).
5. L. D. Landau, *Sov. Phys. JETP* **3**, 920 (1957).
6. L. D. Landau, *Sov. Phys. JETP* **5**, 101 (1957).
7. L. D. Landau, *Sov. Phys. JETP* **8**, 70 (1959).
8. J. Feldman et al., *Europhys. Lett.* **24**, 437 (1993).
9. R. Shankar, *Rev. Mod. Phys.* **66**, 129 (1994).
10. I. J. Pomeranchuk, *Sov. Phys. JETP* **8**, 361 (1958).
11. J. Quintanilla and A. Schofield, *Phys. Rev. B* **74**, 115126 (2006).
12. A. V. Chubukov et al., *Phys. Rev. B* **73**, 045128 (2006).
13. I. L. Aleiner and K. B. Efetov, *Phys. Rev. B* **74**, 075102 (2006).
14. D. Belitz et al., *Phys. Rev. B* **55**, 9452 (1997).
15. A. V. Chubukov et al., *Phys. Rev. B* **71**, 205112 (2005).
16. D. S. Greywall, *Phys. Rev. B* **27**, 2747 (1983).
17. K. Kadowaki and S. B. Woods, *Solid State Commun.* **58**, 507 (1986).
18. A. C. Hewson, *The Kondo Problem to Heavy Fermions*, Cambridge University Press, Cambridge, England (1993).
19. J. Kondo, *Prog. Theor. Phys.* **32**, 37 (1964).
20. P. Nozieres, *J. Low. Temp. Phys.* **17**, 31 (1974).
21. P. W. Anderson, *Phys. Rev.* **124**, 41 (1961).
22. J. R. Schrieffer and P. A. Wolff, *Phys. Rev.* **149**, 491 (1966).
23. P. W. Anderson, J. *Phys. C* **3**, 2436 (1970).
24. P. W. Anderson and G. Yuval, *Phys. Rev. B* **1**, 1522 (1970).
25. N. Andrei, *Phys. Rev. Lett.* **45**, 379 (1980).
26. P. B. Wiegmann, *Sov. Phys. JETP* **31**, 392 (1980).
27. N. Andrei et al., *Rev. Mod. Phys.* **55**, 331 (1983).
28. T. A. Costi and A. C. Hewson, *Philos. Mag. B* **65**, 1165 (1992).
29. M. Oshikawa, *Phys. Rev. Lett.* **84**, 3370 (2000).
30. S. Doniach, *Physica B & C* **91**, 231 (1977).
31. M.A. Continentino et al., *Phys. Rev. B.* **39**, 9734(R) (1989).
32. M.A. Continentino, *Phys. Rev. B* **47**, 11587(R) (1993).
33. Z. Fisk et al., *Jpn. J. Appl. Phys.* **26**, Suppl. 182 (1987).
34. D. M. Newns and N. Read, *Adv. Phys.* **36**, 799 (1987).
35. A. Houghton et al., *Phys. Rev. B* **37**, 3782 (1988).
36. S. Burdin et al., *Phys. Rev. Lett.* **85**, 1048 (2000).
37. W. Metzner and D. Vollhardt, *Phys. Rev. Lett.* **62**, 324 (1989).
38. A. Georges et al., *Rev. Mod. Phys.* **68**, 13 (1996).
39. R. Bulla, *Phys. Rev. Lett.* **83**, 136 (1999).
40. T. Pruschke et al., *Phys. Rev. B* **61**, 12799 (2000).
41. P. NoziĄères, *J. Phys. Soc. Jpn.* **74**, 4 (2005).
42. P. Nozieres, *Ann. Phys. (Paris)* **10**, 19 (1974).
43. C. L. Lin et al., *Phys. Rev. Lett.* **85**, 1232 (1987).
44. P. Nozieres and A. Blandin, J. Phys. (France) **41**, 193 (1980).
45. N. Andrei and C. Destri, *Phys. Rev. Lett.* **52**, 364 (1984).
46. A. M. Tsvelik and P. B. Wiegmann, *Z. Phys. B: Condens. Matter* **54**, 201 (1984).
47. P. Schlottmann and P. D. Sacramento, *Adv. Phys.* **42**, 641 (1993).
48. I. Affleck and A. W. W. Ludwig, *Nucl. Phys. B* **352**, 849 (1991).
49. I. Affleck and A. W. W. Ludwig, *Nucl. Phys. B* **360**, 641 (1991).
50. C. L. Seaman et al., *Phys. Rev. Lett.* **67**, 2882 (1991).

51. D. L. Cox and A. Zawadowski, *Adv. Phys.* **47**, 599 (1998).
52. P. A. Lee and T. V. Ramakrishnan, *Rev. Mod. Phys.* **57**, 287 (1985).
53. M. A. Paalanen et al., *Phys. Rev. Lett.* **57**, 2061 (1986).
54. M. A. Paalanen and R. N. Bhatt, *Physica B* **169**, 223 (1991).
55. M. Lakner et al., *Phys. Rev. B* **50**, 17064 (1994).
56. M. P. Sarachik, in *Metal-insulator transitions revisited* edited by P. P. Edwards and C. N. R. Rao (Taylor & Francis, London), p. 79 (1995).
57. B. M. McCoy and T. T. Wu, *Phys. Rev.* **176**, 631 (1968).
58. R. B. Griffiths, *Phys. Rev. Lett.* **23**, 17 (1969).
59. E. Miranda and V. Dobrosavljevic, *Rep. Prog. Phys.* **68**, 2337 (2005).
60. S. L. Sondhi et al., *Rev. Mod. Phys.* **69**, 315 (1997).
61. S. Sachdev, *Quantum Phase Transitions* (Cambridge University Press, Cambridge, England) 1999.
62. M. Vojta, *Rep. Prog. Phys.* **66**, 2069 (2003).
63. see, e.g., J. J. Binrey, N. J. Dowsick, A. J. Fisher, and M. E. J. Newman, *The theory of criticl phenomena* (Clarendon Press, Oxford, England) 1992.
64. V. Privman, in *Finite size scaling and numerical simulation of statistical systems*, edited by V. Privman (World Scientific, Singapore), p. 1, 1990.
65. L. Zhu et al., *Phys. Rev. Lett.* **91**, 066404 (2003).
66. T. Moriya, *Spin fluctuations in itinerant electron magnetism* (Springer, Berlin) 1985.
67. S. Sachdev et al., *Phys. Rev. B.* **51**, 14874 (1995).
68. S. Sachdev and E. R. Dunkel, *Phys. Rev. B.* **73**, 085116 (2006).
69. A. Abanov and A. Chubukov, *Phys. Rev. Lett.* **93**, 255702 (2004).
70. I. Fischer and A. Rosch, *Phys. Rev. B* **71**, 184429 (2005).
71. T. Moriya and A. Kawabata, *J. Phys. Soc. Jpn.* **34**, 639 (1973).
72. H. Hasegawa and T. Moriya, *J. Phys. Soc. Jpn.* **36**, 1542 (1974).
73. T. Moriya, *Proc. Jpn. Acad., Ser. B: Phys. Biol. Sci* **82**, 1 (2006).
74. R. Hlubina and T. M. Rice, *Phys. Rev. B* **51**, 9253 (1995).
75. A. Rosch, *Phys. Rev. Lett.* **82**, 4280 (1999).
76. A. Rosch, *Phys. Rev. B* **62**, 4945 (2000).
77. H. Kontani et al., *Phys. Rev. B* **59**, 14723 (1999).
78. S. Mathon, *Proc. R. Soc. London, Ser. A* **306**, 355 (1968).
79. T. Moriya and T. Takimoto, *J. Phys. Soc. Jpn.* **64**, 960 (1995).
80. P. Gegenwart et al., *Phys. Rev. Lett.* **81**, 001501 (1998).
81. R. Hauser et al., *J. Magn. Magn. Mater.* **177-181**, 292 (1998).
82. K. Heuser et al., *Phys. Rev. B* **57**, R4198 (1998).
83. F. M. Grosche et al., *J. Phys.: Condens. Matt.* **12**, L533 (2000).
84. S. R. Julian et al., *J. Phys.: Condens. Matt.* **8**, 9675 (1996).
85. N. D. Mathur et al., *Nature* **394**, 38 (1998).
86. P. Estrala et al., *Phys. Rev. B* **63**, 212409 (2001).
87. P. Estrala et al., *Eur. Phys. J. B* **23**, 449 (2001).
88. A. Rosch et al., *Phys. Rev. Lett.* **79**, 159 (1997).
89. T. Vojta et al., *Z. Phys. B: Condens. Matt.* **103**, 451 (1997).
90. D. Belitz et al., *Phys. Rev. Lett.* **85**, 4602 (2000).
91. D. Belitz et al., *Phys. Rev. B* **63**, 174427 (2001).
92. D. Belitz et al., *Phys. Rev. B* **63**, 174428 (2001).
93. D. Belitz et al., *Phys. Rev. Lett.* **93**, 155701 (2004).
94. D. Belitz et al., *Phys. Rev. Lett.* **94**, 247205 (2005).
95. T. Vojta and R. S'knepnek, *Phys. Rev. B* **64**, 052404 (2001).
96. A. V. Chubukov et al., *Phys. Rev. Lett.* **92**, 147003 (2004).
97. J. R. Schrieffer, *J. Low Temp. Phys.* **99**, 397 (1995).
98. P. W. Anderson, *The theory of superconductivity in the high-T_c cuprates* (Princeton University Press, Princeton, NJ) 1997.
99. A. Rosch, *Phys. Rev. B.* **64**, 174407 (2001).
100. S. J. Yamamoto and Q. Si, *Phys. Rev. Lett.* **99**, 016401 (2007).
101. A. Schröder et al., *Phys. Rev. Lett.* **80**, 5623 (1998).
102. A. Schröder et al., *Nature* **407**, 351 (2000).

103. P. Coleman, *Physica B* **259-261**, 353 (1999).
104. Q. Si et al., *Int. J. Mod. Phys. B* **13**, 2331 (1999).
105. Q. Si et al., *Nature* **413**, 804 (2001).
106. Q. Si et al., *Phys. Rev. B* **68**, 115103 (2003).
107. T. Senthil et al., *Phys. Rev. Lett.* **90**, 216403 (2003).
108. T. Senthil et al., *Phys. Rev. B* **69**, 035111 (2004).
109. P. Coleman et al., *J. Phys.: Condens. Matt.* **13**, R723 (2001).
110. C. Pépin et al., *Phys. Rev. Lett.* **94**, 066420 (2005).
111. P. Coleman et al., *Phys. Rev. B* **72**, 245111 (2005).
112. A. K. McMahan et al., *J. Comput. Aided Mater. Des.* **5**, 131 (1998).
113. S. Paschen et al., *Nature* **432**, 881 (2004).
114. P. Gegenwart et al., *Science* **315**, 969 (2007).
115. H. v. Löhneysen et al., *Acta Phys. Polon. B* **34**, 707 (2002).
116. H. Maebashi et al., *Phys. Rev. Lett.* **95**, 207207 (2005).
117. J. L. Smith and Q. Si, *Phys. Rev. B* **61**, 5184 (2000).
118. D. R. Grempel and Q. Si, *Phys. Rev. Lett.* **91**, 026401 (2003).
119. J.-X. Zhu et al., *Phys. Rev. Lett.* **91**, 156404 (2003).
120. T. Senthil et al., Physica B **359-361**, 9 (2005).
121. O. Stockert et al., *Phys. Rev. Lett.* **80**, 5627 (1998).
122. C. Varma et al., *Phys. Rev. Lett.* **63**, 1996 (1989).
123. S. A. Grigera et al., Science **294**, 329 (2001).
124. A. J. Millis et al., *Phys. Rev. Lett.* **88**, 217204 (2002).
125. K. Andres et al., *Phys. Rev. Lett.* **35**, 1779 (1975).
126. F. Steglich et al., *Phys. Rev. Lett.* **43**, 1892 (1979).
127. N. Grewe and F. Steglich, in *Handbook of Physics and Chemistry of Rare Earths,* edited by K. A. Geschneidner, Jr. and L Eyring (Elsevier, Amsterdam), Vol. 14, p. 343, 1991.
128. B. Lüthi and M. Yoshizawa, *J. Magn. Magn. Mater.* **63-64**, 274 (1987).
129. G. Lapertot et al., *Physica B* **186-188**, 454 (1993).
130. J. S. Gavilano et al., *Phys. Rev. B* **52**, R13106 (1995).
131. A. Amato et al., *J. Low Temp. Phys.* **68**, 371 (1987).
132. Y. Onuki and T. Komatsubara, *J. Magn. Magn. Mater.* **63-64**, 281 (1987).
133. E. Gratz et al., *J. Magn. Magn. Mater.* **63-64**, 312 (1987).
134. E. A. Schuberth et al., *Phys. Rev. B* **51**, 12892 (1995).
135. L. Pollack et al., *Phys. Rev. B* **52**, R15707 (1995).
136. H. Tsujii et al., *Phys. Rev. Lett.* **84**, 5407 (2000).
137. A. Amato et al., *Physica B* **186-188**, 273 (1993).
138. H. G. Schlager et al., *J. Low Temp. Phys.* **90**, 181 (1993).
139. J. Rossat-Mignod et al., *J. Magn. Magn. Mater.* **76-77**,376 (1988).
140. H. v. Löhneysen et al., *Physica B* **186-188**, 590 (1993).
141. H. v. Löhneysen et al., *Physica B* **223-224**, 471 (1996).
142. H. v. Löhneysen et al., *J. Magn. Magn. Mater.* **177-181**, 12 (1998).
143. T. Pietrus et al., *Physica B* **206-207**,317 (1995).
144. A. Germann et al., *J. Phys., Colloq* **49**, C8-755 (1988).
145. T. Chattopadhyay et al., *Z. Phys. B: Condens. Matter* **80**, 159 (1990).
146. A. Schröder et al., *Physica B* **199-200**, 47 (1994).
147. H. v. Löhneysen et al., *Eur. Phys. J. B* **5**, 447 (1998).
148. C. Paschke et al., *J. Low Temp. Phys.* **97**, 229 (1994).
149. K. Grube et al., *Phys. Rev. B* **60**, 11947 (1999).
150. A. Hamann et al., *Physica B* **403**, 1323 (2008).
151. A. Amato et al., *Phys. Rev. B* **52**, 554 (1995).
152. O. Bernal et al., *Phys. Rev. B* **54**, 13000 (1996).
153. H. v. Löhneysen, *J. Phys.: Condens. Matt.* **8**, 9689 (1996).
154. Y. Miyako et al., *Physica B* **230-232**, 1011 (1997).
155. H. v. Löhneysen et al., *Phys. Rev. Lett.* **72**, 3262 (1994).
156. D. Finsterbusch et al., *Ann. Phys.* **5**, 184 (1996).
157. R. E. Walstedt et al., *Phys. Rev. Lett.* **90**, 067601 (2003).

158. A. K. Gangopadhyay et al., *Phys. Rev. B* **38**, 2603 (1988).
159. G. Fraunberger et al., *Phys. Rev. B.* **40**, 4735 (1989).
160. R. Küchler et al, *Phys. Rev. Lett.* **93**, 096402 (2004).
161. E.-W. Scheidt et al., *Physica B* **259-261**, 388 (1999).
162. H. v. Löhneysen et al., *Phys. Rev. B* **63**, 134411 (2001).
163. O. Stockert et al., *Phys. Rev. Lett.* **99**, 237203 (2007).
164. M.Sieck et. al., *Physica B* **223-224**, 325 (1996).
165. L. P. Regnault et al., *Phys. Rev. B* **38**, 4481 (1988).
166. A. Amato et al., *Phys. Rev. B* **50**, 619 (1994).
167. R. A. Fischer, *J. Low Temp. Phys.* **84**, 49 (1991).
168. H. Kadowaki et al., *Phys. Rev. B* **92**, 097204 (2004).
169. P. Haen et al., *J. Low Temp. Phys.* **67**, 391 (1987).
170. S. Holtmeier et al., *Physica B* **204**, 250 (1995).
171. F. S. Tautz et al., *Physica B* **206-207**, 29 (1995).
172. T. Sakakibara et al., *Physica B* **186-187**, 317 (1993).
173. R. Djerbi et al., *J. Magn. Magn. Mater.* **76-77**, 260 (1988).
174. L. P. Regnault et al., *J. Magn. Magn. Mater.* **90-91**, 398 (1990).
175. S. Quezel et al., *J. Magn. Magn. Mater.* **76-77**, 403 (1988).
176. J. Flouquet et al., *Physica B* **215**, 77 (1995).
177. S. Kambe et al., *J. Phys. Soc. Jpn.* **65**, 3294 (1996).
178. S. Raymond et al., J. *Low Temp. Phys.* **109**, 205 (1997).
179. W. Knafo et al., *Phys. Rev. B* **70**, 174401 (2004).
180. S. Raymond et al., *J. Phys.: Condens. Matt.* **13**, 8303 (2001).
181. A. Böhm et al., *J. Magn. Magn. Mater.* **76-77**, 150 (1988).
182. M. B. Fontes et al., *Phys. Rev. B* **53**, 11678 (1996).
183. S. Süllow et al., *Phys. Rev. Lett.* **82**, 2963 (1999).
184. H. Wilhelm et al., *Phys. Rev. B* **59**, 3651 (1999).
185. S. Quezel et al., *Solid State Commun.* **49**, 685 (1984).
186. J. D. Thompson et al., *J. Magn. Magn. Mater.* **54-57**, 377 (1986).
187. R. Movshovich et al., *Physica B* **223-224**, 126 (1996).
188. C. Pfleiderer et al., *Phys. Rev. B* **55**, 8330 (1997).
189. T. Graf et al., *Phys. Rev. Lett.* **78**, 3769 (1997).
190. B. Lloret et al., *J. Magn. Magn. Mater.* **63-64**, 85 (1987).
191. R. Calemczuk et al., *J. Magn. Magn. Mater.* **90-91**, 477 (1990).
192. S. Kawarazaki et al., *Physica B* **206-207**, 298 (1995).
193. B. H. Grier et al., *Phys. Rev. B* **29**, 2664 (1984).
194. Y. Miyako et al., *Z. Phys. B: Condens. Matt.* **101**, 339 (1996).
195. H. Kadowaki et al., *Phys. Rev. Lett.* **96**, 016401 (2006).
196. S. Muramaya et al., *Phys. Rev. B* **56**, 11092 (1997).
197. T. Taniguchi et al., *J. Magn. Magn. Mater.* **177-181**, 419 (1998).
198. H. Gu et al., *Physica B* **312-313**, 248 (2002).
199. A. Germann and H. v. Löhneysen, *Euro. Phys. Lett.* **9**, 367 (1989).
200. F. Steglich et al., *J. Phys.: Condens. Matt.* **8**, 9909 (1996).
201. F. Steglich et al., *Z. Phys. B: Condensed Matter* **103**, 235 (1997).
202. O. Stockert et al., *Phys. Rev. Lett.* **92**, 136401 (2004).
203. F. R. deBoer et al., *J. Magn. Magn. Mater.* **63-64**, 91 (1987).
204. D. Jaccard et al., *Phys. Lett. A* **163**, 475 (1992).
205. E. Vargoz and D. Jaccard, *J. Magn. Magn. Mater.* **177-181**, 294 (1998).
206. D. Jaccard et al., *Physica B* **259-261**, 1 (1999).
207. G. Bruls et al., *Phys. Rev. Lett.* **72**, 1754 (1994).
208. O. Stockert et al., *Physica B* **374-375**, 167 (2006).
209. P. Gegenwart et al., *Phys. Rev. Lett.* **82**, 1293 (1999).
210. R. Küchler et al., *Phys. Rev. Lett.* **91**, 066405 (2003).
229. C. Petrovic et al., *J. Phys.: Condens. Matt.* **13**, L337 (2001).
212. H. Hegger et al., *Phys. Rev. Lett.* **84**, 4986 (2000).
213. P. G. Pagliuso et al., *Physica B* **312-313**, 129 (2002).

214. J. R. Jeffries et al., *Phys. Rev. B* **72**, 024551 (2005).
215. S. Nakatsujii et al., *Phys. Rev. Lett.* **89**, 106402 (2002).
216. A. Bianchi et al., *Phys. Rev. Lett.* **91**, 257001 (2003).
217. B. Andraka and A. M. Tsvelik, *Phys. Rev. Lett.* **67**, 2886 (1991).
218. O. Trovarelli et al., *Phys. Rev. Lett.* **85**, 626 (2000).
219. K. Heuser et al., *Phys. Rev. B* **58**, R15959 (1998).
220. J. Paglione et al., *Phys. Rev. Lett.* **91**, 246405 (2003).
221. E.D. Bauer et al., *Phys. Rev. Lett.* **94**, 047001 (2005).
222. F. Ronning et al., *Phys. Rev. Lett.* **73**, 064519 (2006).
223. M. Nicklas et al., *J. Phys.: Condens. Matt.* **13**, L905 (2001).
224. R. A. Fisher et al., *Phys. Rev. B* **65**, 224509 (2002).
225. W. Bao et al., *Phys. Rev. B* **65**, 100505(R) (2002).
226. T. Muramatsu et al., *J. Phys. Soc. Jpn.* **70**, 3362 (2001).
227. T. Park et al., *Nature* **440**, 65 (2006).
228. G. Knebel et al., *Phys. Rev. B* **74**, 020501(R) (2006).
229. C. Petrovic et al., *Europhys. Lett.* **53**, 354 (2001).
230. C. Capan et al., *Phys. Rev. B* **70**, 180502(R) (2004).
231. P. Gegenwart et al., *Phys. Rev. Lett.* **89**, 056402 (2002).
232. P. Gegenwart et al., *Phys. Rev. Lett.* **94**, 076402 (2005).
233. J. Custers et al., *Nature* **424**, 524 (2003).
234. J. Ferstl et al., *Physica B* **359-361**, 26 (2005).
235. A. Fert and P. M Levy, *Phys. Rev. B* **36**, 1907 (1987).
236. M. R. Norman, *Phys. Rev. B* **71**, 220405(R) (2005).
237. K. Ishida et al., *Phys. Rev. Lett.* **89**, 107202 (2002).
238. J. Sichelschmidt et al., *Phys. Rev. Lett.* **91**, 156401 (2003).
239. G. G. Lonzarich and L. Taillefer, *J. Phys. C.* **18**, 4339 (1985).
240. M. Nicklas et al., *Phys. Rev. Lett.* **82**, 4268 (1999).
241. P. Niklowitz et al., *Phys. Rev. B* **72**, 024424 (2005).
242. T. Freltoft et al., *Phys. Rev. B* **37**, 3454 (1988).
243. A. Block et al., *Phys. Rev. B* **49**, 12365 (1994).
244. A. Yeh et al., *Nature* **419**, 459 (2002).
245. M. Lee et al., *Phys. Rev. Lett.* **92**, 187201 (2004).
246. M. Norman et al., *Phys. Rev. Lett.* **90**, 116601 (2003).
247. J. R. Stewart et al., *Phys. Rev. Lett.* **89**, 186403 (2002).
248. P. Bak and M. H. Jenson, *J. Phys. C***13**, L881 (1980).
249. O. Nakanishi et al., *Solid State Commun.* **35**, 995 (1980).
250. F. P. Mena et al., *Phys. Rev. B* **67**, 241101(R) (2003).
251. E. Fawcett et al., *Int. J. Magn.* **1**, 29 (1970).
252. Y. Ishikawa et al., *Phys. Rev. B* **31**, 5884 (1985).
253. L. Taillefer et al., *J. Magn. Magn. Mater.* **54-57**, 957 (1986).
254. J. D. Thompson et al., *Physica B* **161**, 317 (1989).
255. C. Pfleiderer et al., *Nature* **414**, 427 (2001).
256. N. Doiron-Leyraud et al., *Nature* **425**, 595 (2003).
257. P. Pedrazzini et al., *Physica B* **378-380**, 165 (2006).
258. C. Pfleiderer et al., *Nature* **427**, 227 (2004).
259. J. Schmalian and M. Turlakov, *Phys. Rev. Lett.* **93**, 036405 (2004).
260. W. Yu et al., *Phys. Rev. Lett.* **92**, 086403 (2004).
261. B. Binz et al., *Phys. Rev. Lett.* **96**, 207202 (2006).
262. U. K. Rößler et al., *Nature* **442**, 797 (2006).
263. I. Fischer et al., *condmat/0702287* (2007).
264. D. Belitz et al., *Phys. Rev. B* **73**, 054431 (2006).
265. D. Belitz et al., *Phys. Rev. B* **74**, 024409 (2006).
266. I. Fischer and A. Rosch, *Europhys. Lett.* **68**,93 (2004).
267. Y. J. Uemura et al., *Nature Physics* **3** , 29 (2007).
268. J. G. Huber et al., *Solid State Commun.* **16**, 211 (1975).
269. D. Fay and J. Appel, *Phys. Rev. B* **22**, 3173 (1980).

270. C. Pfleiderer et al., *Nature* **412**, 58 (2001).
271. E. A. Yelland et al., *Phys. Rev. B* **72**, 214523 (2005).
272. M. Uhlarz et al., *Phys. Rev. Lett.* **93**, 256404 (2004).
273. N. Kimura et al., *Phys. Rev. Lett.* **92**, 197002 (2004).
274. K. G. Sandeman et al., *Phys. Rev. Lett.* **90**, 167005 (2003).
275. E. Vargoz et al., *Physica B* **230-232**, 182 (1997).
276. H. Q. Yuan et al., *Science* **302**, 2104 (2003).
277. H. Q. Yuan et al., *Phys. Rev. Lett.* **96**, 047008 (2006).
278. F. M. Grosche et al., *Physica B* **239**, 62 (1997).
279. S. Julian et al., *J. Magn. Magn. Mater.* **177-181**, 265 (1998).
280. A. Bianchi et al., *Phys. Rev. Lett.* **91**, 187004 (2003).
281. P. Fulde and R. A. Ferrell, *Phys. Rev.* **135**, A550 (1964).
282. A. I. Larkin and Y. N. Ovchinnikov, *Sov. Phys. JETP* **20**, 762 (1965).
283. F. Steglich, *Physica B* **359-361**, 326 (2005).
284. S. S. Saxena et al., *Nature* **406**, 587 (2000).
285. D. Aoki et al., *Nature* **413**, 613 (2001).
286. C. Pfleiderer and A. D. Huxley, *Phys. Rev. Lett.* **89**, 147005 (2002).
287. F. Levy et al., *Science* **309**, 1343 (2005).
288. E. Bauer et al., *Phys. Rev. Lett.* **92**, 027003 (2004).
289. D. Belitz and T. R. Kirkpatrick, *Rev. Mod. Phys.* **66**, 261 (1994).
290. A. L. Efros and B. I. Shklovskii, *J. Phys. C* **8**, L49 (1975).
291. R. N. Bhatt and P. A. Lee, *Phys. Rev. Lett.* **48**, 344 (1982).
292. Y. Ootuka and N. Matsunaga, *J. Phys. Soc. Jpn.* **59**, 1801 (1990).
293. H. G. Schlager and H. v. Löhneysen, *Europhys. Lett.* **40**, 661 (1997).
294. H. v. Löhneysen, in: *Festkörperprobleme*, Adv. in Solid State Phys. **40**, 143 (2000).
295. E. Abrahams et al., *Rev. Mod. Phys.* **73**, 251 (2001).
296. T. Trappmann et al., *Europhys. Lett.* **38**, 177 (1997).
297. T. Trappmann et al., *Appl. Phys. A* **68**, 167 (1999).
298. M. Schöck et al., *Phys. Rev. B* **61**, 7622 (2000).
299. M. C. M. M. van der Wieden et al., *Phys. Rev. Lett.* **76**, 1075 (1996).
300. C. Wittneven et al., *Phys. Rev. Lett.* **81**, 5616 (1998).
301. A. Roy and M. P. Sarachik, *Phys. Rev. B* **37**, 5531 (1988).
302. H. Alloul and P. Dellouve, *Phys. Rev. Lett.* **59**, 578 (1978).
303. J. R. Marko et al., *Phys. Rev. B* **10**, 2448 (1974).
304. N. Kobayashi et al., *Solid State Commun.* **24**, 67 (1977).
305. M. Lakner and H. v. Löhneysen, *Phys. Rev. Lett.* **83**, 648 (1989).
306. S. Wagner, M. Lakner, H. v. Löhneysen, *Phys. Rev. B* **55**, 4219 (1997).
307. M. Milovanović, S. Sachdev, R. N. Bhatt, *Phys. Rev. Lett.* **63**, 82 (1989).
308. R. N. Bhatt and D. S. Fisher, *Phys. Rev. Lett.* **68**, 3072 (1992).
309. V. Dobrosavljević and G. Kotliar, *Phys. Rev. Lett.* **71**, 3218 (1993).
310. A. Langenfeld and P. Wölfle, *Ann. Phys. (Leipzig)* **4**, 43 (1995).
311. T. F. Rosenbaum et al., *Phys. Rev. B* **27**, 7509 (1983).
312. H. v. Löhneysen and M. Welsch, Phys. Rev. B **44**, 9045 (1991).
313. A. Blaschette et al., *Europhys. Lett.* **36**, 527 (1996).
314. M. Lakner and H. v. Löhneysen, *Phys. Rev. Lett.* **70**, 3475 (1993).
315. P. Ziegler et al., *Europhys. Lett.* **33**, 285 (1996).
316. K. Maki, *Prog. Theor. Phys.* **41**, 586 (1969).
317. A. Gaymann et al., *Phys. Rev. Lett.* **71**, 3681 (1993).
318. A. Gaymann et al., *Phys. Rev. B* **52**, 16486 (1995).
319. N. F. Mott, *Metal-insulator transitions*, London Taylor & Francis (1990).
320. X. Liu et al., *Phys. Rev. Lett.* **77**, 3395 (1996).
321. N. F. Mott and E. A. Davis, *Electronic processes in non-crystalline materials*, 2nd edn. Oxford: Clarendon (1979).
322. P. Norton, *Phys. Rev. Lett.* **37**, 164 (1976).
323. G. A. Thomas et al., *Phys. Rev. B* **23**, 5472 (1981).
324. G. Kotliar et al., *Rev. Mod. Phys.* **68**, 13 (1996).

325. Th. Maier et al., *Eur. Phys. J. B* **13**, 613 (2008).
326. E. Abrahams et al., *Phys. Rev. Lett.* **42**, 673 (1979).
327. D. Vollhardt and P. Wölfle, in: *Electronic phase transitions*, ed. by W. Hanke and Yu. V. Kopaev, Elsevier B.V. (1992), 1.
328. B. Kramer et al., *Physica A* **167**, 163 (1990).
329. B. Kramer and A. MacKinnon, *Rep. Prog. Phys.* **56**, 1469 (1993).
330. K. Slevin and T. Ohtsuki, *Phys. Rev. Lett.* **82**, 382 (1999).
331. G. A. Thomas, *Phil. Mag. B* **52**, 479 (1985).
332. G. A. Thomas et al., *Phys. Rev. B* **27**, 3897 (1983).
333. M. A. Paalanen et al., *Phys. Rev. Lett.* **48**, 1284 (1982).
334. H. Stupp et al., *Phys. Rev. Lett.* **71**, 2634 (1993).
335. S. Waffenschmidt et al., *Phys. Rev. Lett.* **83**, 3005 (1999).
336. R. N. Bhatt, *Phys. Rev. B* **26**, 1082 (1982).
337. S. Bogdanovich et al., *Phys. Rev. Lett.* **82**, 137 (1999).
338. H. Stupp et al., *Phys. Rev. Lett.* **72**, 2122 (1994).
339. M. Hornung et al., *phys. stat. sol. (b)* **218**, 75 (2000).
340. P. Dai, Y. Zhang and M. P. Sarachik, *Phys. Rev. Lett.* **66**, 1914 (1991).

Correlated thermoelectrics

V. Zlatić

Institute of Physics, Bijenicka c. 46, P. O. B. 304,1000 Zagreb,Croatia

Abstract. In the first part of these lecture notes we introduce the phenomenological equations for describing the heat and charge transport in thermoelectric samples. We discuss the solution obtained for various boundary conditions that are appropriate for the homogeneous and inhomogeneous thermoelectrics. In the second part we develop the formalism for a linear-response many-body description of the transport properties of correlated electrons. By properly determining the local heat-current operator we show that the Jonson-Mahan theorem applies to the Hamiltonians that are commonly used for the intermetallic compounds with Cerium, Europium and Ytterbium ions, so the various thermal-transport coefficient integrands are related by powers of frequency. We illustrate how to use this formalism by calculating the thermoelectric properties of the periodic Anderson model and, then, show that these results explain the experimental data on heavy fermions and valence fluctuators. Finally, we calculate the thermoelectric properties of the Falicov-Kimball model and use the results to explain the anomalous features of the intermetallic compounds in which one observes the valence-change transition.

Keywords: thermoelectricity, electron correlations
PACS: 71.27.+a,72.15Jf, 72.20Pa

INTRODUCTION

We are interested in the processes that occur in a thermoelectric sample connected to two reservoirs on different temperatures, chemical potentials, and voltage. The temperature difference, ΔT, and the difference in the voltage and chemical potentials, ΔV and $\Delta\mu$, give rise to the charge and energy transport through the sample that would bring, eventually, both reservoirs into equilibrium. For a given thermoelectric material and the boundary condition defined by ΔT, $\Delta\mu$, and ΔV we ask what is the distribution of temperature, electrical and chemical potential in the sample and how are the current densities related to the gradients $\nabla T(\mathbf{x})$, $\nabla\phi(\mathbf{x})$ and $\nabla\mu(\mathbf{x})$, where $\phi(\mathbf{x})$ and $\mu(\mathbf{x})$ are the electrical and chemical potential, respectively. These questions can be answered by solving the transport equations, together with the continuity equations and the Maxwell equation. The transport equations relate the currents to the driving forces, which are assumed to be given by a linear combinations of the gradients $\nabla T(\mathbf{x})$, $\nabla\phi(\mathbf{x})$ and $\nabla\mu(\mathbf{x})$. The transport coefficients are specified by the thermoelectric properties of the sample and can either be obtained from the measurements or calculated from a microscopic model. The choice of the currents and the forces is not unique, i.e., the transport equations can be written in a number of equivalent ways. Here, we consider only the pairs that lead to the symmetric transport coefficients. In what follows, we derive the compleat set of equations for the thermoelectric problem and discuss the solution. We neglect any changes in the thermodynamic state of the reservoirs and assume that each elementary (macroscopic) volume of the sample can be described by local thermodynamic functions. In our case, the lo-

CP1014, *Lectures on the Physics of Strongly Correlated Systems XII: Twelfth Training Course*, edited by A. Avella and F. Mancini

cal quantities describe a system of interacting particles in the presence of an external potential.

The currents generated by the driving forces give rise to irreversible processes that increase the entropy of the thermoelectric sample. The rate of change of the entropy is obtained from the integral[1, 2]

$$S=\int d^3\mathbf{x}\, s(u,n), \tag{1}$$

where $s(\mathbf{x})$ is the local entropy density, $n(\mathbf{x})$ particle density, $u(\mathbf{x})$ the energy density, and the integration is over the sample. Using the first law of thermodynamics,

$$ds=\frac{\partial s}{\partial u}du+\frac{\partial s}{\partial n}dn, \tag{2}$$

where $\partial s/\partial u=1/T$ and $\partial s/\partial n=-\mu/T$, we can write the rate of change of entropy as

$$\frac{dS}{dt}=\int d^3x\left\{\frac{1}{T}\frac{du}{dt}-\frac{\mu}{T}\frac{dn}{dt}\right\}. \tag{3}$$

The conservation of charge and energy provide the continuity equations that relate the time-derivatives of the charge and energy densities to the divergencies of the corresponding current vectors,

$$\frac{dn}{dt}=-\frac{1}{e}\nabla\cdot\mathbf{J} \tag{4}$$

and

$$\frac{du}{dt}=-\nabla\cdot\mathbf{J}_u, \tag{5}$$

where $\mathbf{J}(\mathbf{x})$ and $\mathbf{J}_u(\mathbf{x})$ are the charge current density and the energy current density, respectively. The continuity equation can also be written in terms of the internal energy current density, $\mathbf{J}_E=\mathbf{J}_u-\phi\mathbf{J}$,

$$\frac{du}{dt}+\nabla\cdot\mathbf{J}_E=\mathbf{E}\cdot\mathbf{J}. \tag{6}$$

where $\mathbf{E}=-\nabla\phi$ is the external field and $\mathbf{E}\cdot\mathbf{J}$ the work done in the system by this field per unit volume and time. Making the substitution $\nabla\cdot\mathbf{J}_E=T\nabla(\mathbf{J}_E/T)-T\mathbf{J}_E\nabla(1/T)$ and $(\mu/T)\nabla\cdot\mathbf{J}=\nabla(\mu\mathbf{J}/T)-\mathbf{J}\nabla(\mu/T)$ we rewrite Eq.(3) in the form

$$\frac{dS}{dt}=\int d^3x\frac{1}{T}\left\{\mathbf{J}\cdot(\mathbf{E}-\frac{T}{e}\nabla\frac{\mu}{T})+\mathbf{J}_E\cdot T\nabla\frac{1}{T}\right\}-\int d^3x\nabla\cdot\frac{\mathbf{J}_E-\mu\mathbf{J}/e}{T} \tag{7}$$

that allows a simple interpretation[1, 2]. The second term can be trqansformed into a surface integral which gives the entropy change arising from an entropy current flowing across the boundary of the sample. Inroducing the heat current density $\mathbf{J}_Q=\mathbf{J}_E-\mu\mathbf{J}/e$ and the entropy current density $\mathbf{J}_S=\mathbf{J}_Q/T$, the rate of change of the entropy due to these boundary currents can be written as,

$$-\int d^3x\nabla\cdot\frac{\mathbf{J}_Q}{T}=-\int_{surface}d\mathbf{A}\cdot\mathbf{J}_S. \tag{8}$$

The first term in Eq. (7) gives the volume contribution which is due to change of the local entropy density at each point. The rate of chage of this local entropy density is defined by the expression,

$$T\frac{ds_V}{dt} = \mathbf{J}\cdot(\mathbf{E} - \frac{T}{e}\nabla\frac{\mu}{T}) + \mathbf{J}_E \cdot T\nabla\frac{1}{T}. \tag{9}$$

In a stationary state, the forces and currents are time-independent and all the entropy produced locally is transported out of the sample by the entropy current.

The charge and energy conservation allow us to express the rate of change of the local entropy density as a scalar product

$$T\frac{ds_V}{dt} = \mathbf{J}\cdot\mathbf{x}_\phi^L + \mathbf{J}_E\cdot\mathbf{x}_E^L, \tag{10}$$

where $\mathbf{J}$ and $\mathbf{J}_E$ are the currents corresponding to the generalized forces,

$$\begin{aligned} \mathbf{x}_\phi^L &= -\nabla\phi - \frac{T}{e}\nabla\frac{\mu}{T} = -\nabla(\phi + \frac{\mu}{e}) - \frac{\mu T}{e}\nabla\frac{1}{T}, \\ \mathbf{x}_T^L &= T\nabla\frac{1}{T}. \end{aligned} \tag{11}$$

Following Onsager, we introduce the generalized current vector

$$\hat{\mathbf{J}} = (J^x, J^y, J^z, J_E^x, J_E^y, J_E^z) = (\hat{J}_1, \ldots, \hat{J}_6), \tag{12}$$

and the generalized force vector

$$\hat{\mathbf{x}} = (x_x^\phi, x_y^\phi, x_z^\phi, x_x^T, x_y^T, x_z^T) = (\hat{x}_1, \ldots, \hat{x}_6), \tag{13}$$

such that

$$\frac{ds_V}{dt} = \frac{\hat{\mathbf{J}}\cdot\hat{\mathbf{x}}}{T}. \tag{14}$$

Finally, assuming that the system is close to an overall equilibrium, and that the generalized forces are small, we approximate the current vectors by a linearized form,

$$\hat{J}_a = \sum_b L_{ab}\hat{x}_b. \tag{15}$$

The general arguments of irreversible statistical physics show that one can define the currents and the forces in a number of ways but as long as the scalar product in Eq. (14) is conserved the transport coefficients are symmetric and satisfy the Onsager relations [1, 2, 3].

$$L_{ab} = L_{ba}. \tag{16}$$

Note, the transport equations and the continuity equation are not sufficient for a unique determination of the current vectors and the scalar fields $T(\mathbf{x})$, $\phi(\mathbf{x})$ and $\mu(\mathbf{x})$ for a given set of boundary conditions. To obtain the complete solution we have to relate the electrical potential to the charge and current densities by the Maxwell equations. For a non-magnetic sample without the displacement currents these are

$$\nabla \cdot \mathbf{E} = 4\pi\rho_T \qquad \nabla \times \mathbf{E} = -\frac{1}{c}\frac{\partial \mathbf{B}}{\partial t}, \tag{17}$$

$$\nabla \cdot \mathbf{B} = 0 \qquad \nabla \times \mathbf{B} = \frac{4\pi}{c}\mathbf{J}^T, \tag{18}$$

where $\rho^T = \rho + \rho_{ext}$ and $\mathbf{J}^T = \mathbf{J} + \mathbf{J}_{ext}$ are the total charge and current densities, ρ_{ext} and $\mathbf{J}_{ext}$ are the external charged and current densities, and $\rho = e(n - n_0)$, where n_0 is the equilibrium number density of electrons in the sample.

We now have a complete set of equations and can find the solution of the thermoelectric problem by the following procedure. For an arbitrary choice of $T(\mathbf{x})$, $\phi(\mathbf{x})$ and $\mu(\mathbf{x})$ find the local values of $n(\mathbf{x})$ from the partition function and obtain the current densities $\mathbf{J}$ and $\mathbf{J}_u$ from the transport equations. Use Maxwell equations to obtain a relationship between $\mathbf{E}(\mathbf{x})$, $\mathbf{B}(\mathbf{x})$, $n(\mathbf{x})$ and $\mathbf{J}(\mathbf{x})$, and solve these equations to determine the electrical field for a given value of $T(\mathbf{x})$ and $\mu(\mathbf{x})$. Alternatively, if no current is flowing, determine the electrical potential for a given charge distribution which depends on $T(\mathbf{x})$ and $\mu(\mathbf{x})$ from the Poisson equation. Finally, use the continuity equations to find the distribution of temperatures and the chemical potential in the sample. The particular solution obtained in such a way depends on the choice of transport coefficients (they describe the sample) and on the boundary conditions (they describes the reservoirs).

PHENOMENOLOGICAL EQUATIONS

The irreversible thermodynamics[1, 3] deals with the systems subject to a thermal gradient, and the inhomogeneities in the electrical and chemical potential. The inhomogeneities of $\mu(\mathbf{x})$ are due to it deviation from the uniform chemical potential μ_0 that we use to control the total number of particles in the sample. The electro-chemical potential which drives the currents in the absence of the temperature gradient is $\bar{\phi} = \phi + \mu/e$; the electro-chemical field is $\bar{E} = -\nabla(\phi + \mu/e)$.

The pair $(\mathbf{J}, \mathbf{J}_E)$ and $(\mathbf{x}_\phi^L, \mathbf{x}_T^L)$ which we introduced before is convenient because the coefficients L_{ab} can be calculated for a given Hamiltonian by the linear response theory. The phenomenological transport equations read,

$$\begin{aligned} \mathbf{J}(x) &= L_{11}\mathbf{x}_\phi^L + L_{12}\mathbf{x}_T^L, \\ \mathbf{J}_E(x) &= L_{21}\mathbf{x}_\phi^L + L_{22}\mathbf{x}_T^L, \end{aligned} \tag{19}$$

and the corresponding continuity equation is

$$\nabla \cdot \mathbf{J}_E = \mathbf{E} \cdot \mathbf{J}, \tag{20}$$

which shows that the Joule heat is the "source" of the energy current. The transport coefficients in these equations have been used by Lutinger [2] to relate the phenomenological description of heat and charge transport to microscopic calculations.

Mahan[4] redefines the coefficients and the forces as $\bar{L}_{ij} = TL_{ij}$ and $\mathbf{x}^L = \mathbf{x}^M/T$, such that

$$\begin{aligned} \mathbf{x}_c^M &= -\frac{\nabla\phi}{T} - \frac{1}{e}\nabla\frac{\mu}{T}, \\ \mathbf{x}_T^M &= \nabla\frac{1}{T}. \end{aligned}$$

Since we have $\bar{L}_{ij}\mathbf{x}^M = L_{ij}\mathbf{x}^L$, the scalar product $\bar{\mathbf{J}} \cdot \bar{\mathbf{x}}$ is obviously conserved and the transport equations read

$$\begin{aligned} \mathbf{J}(x) &= -\bar{L}_{11}[\frac{\nabla\phi}{T} + \frac{1}{e}\nabla\frac{\mu}{T}] + \bar{L}_{12}\nabla\frac{1}{T}, \\ \mathbf{J}_E(x) &= -\bar{L}_{21}[\frac{\nabla\phi}{T} + \frac{1}{e}\nabla\frac{\mu}{T}] + \bar{L}_{22}\nabla\frac{1}{T}. \end{aligned} \tag{21}$$

The coefficients $\bar{L}_{ij}$ are also convenient for relating the phenomenological theory to microscopic calculations[4].

In addition to the charge current density $\mathbf{J}$ and the energy current density $\mathbf{J}_E$ one often considers other pairs of currents, like the number current density $\mathbf{J}_N = \mathbf{J}/e$ and the heat current density $\mathbf{J}_Q = \mathbf{J}_E - (\mu/e)\mathbf{J} = \mathbf{J}_u - \bar{\phi}\ \mathbf{J}$ or the entropy current density $\mathbf{J}_S = \mathbf{J}_Q/T$ or the total energy current density $\mathbf{J}_u$. In a non-interacting system, $\mathbf{J}_Q = \mathbf{J}_u - \bar{\phi}\ \mathbf{J}$ is field-free, i.e., $\mathbf{J}_Q = \mathbf{J}_K$ describes the kinetic energy current. 'This agrees with the intuitive notion of the heat (current) as a kinetic energy current' [1].

Domenicali[3] writes the transport equations in terms of the number current $\mathbf{J}_N(x)$, the entropy current $\mathbf{J}_S(x)$, and the generalized forces

$$\begin{aligned} \mathbf{x}_C &= -\nabla\bar{\mu}, \\ \mathbf{x}_Q &= -\nabla T. \end{aligned} \tag{22}$$

The corresponding transport equations read,

$$\begin{aligned} \mathbf{J}_N(x) &= -D_{11}\nabla\bar{\mu}(x) - D_{12}\nabla T(x), \\ \mathbf{J}_S(x) &= -D_{21}\nabla\bar{\mu}(x) - D_{22}\nabla T(x). \end{aligned} \tag{23}$$

This choice conserves the scalar product

$$-\mathbf{J}_N(x)\nabla\bar{\mu}(x) - \mathbf{J}_S(x)\nabla T(x)$$

and ensures the Onsager relation $D_{12} = D_{21}$.

Most applications use transport equations with the charge current $\mathbf{J}$ and the heat current $\mathbf{J}_Q = T\mathbf{J}_S$. The condition

$$e\mathbf{J}_N(x)\mathbf{x}_c^G + T\mathbf{J}_s(x)\mathbf{x}_Q^G = -\mathbf{J}_N(x)\nabla\bar{\mu}(x) - \mathbf{J}_S(x)\nabla T(x),$$

gives the generalized forces

$$\begin{aligned}
\mathbf{x}_c^G &= -\frac{\nabla\bar{\mu}}{e} = -\nabla\bar{\phi}, \\
\mathbf{x}_Q^G &= -\frac{1}{T}\nabla T,
\end{aligned} \tag{24}$$

where $\bar{\mu} = e\bar{\phi}$. The transport equations are

$$\begin{aligned}
\mathbf{J}(x) &= -G_{11}\nabla\bar{\phi}(x) - G_{12}\frac{\nabla T(x)}{T}, \\
\mathbf{J}_Q(x) &= -G_{21}\nabla\bar{\phi}(x) - G_{22}\frac{\nabla T(x)}{T}
\end{aligned} \tag{25}$$

and the coefficients G_{ij} satisfy Onsager relation. The continuity equation for the heat current density $q(x,t)$ is

$$\frac{dq}{dt} + \nabla\cdot\mathbf{J}_Q = \bar{\mathbf{E}}\cdot\mathbf{J}. \tag{26}$$

The connections between the coefficients in (25) and (23)) is given by $G_{11} = e^2D_{11}$, $G_{12} = eTD_{12}$ and $G_{22} = TD_{22}$. Equating the currents in Eqs.(25) and (19) we find $G_{11} = L_{11}$ and $G_{12} = L_{12} - (\mu/e)L_{11}$. Equating $\mathbf{J}_E(x) = \mathbf{J}_Q(x)$ for $\mathbf{J} = 0$, and using $\nabla\bar{\phi} = (G_{12}/G_{11})\nabla T/T)$, gives $G_{22} = L_{22} - 2(\mu/e)L_{12} - (\mu/e)^2L_{11}$.

Callen[1] redefines the coefficients in (25) as $N_{ij} = TG_{ij}$ and the forces as

$$\begin{aligned}
\mathbf{x}_c^N &= -\frac{\nabla\bar{\phi}}{T}, \\
\mathbf{x}_Q^N &= -\frac{1}{T^2}\nabla T = \nabla\frac{1}{T}.
\end{aligned} \tag{27}$$

The transport equations now read

$$\begin{aligned}
\mathbf{J}(x) &= -\frac{N_{11}}{T}\nabla\bar{\phi} - \frac{N_{12}}{T^2}\nabla T, \\
\mathbf{J}_Q(x) &= -\frac{N_{21}}{T}\nabla\bar{\phi} - \frac{N_{22}}{T^2}\nabla T.
\end{aligned} \tag{28}$$

Another pair[4] that is often used for a system of particles moving in an external field $-\nabla\phi(\mathbf{x})$ is the charge current $\mathbf{J}(x)$ and the total energy current $\mathbf{J}_u(x) = \mathbf{J}_Q(x) + \bar{\phi}\mathbf{J}(x)$. The invariance of the scalar product

$$\mathbf{J}\mathbf{x}_c^P + \mathbf{J}_u\mathbf{x}_u^P = -\mathbf{J}\frac{\nabla\bar{\phi}}{T} + \mathbf{J}_Q\nabla\frac{1}{T}$$

gives the generalized forces

$$\begin{aligned}
\mathbf{x}_c^P &= -\nabla\frac{\bar{\phi}}{T} \\
\mathbf{x}_u^P &= \nabla\frac{1}{T}.
\end{aligned}$$

The transport equations are

$$\begin{aligned}
\mathbf{J}(x) &= -P_{11}\nabla\frac{\bar{\phi}}{T} + P_{12}\nabla\frac{1}{T}, \\
\mathbf{J}_u(x) &= -P_{21}\nabla\frac{\bar{\phi}}{T} - P_{22}\nabla\frac{1}{T}.
\end{aligned} \tag{29}$$

The coefficients P_{ij} satisfy Onsager's relations and are related to L_{ij} or $\bar{L}_{ij}$ in an obvious way.

PHYSICAL INTERPRETATION

Transport equations describe the properties of a sample connected to the reservoirs at different temperatures and electro-chemical potentials, such that the charge and heat currents flow from one reservoir to the other. The currents flow in response to the external perturbation, so as to bring the system in the equilibrium. Since the driving fields are given by the gradients of potentials, the particles move towards the potential minimum.

Physical interpretation of the transport equations is particularly simple if we define a new set of transport coefficients in the following way: $\sigma = C_{11}$, $\sigma\alpha = C_{12}/T$, $\beta = C_{21}$, and $\bar{\kappa} = C_{22}/T$. The transport equations then read[5],

$$\begin{aligned}
\mathbf{J}(x) &= -\sigma\nabla\bar{\phi} - \sigma\alpha\nabla T, \\
\mathbf{J}_Q(x) &= -\beta\nabla\bar{\phi} - \bar{\kappa}\nabla T,
\end{aligned} \tag{30}$$

where σ can be interpreted as the electrical conductivity and α as the Seebeck coefficient. However, $\nabla\bar{\phi}$ and ∇T are not the correct generalized forces for the currents $\mathbf{J}(x)$ and $\mathbf{J}_Q(x)$ and the symmetry between transport coefficients is lost, i.e., $\beta \neq \sigma\alpha$ ($\beta = \sigma\alpha T$). The analysis of thermoelectric effects simplifies further if we write the charge and heat currents as[5],

$$\mathbf{J}(x) = \sigma(\bar{\mathbf{E}} - \alpha\nabla T) \tag{31}$$

$$\mathbf{J}_Q(x) = \Pi\mathbf{J}(x) - \kappa\nabla T, \tag{32}$$

where $\Pi = \alpha T$ is the Peltier coefficient and $\kappa = \alpha\beta - \bar{\kappa}$ is the thermal conductivity. The first part of the heat current in Eq.(30), due to $\mathbf{E}(x)$, arises because the electrons experience a change in the velocity when there are local inhomogeneities in the electrical potential. In a stationary flow, this additional kinetic energy is converted into reversible thermoelectric heat. The second term of the heat current is due to the diffusion of particles from the high to low temperatures, which gives rise to an energy current. The overall entropy production can be written as[5],

$$\frac{dS}{dt} = \int\left[\frac{\kappa(\nabla T)^2}{T^2} + \frac{\mathbf{J}^2}{\sigma T}\right]dV \geq 0, \tag{33}$$

where the thermoelectric term has been eliminated by partial integration. The thermoelectric effects are reversible and do not change the total entropy, even though they produce the local heat and transfer the entropy from one part of the system to another. The overall entropy production is manifestly positive-definite.

Transport Eqs.(31) and (32) together with Maxwell equation (18) and the continuity equations (4) and (5) determine completely the electrodynamics of a conductor in which the currents flow due to the electrical and thermal fields. The solution provides the distribution of electrical fields, temperature, current and heat current in the sample for any value of kinetic coefficients σ, α, Π and κ, and for any value of the external potential drop ΔV and temperature difference ΔT. These coefficients can either be calculated from the Hamiltonian or taken as material-specific constants.

Expressing the coefficients in Eqs.(31) and (32) in terms of Luttinger's coefficient gives $\sigma = L_{11}$ for the conductivity,

$$\kappa = \frac{1}{T}\left(L_{22} - \frac{L_{12}^2}{L_{11}}\right)$$

for the thermal conductivity, and

$$\alpha = \frac{1}{eT}\left(\frac{L_{12}/e}{L_{11}/e^2} - \mu\right)$$

for the Seebeck coefficient.

Thermoelectric heat. The total energy generated per unit volume and unit time is $\dot{u} = -\nabla \cdot \mathbf{J}_u$. Using $\mathbf{J}_u = \mathbf{J}_Q + \bar{\phi}\mathbf{J}$, together with Eqs.(32) and (31), gives

$$\dot{u} = \frac{\mathbf{J}^2}{\sigma} + \mathrm{div}(\kappa \nabla T) - T\mathbf{J}\nabla\alpha. \tag{34}$$

The first term is the Joule heat, the second term is due to the heat flowing down the temperature gradient, and the third part is the thermoelectric part $\dot{u}_{TE} = \mathbf{j}T\nabla\alpha$.

Ohm's law. In the absence of the temperature gradient, the current flowing in a homogeneous sample, such that $\nabla\alpha = 0$, obeys the Ohm's law $\mathbf{J} = -\sigma\nabla\phi$ and dissipates the Joule's heat $\dot{q} = \mathbf{J}^2/\sigma$.

Thermal conductivity in an open circuit. I the absence of current, Eq.(32) gives

$$\mathbf{J}_Q(x) = -\kappa \nabla T(x),$$

which is the Fourier's law for the heat flow due to temperature gradient. In a homogeneous sample, this flow generates the energy dissipation $\mathrm{div}(\kappa\nabla T)$. For a linear temperature profile in a uniform sample, such that $\nabla^2 T = 0$, and $\dot{q} = \mathrm{div}(\kappa\nabla T) = 0$. linear temperature profile leads to the heat conduction which does not change the entropy of the electron gas.

Thomson heat. In a uniform conductor the coefficient α depends on x only through temperature, so that $\nabla\alpha = (d\alpha/dT)\nabla T$. Thus, the current flowing through a uniform conductor in the presence of a temperature gradient ∇T generates the Thomson heat,

$$\dot{q}_T = -\mathbf{J}\rho\nabla T. \tag{35}$$

where the Thomson coefficient $\rho = -Td\alpha/dT$. Contrary to the irreversible Joule heat, the Thomson heat is linear in $\mathbf{J}$ and can be either positive or negative.

Peltier heat. The Peltier effect is produced when the current crosses an interface between two different materials, say 1 and 2. In the absence of a temperature gradient Eq.(34) shows that the heat dissipated on the interface is

$$\dot{q} = -T\mathbf{J}\cdot\nabla\alpha, \tag{36}$$

where we neglected $\mathbf{J}^2$ with respect to $\mathbf{J}$. Integrating over an infinitesimally thin volume element that contains the interface, and using Gauss theorem, gives the heat density generated by the current passing through the interface,

$$\mathbf{J}_Q = \mathbf{J}\Pi_{12} \tag{37}$$

where

$$\Pi_{12} = -T(\alpha_2 - \alpha_1) \tag{38}$$

is the Peltier coefficient, and $\alpha(2)$ and $\alpha(1)$ are the Seebeck coefficients of the material to the left and to the right of the interface. Since the sign of $\mathbf{J}$ is arbitrary, the current flowing through a thermoelectric element at a constant temperature will dissipate or absorb the heat as it crosses the interface. Relation (38) follows from the Onsager's relation.

The Peltier heat is due to the fact that the entropy of n electrons which cross the interface is not the same on the two sides. Even for non-interacting electrons, the structure of the energy levels is different in two metals, so that the partition functions and the entropies of n electrons crossing from one metal to the other, in unit time, are different on two sides. Thus, the electrons or holes absorb or emit some heat as they cross the interface and get thermally distributed over available energy levels.

What happens if the heat current flowing through the interface and $\nabla T \neq 0$? Because the temperature, the electrochemical potential, the normal components of the heat current and electrical current, are all continuous at the interface, Eq.(34) gives in a stationary state,

$$J_x T(\alpha_2 - \alpha_1) - \kappa \left.\frac{dT}{dx}\right|_1^2 = 0. \tag{39}$$

If we neglect the Thompson and Joule heat, the heat current entering the interface is $\kappa \frac{dT}{dx}\big|_1$ and the heat current leaving the interface is $\kappa \frac{dT}{dx}\big|_2$. The heat current densities at the two sides of the interface differ by $\kappa(dT/dx|_2 - dT/dx|_1) = j_x\Pi_{12}$, i.e., they differ by the Peltier heat given away or taken at the interface in unit time.

The Peltier and the Thomson coefficients are related,

$$\frac{\rho_2 - \rho_1}{T} = -\frac{d(\Pi_{12}/T)}{dT}. \tag{40}$$

Thermoelectric power. In the stationary state, the temperature distribution in the sample is obtained from the condition $\mathbf{J}^2/\sigma + \mathrm{div}(\kappa\nabla T) - T\mathbf{J}\nabla\alpha = 0$, the electrical field $\mathbf{E}(x)$ follows from the transport equation, and the total potential drop across the sample is given by the integral

$$\Delta V = \int_0^a dx\, \mathbf{E}(x) = \int_0^a dx\, [\frac{\mathbf{J}}{\sigma} + \alpha\nabla T(x)]. \tag{41}$$

In an open thermoelectric circuit, as the one used for measuring the Seebeck coefficient, we have $\mathbf{J} = 0$ and obtain the electromotive force between the end-points a and b as,

$$\mathscr{E}_T = \int_a^b dx\, \alpha(T)\frac{dT}{dx} = \int_{T(a)}^{T(c)} dT\, \alpha_2(T) - \int_{T(b)}^{T(c)} dT\, \alpha_1(T). \tag{42}$$

$\alpha_1(T)$ and $\alpha_2(T)$ are the Seebeck coefficients of metals 1 and 2, and $T(a)$, $T(b)$, and $T(c)$ are temperatures at the points a, b, and c, respectively. If $T(a) = T(b) = T$ and $\Delta T = T(c) - T(a)$ is small, then

$$\mathscr{E}_T = \int_T^{T+\Delta T} dT\, (\alpha_2 - \alpha_1) \simeq \Delta T(\alpha_2(T) - \alpha_1(T)). \tag{43}$$

The Seebeck coefficient is defined by the ratio $\Delta V/\Delta T$, where $\Delta V = \mathscr{E}_T$. For constant $\alpha(x)$, we have simply $\Delta V = \alpha\Delta T$. In a thermoelectric power generator one has to consider the case $\mathbf{J} \neq 0$.

The Seebeck voltage appears because the charged particles diffuse from the hot to the cold end and the imbalance of charge gives rise to a potential gradient across the sample. In a quasiparticle picture, the electrical energy required to transfer n electrons from the hot end to the cold end against the voltage ΔV is balanced by the change in thermal energy (that is, the heat). Neglecting the shift of the quantum states due to the external potential, we approximate $ne\Delta V \simeq \sigma\Delta T$, where n is the particle density and σ the entropy density of the charge carriers. This gives the approximate relationship between the Seebeck coefficient and the entropy as

$$\alpha(T) = \frac{\sigma(T)}{en}. \tag{44}$$

SOLUTION OF TRANSPORT EQUATIONS

The heat and charge transport in a given thermoelectric device is described by the phenomenological transport equations

$$\mathbf{J} = \sigma\mathbf{E} - \sigma\alpha\nabla T, \tag{45}$$

$$\mathbf{J}_Q = \alpha T\mathbf{J} - \kappa\nabla T, \tag{46}$$

that have to be solved together with the continuity equations for the charge and energy densities,

$$\frac{dn}{dt} = -\nabla\cdot\mathbf{J} \tag{47}$$

$$\frac{du}{dt} = -\nabla \cdot \mathbf{J}_u, \tag{48}$$

and Maxwell equations for the electrical and magnetic fields. The particular solution of these equations describes a thermoelectric device that can operate as a refrigerator or a power generator, depending on the boundary (or the initial) condition.

Simple thermoelectric device. Consider, as an example, the stationary state of a system shown in the figure We want the distribution of temperatures, electrical field, and the currents in the sample of length a and unit cross-section A, which is connected to the heat and charge reservoirs at points $x = 0$ and $x = a$. The first reservoir is at temperature T_0, and the second one at T_A. The sample is characterized by transport coefficients σ, α, and κ. The boundary between the sample and the reservoirs is considered to be 'ideal', i.e., without any effect on thermal transport. Magnetic effects are neglected.

For a given current $\mathbf{J}$, temperature T_A and T_0, we find $T(x)$, $\mathbf{E}(x)$, and the heat current $\mathbf{J}_Q(x)$ in the sample. If we specify the cooling rate at the source (the heat leaving the source in unit time defines $\mathbf{J}_Q$ in Eq.(46), we can determine T_0 for each $\mathbf{J}$ and T_A.

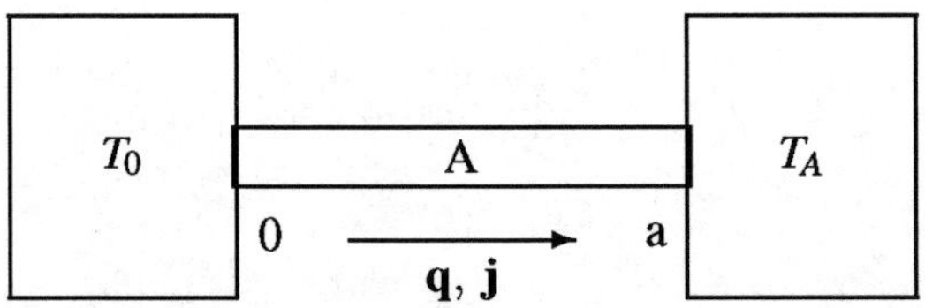

Temperature distribution. In a stationary state, the energy produced in unit time in volume $d^3\mathbf{x}$ around point $\mathbf{x}$ is transported out of that volume by the energy current. The temperature profile of the stationary state is defined by the continuity equation $-\nabla \mathbf{J}_u = -\nabla \mathbf{J}_Q + \mathbf{J} \cdot \mathbf{E} = 0$. This leads to the Domenicali equation,

$$\mathrm{div}(\kappa \nabla T) + \frac{\mathbf{J}^2}{\sigma} - T\mathbf{J}\nabla\alpha = 0, \tag{49}$$

which has to be solved together with the boundary condition $T(0) = T_0$ and $T(a) = T_A$.

In a uniform system $\alpha(x)$ depends on x only through temperature, such that $\nabla\alpha = (d\alpha/dT)\nabla T$, which simplifies Eq.(49) to

$$-\mathrm{div}(\kappa \nabla T) = \frac{\mathbf{J}^2}{\sigma} - \mathbf{J}\,\frac{d\alpha}{dT}\,T\nabla T. \tag{50}$$

The last term is the heat generated in unit time by Thompson effect and $d\alpha/dT$ is the Thompson coefficient. The integration of Eq.(50) gives, for constant σ,

$$-\kappa \nabla T(x) = \frac{\mathbf{J}^2}{\sigma}x + F - \frac{\mathbf{J}}{2}\frac{d\alpha}{dT}T^2(x) \tag{51}$$

and, for constant κ,

$$-\kappa T(x) = \frac{\mathbf{J}^2}{\sigma}\frac{x^2}{2} + F\,x + G - \frac{\mathbf{J}}{2}\frac{d\alpha}{dT}\int_0^x T^2(x), \tag{52}$$

where the integration constants in Eq.(50) are to be determined from the boundary condition for $T(x)$ at $x=0$ and $x=l_a$. In what follows, we consider two limiting cases, which can be solved analytically. First, we keep the Joule heat and neglect the Thompson heat. Second, we neglect the Joule heat and keep the Thompson heat. In the general case, the $T(x)$ can be obtained by iteration, and the lowest order iterative solution is quite simple.

If we neglect the Thompson heat, the boundary values $T(0)=T_0$ and $T(a)=T_A$ give $G=-\kappa T_0$ and

$$F = -K(T_A - T_0) - R\frac{\mathbf{J}^2}{2},$$

where $K=\kappa/l_a$ and $R=l_a/\sigma$ are the thermal conductance and electrical resistance, respectively. (The sample is of a unit crossection.) From Eqs.(46) and (51), we can find the cooling rate at the source for given $\mathbf{J}$ and $\Delta T = T_0 - T_a$.

Alternatively, rather that using the boundary value for $T(x)$, we can use the initial condition for the thermal current, $\mathbf{J}_Q(x)|_{x=0} = \mathbf{q}_0$, and find F from Eqs.(46) and (51) at $x=0$,

$$F = -\,\kappa\nabla T|_{x=0} = (\mathbf{q}_0 - \alpha T_0\mathbf{J}).$$

Together with $T(x)|_{x=a} = T_a$ and Eq.(52) this provides temperature T_0 at the source.

The slope of $T(x)$, for $0\le x\le l_a$, is given by the expression,

$$\kappa\nabla T(x) = -R\,\mathbf{J}^2\,\frac{x}{l_a} + \frac{R}{2}\,\mathbf{J}^2 + K(T_A - T_0), \tag{53}$$

which shows that $\nabla T(x)$ is enhanced with respect to $(T_A-T_0)/l_a$ for $0\le x\le l_a/2$ and is reduced for $l_a/2\le x\le l_a$. Since temperature is constant at the source and the sink, and $T(x)$ has a finite slope at T_0 and T_A, temperature is a continuous function with a kink at T_0 and T_A.

Thermal current. Substituting $\nabla T(x)$ given by Eq.(53) into Eq.(46) gives at the source ($x=0$, $T(0)=T_0$),

$$\mathbf{J}_Q|_{x=0} = \alpha T_0\mathbf{J} - K(T_A - T_0) - \frac{R\mathbf{J}^2}{2}. \tag{54}$$

The first term is the Peltier heat generated by the electrical current at point $x=0$ where the sample is attached to the reservoir, the second term is the heat flux diffusing from the source to the sink due to the temperature difference (T_A-T_0), and the last term is 1/2 of the Joule heat generated by constant current $\mathbf{J}$ in the sample of length a. Interestingly, one half of Joule heat generated in the sample is returned to the source. For $T_A > T_0$, the heat is taken out of the source, provided the Peltier heat overcomes the diffusion heat and 1/2 of the Joule heat.

The thermal current arriving at the sink, is

$$\mathbf{J}_Q|_{x=a} = \alpha T_A \mathbf{J} - K(T_A - T_0) + \frac{R\mathbf{J}^2}{2}, \tag{55}$$

which shows that one half of Joule heat generated in the sample is taken away by the outgoing heat current and dumped into sink.

For given temperatures of the source and the sink, the heat current $\mathbf{J}_Q(T_0)$, defined by (54), is pumped out of the source in order to maintains the stationary state. The parasitic heat which enters the source, say, because of poor thermal insulation, is removed by the thermoelectric device which operates between T_0 and T_A at current $\mathbf{J}$. The accompanying thermal field $T(x)$ is defined by Eq.(52). If the current is switched on when the source and sink temperature are equal, the heat current $\mathbf{J}_Q = \alpha \mathbf{J} T_A - R\mathbf{J}^2/2$ starts flowing out of the source. As long as the parasitic heat which is entering the source is less than $\mathbf{J}_Q(T_0)$ (the heat current pumped out of the source in a stationary state), heat will be removed from the source and its temperature will drop. The stationary state is established at T_0, when the heat removed from the source matches the parasitic heat.

Figure of merit. For a given T_A and T_0, the heat current pumped out of the source in a stationary state depends on $\mathbf{J}$, as shown by Eq.(54). The maximum cooling power at the source is obtained for $dJ_Q/dJ = 0$, which defines the optimal current,

$$J_{opt} = \frac{\alpha T_0}{R},$$

and gives,

$$J_Q^{max} = \frac{\alpha^2 T_0^2}{2R} - K(T_A - T_0),$$

But running the device with J_{opt} and removing the heat at the rate J_Q^{max} is not the most efficient mode of operation.

If the source is thermally isolated, the device will keep removing the heat and reducing T_0, as long as $\mathbf{J}_Q(T_0) \geq 0$. For given temperature of the sink T_A, the lowest T_0 (maximum of $\Delta T = T_A - T_0$) is reached when the thermoelectric reaches the state $\mathbf{J}_Q = 0$, which gives,

$$(T_A - T_0) = \frac{\alpha T_0 \mathbf{J}}{K} - \frac{R\mathbf{J}^2}{2K}.$$

The maximum ΔT which one can achieve using a given thermoelectric material is obtained for $\mathbf{J} = j_{opt}$, such that

$$(T_A - T_0)_{Max} = \frac{1}{2} Z T_0^2,$$

where, $Z = \sigma\alpha^2/2\kappa$ is the figure-of-merit of a given thermoelectric.

Coefficient of performance. When the device operates as a fridge, the coefficient of performance is defined by the ratio of the heat taken from the source and the power supplied by the battery. This power has to maintain the heat flow from T_0 to $T_A \leq T_0$ and compensate the Joule losses and is given by the difference between the outgoing (source) and the incoming (sink) heat currents, $W = [\mathbf{J}_Q]_{x=0} - [\mathbf{J}_Q]_{x=a}$. Using Eqs. (54) and (55) we find,

$$W = \alpha(T_A - T_0)\mathbf{J} + R\mathbf{J}^2, \tag{56}$$

so that the coefficient of performance is

$$\Phi = \frac{q_c}{W} = \frac{\alpha T_0 \mathbf{J} - K(T_A - T_0) - R\mathbf{J}^2/2}{\alpha(T_A - T_0)\mathbf{J} + R\mathbf{J}^2}. \tag{57}$$

The optimal performance follows from $d\Phi/dJ = 0$, which gives the optimal current

$$J_\Phi = \frac{\alpha(T_A - T_0)}{R[(1 + ZT_M)^{1/2} - 1}, \tag{58}$$

where $T_M = (T_1 + T_2)/2$. The best efficiency is,

$$\Phi_{max} = \frac{T_0}{T_A - T_0} \frac{(1 + ZT_M)^{1/2} - T_A/T_0}{1 + ZT_M)^{1/2} + 1}. \tag{59}$$

Φ approaches the Carnot efficiency for $Z \to \infty$.

Efficiency coefficient. When the device is operating as a generator, the efficiency coefficient is defined by the ratio of the useful power of the external load, $W = \mathbf{J}^2 R_{Load}$, and the heat taken from the source. (We now have, $T_A < T_0$.) The efficiency η is defined as,

$$\eta = \frac{W}{\mathbf{J}} Q = \frac{\mathbf{J}^2 R_{Load}}{\alpha T_0 \mathbf{J} - K(T_0 - T_A) - R\mathbf{J}^2/2} \tag{60}$$

The current in the circuit is driven by the Seebeck voltage and is given by $\mathbf{J} = \alpha(T_0 - T_A)/R_{Load}$, where we neglected the resistance of the thermoelectric with respect to R_{Load}. As shown by Ioffe[6], maximizing with respect to R_{Load}/R gives,

$$\eta = \frac{(T_0 - T_A)}{T_0} \frac{\sqrt{1 + ZT_M} - 1}{\sqrt{1 + ZT_M} + T_A/T_0} \tag{61}$$

Hetero-structures

Suppose we consider the device consisting of a thermoelectric B of length b connected to the thermoelectric A of length a. The transport coefficients are σ_A, α_A, κ_A, and σ_B, α_B, κ_B, and the cross-sections are A_a and A_b. The thermoelectric A extends from $x = 0$ to $x = a$, where it connects to the sink at temperature T_A, and the thermoelectrics B extends from $x = 0$ to $x = -b$, where it connects to the source at T_B. The contact between the two thermoelectrics, at $x = 0$, does not impede the heat and charge transport, and temperature at the contact is T_0.

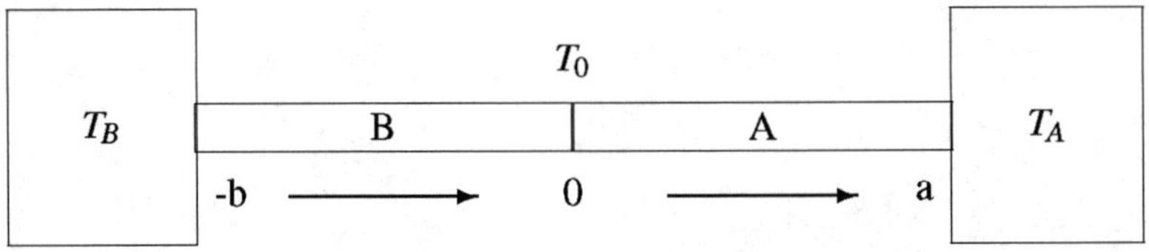

We consider the stationary state with constant electrical current, with the sink at temperature T_A, and the source at T_B. The heat $\mathbf{J}_Q^B$ is taken out of the source by the thermoelectric B and the heat $\mathbf{J}_Q^A$ dumped in the sink by A. The previous analysis shows that the cooling power of B depends on T_B and T_0, and of A on T_A and T_0. Assuming a stationary state and constant current we can find T_0, performing the analysis of the heat and charge transport in A and B in the same way as before. The boundary conditions are shown in the figure above.

The Domenicali equations for A and B are the same as before, and all the results are obtained by using Eqs. (49) to (53) with σ_A, α_A, κ_A for A, and by σ_B, κ_A, κ_B for B, and by taking care of temperatures at the source, at the sink and at the boundary between A and B. For a general point $x \in (0,a)$ in A and $x \in (-b,0)$ in B the solution can be written as,

$$-\kappa_{A,B}\nabla T(x) = \frac{\mathbf{J}^2}{\sigma_{A,B}} x + F_{A,B}, \tag{62}$$

and

$$-\kappa_{A,B} T(x) = \frac{\mathbf{J}^2}{\sigma_{A,B}} \frac{x^2}{2} + F_{A,B}\, x + G_{A,B}, \tag{63}$$

where the integration constants follow form the boundary condition for the thermal field. From Eq.(63) and $T(-b) = T_B$, $T(0) = T_0$, which holds in B, we find $G_B = -\kappa_B T_0$, and

$$F_B = -K_B(T_0 - T_B) + R_B\, \mathbf{J}^2/2,$$

while from $T(0) = T_0$, and $T(a) = T_A$, which holds in A, we find $G_A = -\kappa_A T_0$, and

$$F_A = -K_A(T_A - T_0) - R_A\, \mathbf{J}^2/2,$$

where $K_{A,B}$ and $R_{A,B}$ are the thermal conductance and the electrical resistance in A and B, respectively. Note, the opposite sign of the Joule term in F_A and F_B. However, the problem is not solved unless we specify T_0.

Temperature at the boundary follows from the condition that the heat current us continuous from A to B. Using Eq.(62) for $\nabla T(x)$ we calculate $\mathbf{J}_Q(x) = \alpha T \mathbf{J} - \kappa \nabla T(x)$, at $x = 0$ in B and A, which gives,

$$\mathbf{J}_Q^B\big|_{x=0} = \alpha_B T_0 \mathbf{J} + F_B, \tag{64}$$

and

$$\mathbf{J}_Q^A\Big|_{x=0} = \alpha_A T_0 \mathbf{J} + F_A, \tag{65}$$

where F_A and F_B are the integration constants of Deminicalli equations in segments A and B, respectively. Matching the heat currents gives an equation for T_0,

$$\alpha_B T_0 \mathbf{J} - K_B(T_0 - T_B) + \frac{R_B \mathbf{J}^2}{2} = \alpha_A T_0 \mathbf{J} - K_A(T_A - T_0) - \frac{R_A \mathbf{J}^2}{2}. \tag{66}$$

which takes into account that 1/2 of the Joule heat generated in B is transferred at the boundary to A, and that 1/2 of the Joule heat generated in A is returned to B. In the absence of electrical current, we have $K_B(T_0 - T_B) = K_A(T_A - T_0)$, which gives

$$T_0 = \frac{K_A T_A + K_B T_B}{K_A + K_B},$$

so that $\nabla T(x)$ has a break at T_0 unless $K_A = K_B$. If $\mathbf{J} = 0$, the heat flows in the stationary state without any losses. But if $\mathbf{J} \neq 0$, the temperature profile is parabolic in each segment and the difference in the heat current entering at B and leaving at A is $R_A^2/2 + R_B^2/2$.

Similarly, using the condition that the heat current leaving the source is equal to the heat current in the thermoelectric B at point $x = -b$, $\mathbf{J}_Q^B = \alpha_B T_B \mathbf{J} - \kappa_B \nabla T|_{x=-b}$, we get an equation for T_B,

$$\mathbf{J}_{Q0}^B = \alpha_B T_B \mathbf{J} - K_B(T_0 - T_B) - \frac{\mathbf{J}^2 R_B}{2}, \tag{67}$$

if we specify the cooling rate at the source. Eqs.(66) and (67) provide a set of linear equations for T_B and T_0.

We can now calculate the maximum temperature difference, ΔT, the coefficient of performance and the efficiency coefficient, and find the currents which maximize these coefficients. We can then express these maximal values in terms of effective figures of merit Z_Φ^* and Z_η^*, which are defined by expressions (59) and (61). If Z_Φ^* and Z_η^* are not too different, they can be used to specify the heterostructure in terms of material constants, and initial conditions.

N segments. The generalization to N segments, shown in the figure below, is straightforward. In the i-th segment of length l_i, the solution of the Domenicali equation for $\nabla T(x)$ and $T(x)$ is given by Eqs.(63) and (62), where, say, T_B, α_B, σ_B, κ_B and T_0 are replaced by T_{i-1}, α_{i-1}, σ_{i-1}, κ_{i-1} and T_i, and x is measured with respect to the initial point of the segment x_{i-1}. That is,

$$-\kappa_i \nabla T(x) = \frac{\mathbf{J}^2}{\sigma_i} x + F_i \qquad \text{for } x_{i-1} \leq x \leq i \tag{68}$$

and

$$-\kappa_i T(x) = \frac{\mathbf{J}^2}{\sigma_i} \frac{x^2}{2} + F_i x + G_i \qquad \text{for } x_{i-1} \leq x \leq i, \tag{69}$$

and the corresponding integration constants are $G_i = -T_i$ and

$$F_i = -K_i(T_i - T_{i-1}) + R_i\,\mathbf{J}^2/2,$$

The solution has to satisfy the condition that temperature, electrochemical potential, and normal components of heat current and electrical current, are continuous at every point, which gives,

$$\lim_{\delta\to 0}\,[\alpha_{i-1}T_{i-1}\mathbf{J} - \kappa_{i-1}\nabla T_{i-1}|_{x_i-\delta} = \lim_{\delta\to 0}\,[\alpha_i T_i\mathbf{J} - \kappa_i\nabla T_i]_{x_i+\delta} \tag{70}$$

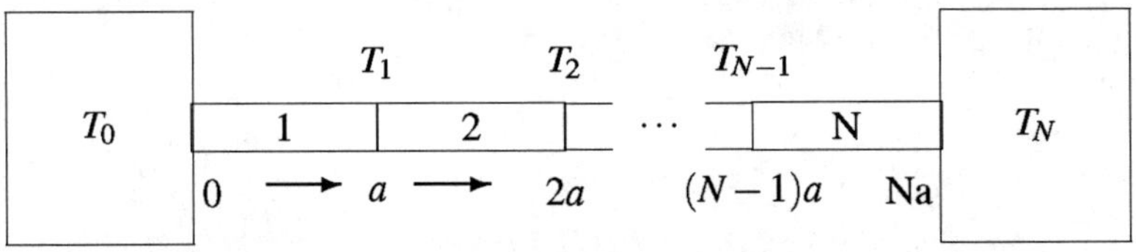

The boundary conditions at $x = 0, \ldots, x = N(a-1)$, provide N equations for temperature field defined at points $\mathbf{t} = \{T_0, \ldots, T_{N-1}\}$. Assuming we know T_N and $\mathbf{J}_Q^0$, Eq.(70) give the set of equations,

$$\begin{aligned}
-\alpha_1 T_0\mathbf{J} + K_1(T_1 - T_0) &= -(\mathbf{J}_Q^0 + \frac{\mathbf{J}^2}{2}R_1^2) \\
\alpha_1 T_1\mathbf{J} - K_1(T_1 - T_0) - \alpha_2 T_1\mathbf{J} + K_2(T_2 - T_1) &= -\frac{\mathbf{J}^2}{2}(R_1^2 + R_2^2) \\
\alpha_2 T_2\mathbf{J} - K_2(T_2 - T_1) - \alpha_3 T_2\mathbf{J} + K_3(T_3 - T_2) &= -\frac{\mathbf{J}^2}{2}(R_2^2 + R_3^2) \\
\cdots \quad \cdots & \quad \cdots \\
\alpha_{i-1}T_{i-1}\mathbf{J} - K_{i-1}(T_{i-1} - T_{i-2}) - \alpha_i T_{i-1}\mathbf{J} + K_i(T_i - T_{i-1}) &= -\frac{\mathbf{J}^2}{2}(R_{i-1}^2 + R_i^2) \\
\cdots \quad \cdots & \quad \cdots \\
\alpha_{N-1}T_{N-1}\mathbf{J} - K_N(T_{N-1} - T_{N-2}) - \alpha_N T_{N-1}\mathbf{J} + K_N(T_N - T_{N-1}) &= -\frac{\mathbf{J}^2}{2}(R_{N-1}^2 + R_N^2)
\end{aligned}$$

which is an array with elements on the main diagonal and two principal sub-diagonals,

$$\begin{aligned}
-(\alpha_1\mathbf{J} - K_1)T_0 + K_1 T_1 &= -(\mathbf{J}_Q^0 + \frac{\mathbf{J}^2}{2}R_1^2) \\
K_1 T_0 + [(\alpha_1 - \alpha_2)\mathbf{J} - (K_1 + K_2)]T_1 + K_2 T_2 &= -\frac{\mathbf{J}^2}{2}(R_1^2 + R_2^2) \\
K_2 T_1 + [(\alpha_2 - \alpha_3)\mathbf{J} - (K_2 + K_3)]T_2 + K_3 T_3 &= -\frac{\mathbf{J}^2}{2}(R_2^2 + R_3^2) \\
\cdots \quad \cdots & \quad \cdots \\
K_{i-1}T_{i-2} + [(\alpha_{i-1} - \alpha_i)\mathbf{J} - (K_{i-1} + K_i)]T_{i-1} + K_i T_i &= -\frac{\mathbf{J}^2}{2}(R_{i-1}^2 + R_i^2)
\end{aligned} \tag{71}$$

$$\dots \quad \dots \quad \dots$$

$$K_{N-1}T_{N-2} + [(\alpha_{N-1} - \alpha_N)\mathbf{J} - (K_{N-1} + K_N)]T_{N-1} = -\frac{\mathbf{J}^2}{2}(R_{N-1}^2 + R_N^2) - K_N T_N$$

or in the matrix form

$$\mathbf{Mt} = -\mathbf{c} - \frac{\mathbf{J}^2}{2}(\mathbf{r} + \mathbf{R})$$

where $\mathbf{c} = \{\mathbf{q_0J}^2/2, 0, \dots, 0, K_N T_N\}$, $\mathbf{r} = \{R_0^2, R_1^2, \dots, R_{N-1}^2\}$, and $\mathbf{R} = \{R_1^2, R_2^2, \dots, R_N^2\}$.

The matrix elements of $\mathbf{M}$ are defined as $M_{11} = -\alpha_1\mathbf{J} + K_1$, and $M_{ij} = [(\alpha_{i-1} - \alpha_i)\mathbf{J} - (K_{i-1} + K_i)]\delta_{ij} + K_i\delta_{i,j-1} + K_{i-1}\delta_{i,j+1}$ for $i, j \geq 1$.

Provided $\det\mathbf{M} \neq 0$ we can find the solution

$$\mathbf{t} = \mathbf{M}^{-1}[-\mathbf{c} - \frac{\mathbf{J}^2}{2}(\mathbf{r} + \mathbf{R})]$$

for a given set of material constants, $\mathbf{A} = \{\alpha_1, \dots, \alpha_N\}$, $\mathbf{K} = \{\kappa_1, \dots, \kappa_N\}$, $\mathbf{R}$, and for a given value of T_N, $\mathbf{J}$, and the heat current at the source $\mathbf{q_0}$.

Alternatively, we can rewrite the first equation as

$$\mathbf{J}_Q^0 + K_1 T_1 = -\frac{\mathbf{J}^2}{2}R_1^2) + -(\alpha_1\mathbf{J} - K_1)T_0$$

and invert new matrix matrix $\mathbf{M}'$ to find $\mathbf{J}_Q^0$ and $\{T_1, \dots, T_{N-1}\}$ in terms of T_0 and T_N. If the material-specific constants are strongly temperature dependent, the calculations have to be performed self-consistently. That is, we start with some initial value of $\mathbf{A}$, $\mathbf{K}$, and $\mathbf{R}$, calculated for the linear temperature dependence of $T(x)$ between T_0 and T_N. We then solve Eq.(72) to find $T(x)$, $\nabla T(x)$ and the heat current in each segment. Then we recalculate $\mathbf{A}$, $\mathbf{K}$, and $\mathbf{R}$, and iterate unless the result converges.

The same procedure has to be performed for the Seebeck effect, when $\mathbf{J} = 0$ and we know $\Delta T = T_0 - T_a$. Eventually, we obtain the electrical field $\mathbf{E}(x) = \alpha(x)\nabla T(x)$, evaluate $\Delta V = \int_0^l dx\, \mathbf{E}(x)$, and define the effective Seebeck coefficient of the heterostructure at $(T_0 + T)N)/2$ as $S^* = \Delta V/\Delta T$.

LINEAR RESPONSE THEORY

The coefficients appearing in the transport equations can either be treated as phenomenological parameters taken from the experiment on a given sample or calculated for a microscopic model that we use to describe this sample. Here, we follow Luttinger's approach[2] to calculate the transport coefficients for a given Hamiltonian

The macroscopic values of the charge and heat currents densities are defined as the statistical averages

$$\mathbf{J} = \mathrm{Tr}\{\rho_\phi \hat{\mathbf{j}}\}, \tag{72}$$

$$\mathbf{J}_u = \mathrm{Tr}\{\rho_\phi \hat{\mathbf{j}}_u\}, \tag{73}$$

where $\hat{\mathbf{j}}$ and $\hat{\mathbf{j}}_u$ are the charge current density and the energy current density operators, respectively. The Hamiltonian is

$$\hat{H}_\phi = \hat{H} + \Delta H_\phi(t) = \hat{H} + e^{\omega t}\hat{F}, \tag{74}$$

where $\hat{H}$ defines the dynamics of interacting particles in the absence of the external field and $\Delta H_\phi(t) = e^{\omega t}\hat{F}$ is the perturbation. The operator

$$\hat{F} = \int d^3x\,\phi(x)\,e\hat{n}(x)$$

provides the coupling of the charge density $e\hat{n}(x)$ to the external potential which is gradually turned on, as described by the exponential factor $e^{\omega t}$. The coupling is absent at $t = -\infty$ and is fully turned on at $t = 0$. Note, a completely different behavior is obtained for 'slow' and 'rapid' switching-on of the perturbation. (The time-scale for the 'slow' and 'rapid' perturbation is set by the characteristic time-scales of the unperturbed system.) In the 'slow' case, the electron wavefunctions adjust to the new potential and the stationary state corresponds to thermal equilibrium, where no currents flow. In the 'rapid' case, the wave functions are not modified by the perturbation; the additional potential generates the currents which flow so as to establish the equilibrium. For a potential characterized by a single spatial Fourier component ϕ_q, the switching is rapid if $q/\omega \to 0$ which describes a uniform system perturbed by a time-dependent potential (the limit $q \to 0$ is taken before $\omega \to 0$). The switching is slow if $\omega/q \to 0$, which describes a static perturbation of a spatially modulated system without any currents (the limit $\omega \to 0$ is taken before $q \to 0$).

The density matrix ρ_ϕ is defined by the equation

$$i\frac{\partial \rho_\phi}{\partial t} = [\hat{H}_\phi, \rho_\phi] \tag{75}$$

and the initial condition $\rho_\phi(-\infty) = \rho_0$. The unperturbed density matrix is ρ_0 and we want the lowest order correction to ρ_0 due to $\Delta H(t)$. Writing

$$\rho_\phi(t) = \rho_0 + f(t)$$

we obtain from the equation of motion

$$i\frac{\partial f(t)}{\partial t} = [\hat{H}, \rho_0] + [\hat{H}, f] + [\Delta\hat{H}_\phi, \rho_0] + [\Delta\hat{H}_\phi, f], \tag{76}$$

which has to be solved for the boundary condition $f(-\infty) = 0$. Using $[\hat{H}, \rho_0] = 0$ and neglecting the non-linear term gives

$$i\frac{\partial f(t)}{\partial t} - [\hat{H}, f] = [\Delta\hat{H}_\phi, \rho_0]. \tag{77}$$

The identity

$$i\frac{\partial f(t)}{\partial t} - [\hat{H}, f] = e^{-Ht} i\left\{\frac{\partial}{\partial t} e^{Ht} f(t) e^{-Ht}\right\} e^{Ht}, \tag{78}$$

allows us to write

$$i\frac{\partial}{\partial t}e^{Ht}f(t)e^{-Ht} = e^{Ht}[\Delta\hat{H}_\phi,\rho_0]e^{-Ht} = [\Delta\hat{H}_\phi(t),\rho_0], \tag{79}$$

such that

$$e^{Ht}f(t)e^{-Ht} = f(-\infty) - i\int_{-\infty}^{t} dt'[\Delta\hat{H}_\phi(t'),\rho_0]. \tag{80}$$

Using $f(-\infty) = 0$ obtains and

$$f(t) = -i\int_{-\infty}^{t} dt'[\Delta\hat{H}_\phi(t'-t),\rho_0] = -i\int_{0}^{\infty} dt' e^{-\omega t'}[F(-t'),\rho_0]. \tag{81}$$

In the last equation we shifted the integration variable by t and used $F(z) = e^{iHz}F(t)e^{-iHz}$ for z in the complex plane. We now show that $f(t)$ is proportional to the gradient of the external. The identity

$$[F(-t'),\rho_0] = -i\rho_0\int_0^\beta d\beta'\frac{\partial F(-t'-i\beta')}{\partial t'} \tag{82}$$

gives

$$f = -\rho_0\int_0^\infty dt' e^{-\omega t'}\int_0^\beta d\beta'\frac{\partial F(-t'-i\beta')}{\partial t} \tag{83}$$

and the definition of the operator $\hat{F}$ and the continuity equation for the number density operator obtain

$$\frac{\partial F(-t-i\beta)}{\partial t} = \int d^3x\phi(x)e^{\beta H}\frac{\partial\hat{n}(-t)}{\partial t}e^{-\beta H} = \int d^3x\phi(x)\nabla\cdot\hat{\mathbf{j}}(x,-t-i\beta). \tag{84}$$

Finally, the integration by parts gives

$$\frac{\partial F(-t-i\beta)}{\partial t} = -\int d^3x'\hat{\mathbf{j}}(x',-t-i\beta)\cdot\mathbf{E}(x'), \tag{85}$$

where $\mathbf{E}(x) = -\nabla\phi(x)$. Thus, the lowest order correction to $\rho_\phi(t)$ is proportional to the applied field and given by

$$f = +\rho_0\sum_\gamma\int d^3x'\int_0^\infty dt' e^{-\omega t'}\int_0^\beta d\beta'\mathbf{E}_\gamma(x')\hat{\mathbf{j}}_\gamma(x',-t'-i\beta'). \tag{86}$$

Since no current flows for $\mathbf{E}(x) = 0$, the current density induced by the field is

$$\mathbf{J}_\alpha(x) = \mathrm{Tr}\{f\,\hat{\mathbf{j}}_\alpha\} = \sum_\gamma\int d^3x'\mathbf{E}_\gamma(x')\int_0^\infty dt' e^{-\omega t'}\int_0^\beta d\beta' < \hat{\mathbf{j}}_\gamma(x',-t'-i\beta')\hat{\mathbf{j}}_\alpha(x) >_0, \tag{87}$$

where $\langle\cdots\rangle_0$ denotes the thermodynamic average with respect to ρ_0 and the α,γ label the spatial directions.

The above expression gives the thermodynamic average of the current density at point $\mathbf{x}$ induced by a spatially varying electrical field. To compare with the phenomenological equations discussed in the first part of these lecure notes we shoud make an additional averaging of the current given by Eq. (87) over a macroscopic region aroud $\mathbf{x}$. Thus, we introduce the Fourier transform of the current density,

$$\mathbf{J}_\alpha(q) = \frac{1}{V}\int d^3x e^{i\mathbf{q}\cdot\mathbf{x}} \mathbf{J}_\alpha(x), \tag{88}$$

denote by $\hat{\mathbf{j}}_\gamma(q)$ the Fourier component of current density operator, and consider the external potential with a single Fourier component,

$$\phi(x) = \phi_q e^{iqx}. \tag{89}$$

The electrical field is

$$\mathbf{E}_\gamma(x) = \mathbf{E}_\gamma(q) e^{i\mathbf{q}\cdot\mathbf{x}}, \tag{90}$$

where

$$\mathbf{E}_\gamma(q) = -iq_\gamma\phi_q. \tag{91}$$

The only non-vanishing Fourier component of the macroscopic current density is then

$$\mathbf{J}_\alpha(q) = \sum_\gamma \mathbf{E}_\gamma(q) \int_0^\infty dt' e^{-\omega t'} \int_0^\beta d\beta' < \hat{\mathbf{j}}_\gamma(-q, -t' - i\beta')\hat{\mathbf{j}}_\alpha(q) >_0 \tag{92}$$

and to compare with the phenomenological expression for the current density we should now take the limit $q, \omega \to 0$. If the limit $\omega \to 0$ is taken with finite q, the above expression describes the electrons moving in a static periodic potential and $\mathbf{J}_\alpha(q)$ should vanish. In general, the correlator in Eq.(92) does not have to vanish in that limit, which means that the above result is unphysical in that limit. The reason is that the physical zero-current state corresponds to a balance between the electrical force, which is due to the potential gradient, and the diffusion force, which is due to the concentration gradient. Since the diffusion force cannot be included in the Hamiltonian, Eq.(92) does not give the correct result, if the limit $\omega \to 0$ is taken before $q \to 0$.) However, if $q \to 0$ is taken before $\omega \to 0$, we find the current that is flowing in response to the external field that has been gradually turned on. The average current obtained by the linear-response theory in such a way should coincide with the phenomenological current written for the sample without the thermal force and the diffusion force. Writing

$$\mathbf{J}_\alpha = \lim_{q\to 0} \mathbf{J}_\alpha(q) = \lim_{q\to 0} \sum_\gamma \sigma_{\gamma\alpha} \mathbf{E}_\gamma(q), \tag{93}$$

we define the static conductivity by the expression

$$\sigma_{\gamma\alpha} = \lim_{\omega\to 0} \int_0^\infty dt' e^{-\omega t'} \int_0^\beta d\beta' < \hat{\mathbf{j}}_{0\gamma}(-t' - i\beta')\hat{\mathbf{j}}_{0\alpha} >_0, \tag{94}$$

where $\mathbf{J}_\alpha$ and $\hat{\mathbf{j}}_{0\gamma}$ denote the $q = 0$ Fourier components of $\mathbf{J}_\alpha(q)$ and $\hat{\mathbf{j}}_\gamma(q)$, respectively. Equating $\mathbf{J}_\alpha$ with the macroscopic current density in Eq. (19) or (21), we relate the phenomenological transport coefficient L_{11} to the correlation function for the static conductivity $\sigma_{\gamma\alpha}$.

In real space, the static conductivity is defined by the expression

$$\mathbf{J}_{\alpha}(x) = \sum_{\gamma} \int d^3x' \mathbf{E}_{\gamma}(x') \sigma_{\gamma,\alpha}(x,x'), \tag{95}$$

where

$$\sigma_{\gamma,\alpha}(x,x') = \lim_{\omega \to 0} \int_0^{\infty} dt' e^{-\omega t'} \int_0^{\beta} d\beta' < \hat{\mathbf{j}}_{\gamma}(x', -t' - i\beta') \hat{\mathbf{j}}_{\alpha}(x) >_0 . \tag{96}$$

This expression should be used for inhomogeneous materials. In translationally invariant systems, the Fourier transform of (96) gives the same results as obtaioned before.

The uniform static conductivity given by expression Eq. (94) agrees with the usual Kubo formula obtained by calculating the $\omega \to 0$ limit of the dynamical conductivity. In the basis which diagonalizes H, we have

$$\begin{aligned} < n|\hat{\mathbf{j}}_{0\gamma}(-t-i\beta)|m > &= < n|e^{iH(-t-i\beta)} \hat{\mathbf{j}}_{0\gamma}(0) e^{-H(-t-i\beta)}|m > \\ &= e^{-i(E_n - E_m)(-t-i\beta)} < n|\hat{\mathbf{j}}_{0\gamma}(0)|m > . \end{aligned}$$

Then, taking the trace with respect to ρ_0 we obtain from Eq. (94) the static conductivity

$$\sigma_{\gamma\alpha} = \mathrm{Re} \left\{ \sum_{mn} e^{-E_m\beta} < n|\hat{\mathbf{j}}_{0\gamma}|m >< m|\hat{\mathbf{j}}_{0\alpha}|n > \right\} \int_0^{\infty} dt e^{it(E_n - E_m + i\omega)} \int_0^{\beta} d\beta e^{(E_n - E_m)\beta)} .$$

The limit

$$\lim_{\omega \to 0} \mathrm{Re} \int_0^{\infty} dt e^{it(E_n - E_m + i\omega)} = \pi \delta(E_n - E_m) \tag{97}$$

and

$$\delta(E_n - E_m) \int_0^{\beta} d\beta e^{(E_n - E_m)\beta)} = \beta \delta(E_n - E_m) \tag{98}$$

provide the uniform conductivity matrix

$$\sigma_{\gamma\alpha} = \pi\beta \sum_{mn} e^{-E_m\beta} < n|\hat{\mathbf{j}}_{0\gamma}|m >< m|\hat{\mathbf{j}}_{0\alpha}|n > \beta \delta(E_n - E_m), \tag{99}$$

that coincides with the spectral representation obtained from the Kubo expression for the dynamical conductivity.

In Kubo formalism, the transport coefficients appearing in the phenomenological equations are expressed in terms of appropriate zero-frequency "polarization operators"[4]. For the coeficients in Eq. (21) we have[4],

$$\bar{L}_{11}^{\gamma\alpha} = T\, \sigma_{\gamma\alpha} \tag{100}$$

such that

$$\bar{L}_{11}^{\gamma\alpha} = \lim_{\nu \to 0} \mathrm{Re} \frac{i}{\nu} \mathscr{L}_{11}^{\gamma\alpha}(\nu),$$

$$\mathscr{L}_{11}^{\gamma\alpha}(i\nu_l), \quad = \quad \pi T \int_0^\beta d\tau e^{i\nu_l \tau} \langle T_\tau j_{0\alpha}^\dagger(\tau) j_{0\gamma}(0)\rangle, \tag{101}$$

where $\nu_l = 2\pi T l$ is the Bosonic Matsubara frequency, the τ-dependence of the operator is with respect to the full Hamiltonian in Eq. (185), and we must analytically continue $\mathscr{L}_{11}^{\gamma\alpha}(i\nu_l)$ to the real axis $\mathscr{L}_{11}^{\gamma\alpha}(\nu)$ before taking the limit $\nu \to 0$.

The derivation of the coefficient $\bar{L}_{21}^{\gamma\alpha}$ that gives the energy current density induced by an applied electrical field, proceeds in the same way and one easily finds[4],

$$\begin{aligned} \bar{L}_{21}^{\gamma\alpha} &= \lim_{\nu\to 0} \mathrm{Re} \frac{i}{\nu} \mathscr{L}_{12}^{\gamma\alpha}(\nu), \\ \mathscr{L}_{21}^{\gamma\alpha}(i\nu_l) &= \pi T \int_0^\beta d\tau e^{i\nu_l \tau} \langle T_\tau j_{0Q\alpha}^\dagger(\tau) j_{0\beta}(0)\rangle. \end{aligned} \tag{102}$$

However, the derivation of $\bar{L}_{12}^{\gamma\alpha}$ and $\bar{L}_{22}^{\gamma\alpha}$ coefficients that describe the currents induced by the thermal gradient is more complicated, because the 'thermal force' (like the 'diffusion force') does not appear in the Hamiltonian. The details are given in Ref. [2] and we just quote the results,

$$\begin{aligned} \bar{L}_{22}^{\gamma\alpha} &= \lim_{\nu\to 0} \mathrm{Re} \frac{i}{\nu} \mathscr{L}_{22}^{\gamma\alpha}(\nu), \\ \mathscr{L}_{22}^{\gamma\alpha}(i\nu_l) &= \pi T \int_0^\beta d\tau e^{i\nu_l \tau} \langle T_\tau j_{0Q\alpha}^\dagger(\tau) j_{0Q\beta}(0)\rangle. \end{aligned} \tag{103}$$

and $\bar{L}_{12}^{\gamma\alpha} = \bar{L}_{21}^{\gamma\alpha}$, in agreement with the Onsager relation. In all of these equations, the subscripts α and β denote the respective spatial index of the current vectors and subscript 0 denotes the $q \to 0$ Fourier component.

In terms of transport coefficients $\bar{L}_{ij}^{\gamma\alpha}$ we then have

$$\alpha = -\frac{k_B}{|e|T} \frac{\bar{L}_{12}^{\gamma\alpha}}{\bar{L}_{11}^{\gamma\alpha}}, \tag{104}$$

for the thermopower and

$$\kappa = \frac{k_B^2}{T^2} \left[\bar{L}_{22}^{\gamma\alpha} - \frac{\bar{L}_{12}^{\gamma\alpha} \bar{L}_{21}^{\gamma\alpha}}{\bar{L}_{11}^{\gamma\alpha}} \right]. \tag{105}$$

for the thermal conductivity.

CURRENT OPERATORS

The linear response theory expresses the transport coefficients in terms of the thermal averages of the current operators. In this Section we calculate these operators for a given Hamiltonian using the conservation laws for the local charge and energy density. That is, we define the divergence of the current densities by the rate of change of the corresponding densities. The current density operators obtained in such a way give the macroscopic current densities that satisfy the continuity equations (EOC),

$$\frac{\partial < e\hat{n}(\mathbf{x},t) >}{\partial t} = - < \mathrm{div}\hat{\mathbf{j}}(\mathbf{x},t) >, \tag{106}$$

and

$$\frac{\partial < \hat{h}_\phi(\mathbf{x},t) >}{\partial t} = - < \mathrm{div}\hat{\mathbf{j}}^{\phi}_{E}(\mathbf{x},t) > . \tag{107}$$

Here, $< \hat{n}(\mathbf{x};t) >= \mathrm{Tr}\{\rho_\phi \hat{n}(\mathbf{x})\}$, $< \hat{h}_\phi(\mathbf{x};t) >= \mathrm{Tr}\{\rho_\phi \hat{h}_\phi(\mathbf{x})\}$ and the averaging is with respect to the Hamiltonian $\hat{H}_\phi = \int d^3x\, \hat{h}_\phi(\mathbf{x},t)$. The energy density operator is $\hat{h}_\phi(\mathbf{x}) = h(\mathbf{x}) + \Delta\hat{h}_\phi(\mathbf{x})$, and the perturbation is $\Delta\hat{h}_\phi(\mathbf{x}) = \phi(x)\, e\hat{n}(x)$.

Charge current. The equation of motion for the density matrix Eqs. (75) and the charge conservation expressed by (106) give

$$\frac{\partial < \hat{n}(\mathbf{x},t) >}{\partial t} = \mathrm{Tr}\left\{\frac{\partial \rho_\phi}{\partial t}\hat{n}(\mathbf{x})\right\} = \left\langle i[\hat{H}_\phi, \hat{n}(\mathbf{x})]\right\rangle, \tag{108}$$

where the last equation follows from the cyclic invariance of the trace. Thus, the current density operator defined by the equation

$$-\mathrm{div}\hat{\mathbf{j}}_\phi(x,t) = i[\hat{H}_\phi, e\hat{n}(\mathbf{x})]. \tag{109}$$

gives the macroscopic current density which satisfies the EOC. In the steady-state, reached at $t = 0$, the flow is divergence-less: div $< \hat{\mathbf{j}}_\phi(x,t) >= 0$.

The $q = 0$ Fourier component of the current density operator $\hat{\mathbf{j}}_\phi = \int d^3x \hat{\mathbf{j}}_\phi(x,t)$ is obtained by commuting the Hamiltonian with the charge polarization operator

$$\mathbf{P}_c(t) = e\int d^3x \mathbf{x}\hat{n}(x,t) \tag{110}$$

We find, using (110), (109) and integrating by parts, the result

$$i[\hat{H}_\phi, \mathbf{P}_c] = \int d^3x \mathbf{x}\, i[\hat{H}_\phi, e\hat{n}(\mathbf{x})] = -\int d^3x \mathbf{x}\, \mathrm{div}\hat{\mathbf{j}}_\phi(x,t) = \int d^3x \hat{\mathbf{j}}_\phi(x,t) = \hat{\mathbf{j}}_\phi. \tag{111}$$

The current density operator is defined by the commutator

$$\hat{\mathbf{j}}_\phi(x,t) = i[\hat{H}_\phi, \mathbf{x}e\hat{n}(\mathbf{x})]. \tag{112}$$

For scalar gauge, such that $\hat{H}_\phi = \hat{H} + e\int d^3x \phi(x)\hat{n}(x)$, the current density operator defined by Eq. (112) is unaffected by the potential term of the Hamiltonian and we have $\hat{\mathbf{j}}_\phi = \hat{\mathbf{j}}$.

Energy current. The energy current density operator follows in the same way. The local energy conservation gives

$$\frac{\partial < \hat{h}_\phi(\mathbf{x},t) >}{\partial t} = \mathrm{Tr}\left\{\frac{\partial \rho_\phi}{\partial t}\hat{h}_\phi(\mathbf{x})\right\} =< i[\hat{H}_\phi,\hat{h}_\phi(\mathbf{x})] > \qquad (113)$$

which we use to define the divergence of the energy current density

$$-\mathrm{div}\,\hat{\mathbf{j}}_u(x,t) = i[\hat{H}_\phi,\hat{h}_\phi(x)]. \qquad (114)$$

The $q = 0$ Fourier component of the energy current density operator $\hat{\mathbf{j}}_u = \int d^3x\hat{\mathbf{j}}_u(x,t)$ is obtained by commuting the Hamiltonian with the energy polarization operator

$$\mathbf{P}_u(t) = \int d^3x\mathbf{x}\hat{h}_\phi(x). \qquad (115)$$

This gives

$$i[\hat{H}_\phi,\mathbf{P}_u] = \int d^3x\mathbf{x}i[\hat{H}_\phi,\hat{h}_\phi(\mathbf{x})] = -\int d^3x\mathbf{x}\,\mathrm{div}\,\hat{\mathbf{j}}_u(x,t) = \int d^3x\hat{\mathbf{j}}_u(x,t) = \hat{\mathbf{j}}_u, \qquad (116)$$

and we take the commutator

$$\hat{\mathbf{j}}_u(x,t) = i[\hat{H}_\phi,\mathbf{x}\hat{h}_\phi(x)] \qquad (117)$$

as the definition of the energy current density operator. This operator has contributions coming from

$$i[\hat{H}_\phi,\hat{h}_\phi(\mathbf{x})] = i[\hat{H}_\phi,\hat{h}(\mathbf{x})] + i[\hat{H},\Delta h_\phi(\mathbf{x})] + i[\Delta H,\Delta h_\phi(\mathbf{x})]. \qquad (118)$$

The last term in (118) vanishes and the contribution of the middle term to the total energy current is

$$\hat{\mathbf{j}}_u^\phi = i[\hat{H},\mathbf{P}_u^\phi(t)] \qquad (119)$$

where we introduced

$$\mathbf{P}_u^\phi(t) = e\int d^3x\mathbf{x}\hat{\phi}(x)n(x,t). \qquad (120)$$

Eqs. (110) and (112) give

$$\hat{\mathbf{j}}_u^\phi = \int d^3x\, e\phi(x)\, i[\hat{H},\hat{n}(\mathbf{x})] = -\int d^3x\mathbf{x}\phi(x)\,\mathrm{div}\,\hat{\mathbf{j}}(x,t). \qquad (121)$$

which becomes by partial integration

$$\hat{\mathbf{j}}_u^\phi = \int d^3x\phi(x)\hat{\mathbf{j}}(x,t), \qquad (122)$$

where we neglected $\nabla\phi$ with respect to ϕ. Thus, $\hat{\mathbf{j}}_u^\phi(x) = \phi(x)\hat{\mathbf{j}}(x)$. Performing similar analysis for the first term in Eq. (118), and neglecting again the gradient terms, gives

$$-\mathrm{div}\,\hat{\mathbf{j}}_E^\phi \simeq -\mathrm{div}\,\hat{\mathbf{j}}_E, \qquad (123)$$

where

$$\hat{\mathbf{j}}_E = i[\hat{H},\mathbf{P}_E], \tag{124}$$

defines the energy current in the absence of the field and

$$\mathbf{P}_E(t) = \int d^3x \mathbf{x} \hat{h}(x,t). \tag{125}$$

is the field-free energy polarization operator. The full energy current density operator which satisfies the continuity equation is

$$\hat{\mathbf{j}}_u(x) = \hat{\mathbf{j}}_E(x) + \phi(x)\hat{\mathbf{j}}(x). \tag{126}$$

In a stationary state, the rate of change of the total energy density around point $\mathbf{x}$ is zero, such that $\dot{u} = -\text{div} < \hat{\mathbf{j}}_u(\mathbf{x}) >= 0$. However, the internal energy is not conserved and $-\text{div} < \hat{\mathbf{j}}_E(\mathbf{x}) > \neq 0$. None-the-less $\hat{\mathbf{j}}_E(\mathbf{x}) = \hat{\mathbf{j}}_u(x) - \phi(x)\hat{\mathbf{j}}(x)$ is convenient for calculations, because the external perturbation appears only in ρ_ϕ, which simplifies the perturbation theory for $\mathbf{J}_E = \text{Tr}\{\rho_\phi \hat{\mathbf{j}}_E(\mathbf{x})\}$. The statistical averaging of the last term in Eq. (126) gives rise to the Joule heat, which ensures the proper temperature distribution in a stationary flow[3]. We cannot neglect this term, because no matter how small $\nabla\phi(x)$, the value of $\phi(x)$ is arbitrary large for $x \to \pm\infty$.

Examples of current operators

As a case study we take the Falicov-Kimball model, Hubbard model and periodic Anderson model with on-site hybridization, $V_{ij} = V\delta_{ij}$. The Hamiltonians for the Falicov-Kimball and Hubbard models can be summarized by the following form:

$$\mathscr{H} = -\sum_{ij} t_{ij} c_i^\dagger c_j - \sum_{ij} t_{ij}^f f_i^\dagger f_j + U \sum_i c_i^\dagger c_i f_i^\dagger f_i. \tag{127}$$

In Equation (127), we have fermionic creation (annihilation) operators $c_i^\dagger$ (c_i) for conduction electrons, and U is the on-site Coulomb repulsion. In the Falicov-Kimball model, we have $t^f = 0$ and the f-electrons are localized on the lattice. In the Hubbard model, we have $t = t^f$ and we identify the c electrons as the up-spin electrons and the f electrons as the down-spin electrons. Equation (127) interpolates continuously between the Falicov-Kimball and Hubbard models.

The periodic Anderson model is written in a standard form,

$$\begin{aligned} \mathscr{H} &= -\sum_{ij\sigma} t_{ij} c_{i\sigma}^\dagger c_{j\sigma} + f_{i\sigma}^\dagger f_{i\sigma}) + E_f \sum_i f_{i\sigma}^\dagger f_{i\sigma} \\ &+ \sum_{i\sigma} (V_i c_{i\sigma}^\dagger f_{i\sigma} + f_{i\sigma}^\dagger c_{i\sigma}) + U \sum_{i\sigma} f_{i\uparrow}^\dagger f_{i\uparrow} f_{i\downarrow}^\dagger f_{i\downarrow}. \end{aligned} \tag{128}$$

The conduction and localized electrons as the same as in the Falicov-Kimball model, except now they have spin (E_f is the localized electron site energy). The parameter V is the on-site hybridization between conduction and localized electrons on the same site, and U is the on-site Coulomb repulsion between f-electrons with different spin.

The Hamiltonians become quite simple in momentum space:

$$\mathscr{H} = \sum_{\mathbf{k}} \varepsilon(\mathbf{k})\, c_{\mathbf{k}}^{\dagger} c_{\mathbf{k}} + \sum_{\mathbf{k}} \varepsilon^{f}(\mathbf{k}) + U \sum_{\mathbf{p},\mathbf{k},\mathbf{q}} c_{\mathbf{k}-\mathbf{q}}^{\dagger} c_{\mathbf{k}} f_{\mathbf{p}+\mathbf{q}}^{\dagger} f_{\mathbf{p}}, \tag{129}$$

for the Falicov-Kimball-Hubbard model, where the fermionic creation and annihilation operators now create or annihilate electrons with well-defined momentum. The periodic Anderson model becomes

$$\begin{aligned}\mathscr{H} &= \sum_{\mathbf{k}\sigma} \varepsilon(\mathbf{k})\, c_{\mathbf{k}\sigma}^{\dagger} c_{\mathbf{k}\sigma} + E_f \sum_{\mathbf{k}\sigma} f_{\mathbf{k}\sigma}^{\dagger} f_{\mathbf{k}\sigma} \\ &+ V \sum_{\mathbf{k}\sigma} (c_{\mathbf{k}\sigma}^{\dagger} f_{\mathbf{k}\sigma} + f_{\mathbf{k}\sigma}^{\dagger} c_{\mathbf{k}\sigma}) + U \sum_{\mathbf{p},\mathbf{k},\mathbf{q}} f_{\mathbf{p}+\mathbf{q}\uparrow}^{\dagger} f_{\mathbf{p}\uparrow} f_{\mathbf{k}-\mathbf{q}\downarrow}^{\dagger} f_{\mathbf{k}\downarrow}.\end{aligned}$$

To find the $q = 0$ component of the charge-current operator we define the polarization operator $\mathbf{P}_c = e\sum_i \mathbf{R}_i c_i^{\dagger} c_i$, with the obvious generalization if we need to include spin, and and evaluate the commutator $[\mathscr{H}, \mathbf{P}_c]$. Since the potential energy depends only on the number operators the charge polarization operator commutes with the U-term and we need evaluate only the commutator with the kinetic-energy piece of the Hamiltonian, which is straightforward in real space. Converting to a momentum-space representation gives

$$\mathbf{j} = e \sum_{\mathbf{k}} \mathbf{v}(\mathbf{k}) c_{\mathbf{k}}^{\dagger} c_{\mathbf{k}}, \tag{130}$$

where $\mathbf{v}(\mathbf{k}) = \nabla \varepsilon(\mathbf{k})$ is the unrenormalized velocity.

The non-local hybridization would generate the additional term,

$$\mathbf{j} = e \sum_{\mathbf{k}} \mathbf{v}(\mathbf{k}) c_{\mathbf{k}}^{\dagger} c_{\mathbf{k}} + e \sum_{\mathbf{pq}} \left[\nabla_{\mathbf{p}} V_{\mathbf{p}} c_{\mathbf{p}+\mathbf{q}}^{\dagger} f_{\mathbf{q}} + \nabla_{\mathbf{p}} V_{\mathbf{p}}^{*} f_{\mathbf{q}}^{\dagger} c_{\mathbf{p}+\mathbf{q}} \right] \tag{131}$$

The energy-current operator is more complicated to construct. The energy-polarization operator is $\mathbf{P}_E = \sum_i \mathbf{R}_i h_i$, and its proper definition requires assigning an energy operator h_i to each lattice site i ($\mathscr{H} = \sum_i h_i$). For a local potential energy, such an assignment is completely straightforward, but the hopping terms in the kinetic energy are more complicated — the standard procedure is to associate one half of the operator that connects sites i and j to the local energy operator at site i and at site j. Once h_i has been defined, we can calculate the commutator $[\mathscr{H}, \mathbf{P}_E]$ and find the energy-current operator. Unlike the charge current operator, which was essentially independent of the model considered, the energy current operator is different for each Hamiltonian considered. So we summarize the results for each case considered here. For the Falicov-Kimball and Hubbard models, the energy-current operator becomes

$$\mathbf{j}_E^{\mathrm{FK}} = \sum_{\mathbf{k}} \varepsilon(\mathbf{k}) \mathbf{v}(\mathbf{k}) c_{\mathbf{k}}^{\dagger} c_{\mathbf{k}} + \frac{U}{2} \sum_{\mathbf{kqq}'} [\mathbf{v}(\mathbf{q}) + \mathbf{v}(\mathbf{q}')] f_{\mathbf{k}}^{\dagger} f_{\mathbf{k}+\mathbf{q}-\mathbf{q}'} c_{\mathbf{q}}^{\dagger} c_{\mathbf{q}'}. \tag{132}$$

For the Hubbard model it is

$$\mathbf{j}_Q^{\rm H} = \sum_{\mathbf{k}\sigma} \varepsilon(\mathbf{k})\mathbf{v}(\mathbf{k})c_{\mathbf{k}\sigma}^\dagger c_{\mathbf{k}\sigma} + \frac{U}{2}\sum_{\mathbf{k}\mathbf{q}\mathbf{q}'}[\mathbf{v}(\mathbf{q})+\mathbf{v}(\mathbf{q}')]c_{\mathbf{q}\uparrow}^\dagger c_{\mathbf{q}'\uparrow} c_{\mathbf{k}\downarrow}^\dagger c_{\mathbf{k}+\mathbf{q}-\mathbf{q}'\downarrow}.$$

For the periodic Anderson model we have to evaluate (in obvious notation)

$$\mathbf{j}_E = i\sum_{ij}\mathbf{R}_j[\mathscr{H}_i^T + \mathscr{H}_i^V + \mathscr{H}_i^U, \mathscr{H}_j^T + \mathscr{H}_j^V + \mathscr{H}_j^U] = \mathbf{j}_E^{TT} + (\mathbf{j}_E^{TV} + \mathbf{j}_E^{VT}), \tag{133}$$

where the last equation holds because $[\mathscr{H}_i^T, \mathscr{H}_j^U] = 0$ and

$$\mathbf{j}_E^{VU} + \mathbf{j}_E^{UV} = i\sum_i \mathbf{R}_i\left\{[\mathscr{H}_i^V, \mathscr{H}_i^U] + [\mathscr{H}_i^U, \mathscr{H}_i^V]\right\} = 0.$$

The kinetic term

$$\mathbf{j}_E^{TT} = i\sum_i \mathbf{R}_i[\mathscr{H}_i^T, \mathscr{H}_i^T]$$

is the same as for the Falicov-Kimbal model and we only have to calculate

$$\mathbf{j}_E^{TV} + \mathbf{j}_E^{VT} = i\sum_i \mathbf{R}_i[\mathscr{H}_i^T, \mathscr{H}_i^V] + i\sum_i \mathbf{R}_i[\mathscr{H}_i^V, \mathscr{H}_i^T].$$

Direct evaluation in the site representation gives

$$\mathbf{j}_E^{TV} + \mathbf{j}_E^{VT} = -i\frac{W}{2}\sum_{i\delta\sigma}\left\{V_{i+\delta}(\mathbf{R}_{i+\delta} - \mathbf{R}_i)c_{i\sigma}^\dagger f_{i+\delta\sigma} + V_{i+\delta}^*(\mathbf{R}_{i+\delta} - \mathbf{R}_i)c_{i\sigma} f_{i+\delta\sigma}^\dagger.\right\}$$

In the momentum representation this term can be written as

$$\mathbf{j}_E^{TV} + \mathbf{j}_E^{VT} = -\frac{1}{2}\sum_{\mathbf{p}\mathbf{q}\sigma}\mathbf{v}_\mathbf{p}\left\{V_{\mathbf{q}-\mathbf{p}}c_{\mathbf{p}\sigma}^\dagger f_{\mathbf{q}\sigma} + V_{\mathbf{p}-\mathbf{q}}^* f_{\mathbf{q}\sigma}^\dagger c_{\mathbf{p}\sigma}\right\}. \tag{134}$$

For the constant hybridization the $q = 0$ Fourier component of the energy density operator is

$$\mathbf{j}_E = \sum_{\mathbf{k}} \varepsilon(\mathbf{p})\mathbf{v}(\mathbf{p})c_\mathbf{p}^\dagger c_\mathbf{p} - \frac{V}{2}\sum_{\mathbf{p}\sigma}\mathbf{v}_\mathbf{p}\left\{c_{\mathbf{p}\sigma}^\dagger f_{\mathbf{p}\sigma} + f_{\mathbf{p}\sigma}^\dagger c_{\mathbf{p}\sigma}\right\} \tag{135}$$

This energy current operator, together with the charge current operator defined by Eq. (130) enters the correlation functions for thermal transport.

MAHAN-JONSON THEOREM

The Jonson-Mahan[7, 8, 9] theorem provides a simple relation between the transport coefficient for the electrical conductivity and that needed for the thermopower and thermal conductivity. The relation is that the integral for the L_{12} coefficient has an extra power of frequency in the integrand than the L_{11} coefficient and that L_{22} has one more power of frequency than L_{21}. This result has been known for many years for a noninteracting system— the Jonson-Mahan theorem generalizes this result for a wide class of many-body systems (including the Falicov-Kimball model, the Hubbard model, and the periodic Anderson model. We provide the proof for infinite dimensional lattices[10].

We begin with the generalized two-particle correlation function

$$F_{\alpha\beta}(\tau,\tau',\tau'',\tau''') = \sum_{qq'\sigma\sigma'} \mathbf{v}_{q\alpha}\mathbf{v}_{q'\beta}\langle T_\tau c^\dagger_{q\sigma}(\tau)c_{q\sigma}(\tau')c^\dagger_{q'\sigma'}(\tau'')c_{q'\sigma'}(\tau''')\rangle, \qquad (136)$$

where $c^\dagger_{q\sigma}$ and $c_{q\sigma}$ denote the creation and the annihilation operators for conduction electrons and the averaging is with respect to the interacting Hamiltonian. This function provides all the relevant "polarization operators", which can be shown by a method based on the Heisenberg equation of motion. Using

$$i\frac{d}{dt}\mathscr{O}_{\mathrm{H}}(t) = [\mathscr{O}_{\mathrm{H}}(t),\mathscr{H}_{\mathrm{H}}(t)] \qquad (137)$$

we find the time-dependence of the fermionic creation and annihilation operators (in the momentum basis and on the imaginary time-axis). The commutators are straightforward to calculate and become

$$[c^\dagger_{\mathbf{k}}(\tau),\mathscr{H}] = -\varepsilon(\mathbf{k})c^\dagger_{\mathbf{k}}(\tau) - U\sum_{\mathbf{p},\mathbf{q}} f^\dagger_{\mathbf{p}+\mathbf{q}}(\tau)f_{\mathbf{p}}(t)c^\dagger_{\mathbf{k}-\mathbf{q}}(\tau), \qquad (138)$$

for the Falicov-Kimball-Hubbard model and

$$[c^\dagger_{\mathbf{k}\sigma}(\tau),\mathscr{H}] = -\varepsilon(\mathbf{k})c^\dagger_{\mathbf{k}\sigma}(\tau) - \sum_{\mathbf{p}} V^*_{\mathbf{p}} f^\dagger_{\mathbf{k}-\mathbf{p}\sigma}(\tau) \qquad (139)$$

for the periodic Anderson model. Note, $[c^\dagger_{\mathbf{k}},H_U]=0$ for PAM. The commutator for the c operators is similar to evaluate. This shows that

$$\lim_{\tau'\to\tau^-}\frac{1}{2}\left(\frac{\partial}{\partial\tau}-\frac{\partial}{\partial\tau'}\right)\sum_{q\sigma}\mathbf{v}_q c^\dagger_{q\sigma}(\tau)c_{q\sigma}(\tau') = \mathbf{j}_Q(\tau). \qquad (140)$$

The Jonson-Mahan theorem will hold for any Hamiltonian that satisfies Eq. (140). In particular, it holds true for the Falicov-Kimball-Hubbard model and the periodic Anderson model[9] and allows us to express the "polarization operators" in terms of the generalized function F and its time-derivatives. We obtain

$$\mathscr{L}_{11} = \pi T e^2\int_0^\beta d\tau e^{i\nu_l\tau}F(\tau,\tau^-,0,0), \qquad (141)$$

for the conductivity,

$$\mathscr{L}_{12} = \pi T e\int_0^\beta d\tau e^{i\nu_l\tau}\frac{1}{2}\left(\frac{\partial}{\partial\tau}-\frac{\partial}{\partial\tau'}\right)F(\tau,\tau',0,0), \qquad (142)$$

(in the limit where $\tau' \to \tau^-$) for the thermopower, and

$$\mathscr{L}_{22} = \pi T \int_0^\beta d\tau e^{i\nu_l \tau} \frac{1}{4}\left(\frac{\partial}{\partial \tau} - \frac{\partial}{\partial \tau'}\right)\left(\frac{\partial}{\partial \tau''} - \frac{\partial}{\partial \tau'''}\right) \times F(\tau, \tau', \tau'', \tau'''), \qquad (143)$$

(in the limit where $\tau' \to \tau^-$, $\tau''' \to \tau''^-$, and $\tau'' \to 0^+$) for the thermal conductivity.

The analytic continuation to the real axis is quite complicated in general[10] and we restrict our considerations to the infinite-dimensional limit. In that case the two-particle correlation function is expressed by just its bare bubble because the irreducible charge vertex has a different symmetry than $\mathbf{v}_q$. Hence, we immediately learn that

$$F_{\alpha\beta}(\tau, \tau', \tau'', \tau''') = -\sum_{q\sigma} \mathbf{v}_{q\alpha}^2 \delta_{\alpha\beta} G_{q\sigma}(\tau''' - \tau) G_{q\sigma}(\tau' - \tau'')., \qquad (144)$$

where $G_{q\sigma}(\tau) = -\langle T_\tau c_{q\sigma}(\tau) c_{q\sigma}^\dagger(0)\rangle$ is the imaginary time Green's function of conduction electrons. Next, we need to determine a spectral representation for the Green's function. Using the fact that

$$G_{q\sigma}(z) = -\frac{1}{\pi}\int d\omega \frac{\mathrm{Im} G_{q\sigma}(\omega)}{z - \omega}, \qquad (145)$$

with z in the upper half plane (which can be shown by using the Lehmann representation), we find that

$$G_{q\sigma}(\tau) = -\frac{1}{\pi}\int d\omega T \sum_n \frac{e^{-i\omega_n \tau}}{i\omega_n - \omega} \mathrm{Im} G_{q\sigma}(\omega). \qquad (146)$$

Now we convert the sum over Matsubara frequencies into a contour integral (that surrounds each Matsubara frequency, but does not cross the real axis—the contour is then deformed into two contours, one running just above and the other just below the real axis), but we must be careful to ensure that the procedure is well-defined. If $\tau < 0$, then

$$\begin{aligned} T\sum_n \frac{e^{-i\omega_n \tau}}{i\omega_n - \omega} &= -\frac{i}{2\pi}\int_C dz \frac{e^{-z\tau}}{z - \omega} f(z), \\ &= -\frac{i}{2\pi}\int_{-\infty}^{\infty} dz e^{-z\tau} f(z) \left[\frac{1}{z + i0^+ - \omega} - \frac{1}{z - i0^+ - \omega}\right], \\ &= -e^{-\omega\tau} f(\omega). \end{aligned} \qquad (147)$$

This result is well-defined because the Fermi factor provides convergence (asymptotically like $\exp[-\beta z]$) for $z \to \infty$ and the $\exp[-z\tau]$ term provides boundedness for $z \to -\infty$ when $\tau < 0$. Since $1 - f(z)$ has the same poles as $f(z)$ on the imaginary axis, with residues that have the opposite sign, and it behaves like $\exp[\beta z]$ for $z \to -\infty$, one finds

$$T\sum_n \frac{e^{-i\omega_n \tau}}{i\omega_n - \omega} = e^{-\omega\tau}[1 - f(\omega)], \qquad (148)$$

for $\tau > 0$. The results in Eqs. (147) and (148) can then be substituted into Eq. (146) to get the final formula for the Green's function

$$G_{q\sigma}(\tau) = \begin{cases} \int d\omega A_{q\sigma}(\omega) e^{-\omega\tau}[1-f(\omega)], & \tau > 0 \\ \int d\omega A_{q\sigma}(\omega) e^{-\omega\tau}[-f(\omega)], & \tau < 0. \end{cases} \tag{149}$$

where $A_{q\sigma}(\omega) = -\text{Im}G_{q\sigma}(\omega)/\pi$ is the spectral function. Substituting into Eq. (144), then yields

$$\begin{aligned} F_{\alpha\beta}(\tau,\tau',\tau'',\tau''') &= \delta_{\alpha\beta}\int d\varepsilon\, \mathbf{v}_\varepsilon^2\, \rho(\varepsilon) \int d\omega \int d\omega' \\ &\times A(\varepsilon,\omega)A(\varepsilon,\omega')e^{\omega(\tau-\tau''')-\omega'(\tau'-\tau'')} f(\omega)[1-f(\omega')], \end{aligned} \tag{150}$$

where we replaced the q-summation by an integral over the non-interacting density of states. Because of Eq. (150), the analytic continuation of the correlation function given by Eq. (141) is trivial (one first converts from imaginary time to Matsubara frequencies and then performs the Wick rotation to the real frequency axis), and if we note the identity

$$f(\omega) - f(\omega+\nu) = -f(\omega)[1-f(\omega+\nu)][e^{-\beta\nu}-1], \tag{151}$$

then we can easily compute that

$$\bar{L}_{11} = Te^2\langle \mathbf{v}_\mathbf{k}^2\rangle_{\text{FS}} \int d\omega \int d\varepsilon \rho(\varepsilon)\left(-\frac{df(\omega)}{d\omega}\right) A^2(\varepsilon,\omega), \tag{152}$$

for the conductivity,

$$\bar{L}_{12} = Te\langle \mathbf{v}_\mathbf{k}^2\rangle_{\text{FS}} \int d\omega \int d\varepsilon \rho(\varepsilon)\left(-\frac{df(\omega)}{d\omega}\right) A^2(\varepsilon,\omega)\omega, \tag{153}$$

for the thermopower, and

$$\bar{L}_{22} = T\langle \mathbf{v}_\mathbf{k}^2\rangle_{\text{FS}} \int d\omega \int d\varepsilon \rho(\varepsilon)\left(-\frac{df(\omega)}{d\omega}\right) A^2(\varepsilon,\omega)\omega^2, \tag{154}$$

for the thermal conductivity. Here, $\langle \mathbf{v}_\mathbf{k}^2\rangle_{FS}$ denotes the square of conduction electron's velocity averaged over the renormalized Fermi surface (FS). This pre-factor has been taken out of the integral because $(-df(\omega)/d\omega)$ restricts the integration region in the q-space to the vicinity of the Fermi surface. This proves Mott's form for the thermal transport.

For the degenerate models with N conduction channels it is convenient to rewrite the above expressions in terms of the Luttinger's coefficients $L_{12} = \bar{L}_{12}/T$ defined by Eq. (21). This gives $\sigma(T) = e^2 N L_{11}$ for the charge conductivity, $\alpha(T)|e|T = -L_{12}/L_{11}$ for the thermopower, and $\kappa(T)T = N(L_{22} - L_{12}^2/L_{11})$ for the electronic contribution to thermal conductivity, where

$$L_{mn} = \langle \mathbf{v}_\mathbf{k}^2\rangle_{\text{FS}} \int d\omega \left(-\frac{df}{d\omega}\right) \omega^{m+n-2}\Lambda(\omega,T). \tag{155}$$

$f(\omega) = 1/[1+exp(\beta\omega)]$ is the Fermi-Dirac distribution function, ω is measured with respect to the chemical potential μ, and $\Lambda(\omega,T)$ is the frequency-dependent correlation function

$$\Lambda(\omega,T) = \int d\varepsilon\rho(\varepsilon)A^2(\varepsilon,\omega) \tag{156}$$

calculated for a single channel. For the infinite-dimensional Falicov-Kimball-Hubbard model and the Anderson model with the Gaussian non-interacting density of states the transport relaxation time can be given an explicit form. Using the Dyson equation for the conduction electrons Green's function

$$G_{q\sigma}(\omega) = \frac{1}{\omega+\mu-\Sigma(\omega)-\varepsilon(q)}$$

and the fact that the self energy is momentum independent we can evaluate the integral in Eq. (156) with the result

$$\Lambda(\omega) = \frac{\mathrm{Im}G(\omega)}{\mathrm{Im}\Sigma(\omega)} + 2 - 2\mathrm{Re}\{[\omega+\mu-\Sigma(\omega)]G(\omega)\}. \tag{157}$$

This expression simplifies at low temperatures, where the frequency integration in Eq. (155) is restricted by $(-df/d\omega)$ to the Fermi window, $|\omega| \leq k_B T$. In the $T \to 0$ limit we can write Eq. (156) as

$$\Lambda(\omega,T) = \mathcal{N}_c(\omega)\tau(\omega,T) \tag{158}$$

where $\mathcal{N}_c(\omega) = \sum_{\mathbf{k}} A_{\mathbf{k}}(\omega)$ is the renormalized DOS of conduction electrons, and $\tau(\omega,T)$ the transport relaxation time

$$\tau(\omega,T) = \frac{-1}{Im\ \Sigma_c(\omega^+,T)} \tag{159}$$

which is given by the momentum-independent self energy of c electrons. In deriving Eq. (158) we included only the divergent term in $\Lambda(\omega,T)$, and dropped the finite term which contributes to the full DMFT expression but does not affect the $T,\omega \to 0$ results. To calculate the renormalized expressions for $v_{k_F}^2$, $\mathcal{N}_c(\omega)$, $\tau(\omega)$, we make the approximations which are consistent with this derivation.

MICROSCOPIC SOLUTION FOR TRANSPORT COEFFICIENTS

We now discus the thermoelectric properties of intermetallic compounds with Cerium, Europium or Ytterbium ions using the theoretical methods explained in the previous sections. We summarize, first, the transport properties of heavy fermions and valence fluctuators and, then, describe the anomalies observed in $YbInCu_4$-like systems which show a valence-change transition. The thermal transport of heavy fermions and valence fluctuators is explained by the periodic Anderson model and that of the $YbInCu_4$-like systems by the Falicov-Kimball model.

Experimental summary

The experimental data on heavy fermions and valence fluctuators can be summarized as follows. The thermoelectric power $\alpha(T)$ exhibits the characteristic features that allow the classification of these compounds into several distinct groups[11, 12, 13]. In the case of Cerium ions, the compounds belonging to the first group (type (a) systems) have the thermopower with a deep negative minimum at low temperatures [11, 12, 14, 15, 16] and a high-temperature maximum, typically between 100 K and 300 K. At the maximum, $\alpha(T)$ could be either positive or negative. At very low temperatures, the type (a) systems order magnetically or become superconducting. The compounds of the second group (type (b) systems) have a negative low-temperature minimum and a positive high-temperature maximum but, in addition, the thermopower shows a smaller positive peak at lowest temperatures.[17, 18, 11, 19, 20] This second peak is sometimes concealed by a low-temperature phase transition; for example, in $CeCu_2Si_2$ it becomes visible only in an external magnetic field which suppresses the superconducting transition,[14] and in $CeRu_2Ge_2$ it shows up when the external pressure suppresses the magnetic transition.[21] The experimental evidence is now accumulating that the initial slope of the thermopower $\alpha(T)/T$ is positive for this class of (heavy fermion) materials, provided the measurements are performed at low enough temperature and with sufficient accuracy.[19, 20, 16, 22] In the third group (type (c) systems), the low-temperature peak is well pronounced and shifted towards the high-temperature peak. The main difference with respect to the type (b) systems is that the sign-change of $\alpha(T)$ does not occur.[23, 24, 25, 26] Finally, in some cases (type (d) systems) the thermopower grows monotonically towards the high-temperature maximum, and the low-temperature structure appears only as a shoulder on a broad peak, or is not resolved at all. [27, 24, 28, 20]

The clue to these various types of behavior comes from the high-pressure[29, 30, 21, 31, 12] and doping studies, [11, 24, 25, 32, 33, 34, 35, 36] which show that the thermopower of Cerium compounds changes continuously from type (a) to type (d). A typical example is provided by the shape of $\alpha(T)$ observed in $CeRu_2Ge_2$at various pressures.[21] At ambient pressure, $CeRu_2Ge_2$ is a type (a) system with a magnetic ground state and negative thermopower below 300 K. An increase of pressure leads to a thermopower with a small positive peak at low temperatures and an enhanced peak at high temperatures. A further increase of pressure enhances both peaks, shifts the low-temperature peak towards the high-temperature one, and makes the thermopower at intermediate temperatures less negative. For large enough pressure, the sign-change does not occur at all and for very high pressure the low-temperature peak merges with the high-temperature one, and transforms into a shoulder. The high-temperature peak grows continuously but its position remains more or less constant, as $\alpha(T)$ changes from type (a) to (c). Eventually, for pressures above 10 GPa, the $\alpha(T)$ assumes the (d) shape. Here, the initial slope of $\alpha(T)$ decreases and the position of the maximum shifts to higher temperatures, but its magnitude does not change as pressure increases. Similar behavior is also seen in the high-pressure data of, $CeCu_2Si_2$,[29] $CeCu_2Ge_2$,[31, 37] or $CePd_2Si_2$.[38] As regards doping, the chemical substitutions which reduce the volume and make Ce ions less magnetic, transform $\alpha(T)$ from type (a) to type (b),[32] from (b) to (c),[24, 25] or from (a) to(c),[33, 34, 35] while the substitutions which expand the volume and make the Ce more magnetic, transform the thermopower from, say, type (d) to type (c) or from type (c) to type (b).[36] This variation of shape is an indication that the local environment plays an important role in determining the magnetic character of Ce and Yb ions. Even at high temperatures, where each 4f ion is an independent scatterer, the thermopower of a sample with a high concentration of 4f ions cannot be obtained by rescaling the low-concentration data.

The Ytterbium intermetallics can be classified using the mirror-image analogy with Cerium systems. This holds because the Yb ions fluctuate between $4f^{13}$ and $4f^{14}$, while the Ce ions fluctuate between $4f^1$ and $4f^0$ configurations, and the dynamics of a single f hole and a single f electron is the same. A well-defined local moment leads in Yb systems to the type (a) behavior, such that the thermopower has a negative minimum at high temperatures and a positive maximum at low temperatures;[39, 40] the size of the minimum is about the same as the size of the maximum. The thermopower of (b)-type Yb systems[41, 42, 43] mirrors the (b)-type Ce systems. Here, one finds two negative minima separated by a small positive maximum. The type (c) Yb systems have a non-monotonic thermopower with a large (negative) minimum at high temperatures and a smaller one at low temperatures, but there is no sign-change.[41, 43, 39, 42] Finally, the thermopower with a single negative peak centered around 100 K [44, 41, 28, 39] mirrors the type (d) behavior of Ce systems. The reduction of volume by pressure or doping[43, 39] stabilizes the magnetic $4f^{13}$ configuration of Yb ions, and transforms $\alpha(T)$ from, say, type (b) to type (a), from (c) to (b), or from (c) to (a).

The experimental results show that 4f systems with similar thermopowers exhibit similarities in other thermodynamic[45] and transport[46, 47, 48] properties, and there is an obvious correlation between the shape of $\alpha(T)$ and the magnetic character of the 4f ions. The thermopower measurements[49] provide a simple and sensitive tool for characterizing the magnetic state of a 4f ion in a given metallic matrix: the shape of $\alpha(T)$ changes from the (a)-type in the case of magnetic Ce (Yb) ions with stable f^1 ($4f^{13}$) configuration to the (d)-type for non-magnetic Ce (Yb) ions which fluctuate between the $4f^1$ ($4f^{13}$) and $4f^0$ ($4f^{14}$) configurations.

At very low temperatures one finds additional interestig features. The data show [22, 50] a striking correlation between the low-temperature Seebeck coefficient α and the specific heat coefficient $\gamma = C_V/T$; the ratio $q = |e|\lim_{T\to 0}\alpha/\gamma T$ is nearly constant, although the absolute values of γ and α/T vary by orders of magnitude. The q ratio increases as the number of conduction electrons n_c decreases and becomes large in bad metals[22, 50]. The Kadowaki-Woods (KW) ratio [51] which is defined as $\rho(T)/(\gamma T)^2$, where $\rho(T)$ is the electrical resistivity, exhibits similar universal features, if one takes into account the effective low-temperature degeneracy of the f states [52, 53]. The near constancy of the KW and q ratios suggests a 'universal law' for the low-temperature power factor $P = \alpha^2/\rho$ and brings to the fore the validity of the Wiedemann-Franz (WF) law, $\kappa\rho/T = \mathscr{L}_0$, and a possibility of enhancing the electronic thermoelectric figure-of-merit in strongly correlated materials, $ZT = \alpha^2 T/\kappa\rho$, where κ is the electronic contribution to the thermal conductivity and $\mathscr{L}_0 = \pi^2 k_B^2/3e^2$ the Fermi liquid (FL) Lorenz number. When the WF law holds, metals must have a thermopower larger than 155 μV/K to achieve $ZT > 1$; to date no metal has been found with so large a thermopower. In the temperature window where the effective Lorenz number ($\mathscr{L} = \kappa\rho/T$) is reduced, one can achieve $ZT > 1$ with substantially lower thermopowers, which might make it possible to find strongly correlated metals that can be used for cooling applications at low temperature.

Finally, we mention a new class of materials in which the transition between the high-entropy phase and the low-entropy does no follow the usual "Kondo route" but is driven by the valence-change transition, as observed in $EuCu_2Ni_2$, $YbIn_{1-x}Ag_xCu_4$ or $YbInCu_4$[54, 55]. The main features of these systems are best revealed by taking $YbIn_{1-x}Ag_xCu_4$ as an example. Indium doping expands the lattice and increases the weight of Yb^{2+} with respect to Yb^{3+} configuration[56] by transferring the electrons from the conduction band to the 4*f*-state. Thus, chemical pressure reduces the number of *f*-holes and increases the Kondo coupling which makes the compound less magnetic and increases charge fluctuations. For $x \geq 0.5$, the compound is a typical heavy fermion in which the substitution of the monovalent Ag by the trivalent In compensates the chemical pressure effects, such that the total number of conduction electrons increases. For for $x \leq 0.5$, the Ag-In substitution brings eventually μ and E_f in the vicinity of the band edge E_c which gives rise to completely new features. In this concentration range $YbIn_{1-x}Ag_xCu_4$ is already a valence fluctuator with temperature dependent $n_f^h(T)$, such that the temperature-induced transfer of *f*-holes in the conduction band can reduce $\mu - E_c$ and $E_f - E_c$ to zero. Once μ is within the gap of the density of states, the effective hybridization is switched off and the magnetic moment of the *f*-ions cannot be quenched by Kondo screening. Thus, the transition from the low-temperature coherent FL state to the high-temperature disordered paramagnetic state cannot follow the usual 'Kondo route', taken by the heavy fermions. The valence fluctuators like $YbIn_{1-x}Ag_xCu_4$ for $x \leq 0.5$, belong to a new class of materials in which the transition between the low- and high-entropy phase is driven by the Falicov-Kimball interaction. This gives rise, at the temperature T_V, to a change in the relative occupancy of the *f* and the conduction states and an abrupt modification of the properties of the system. The valence change transition is clearly seen in the XPS data; above T_V, the spectra indicate a stable $4f^{13}$ configuration of Yb ions and below T_V one has a mixture of $4f^{13}$ and $4f^{14}$ states. The magnetic character of the Yb ions changes at T_V, as indicated by an abrupt change of the susceptibility from Pauli-like to Curie-like. In the high-temperature phase, the Curie constant is close to the free ion value of Yb^{3+}. The conduction states are also modified at T_V, as indicated by a drastic change of the frequency dependence of the optical conductivity and by a large increase in the resistivity[56]. The electrical resistance and the Hall coefficient of the high-temperature phase of $YbIn_{1-x}Ag_xCu_4$ are typical of narrow-band semiconductors, or semi-metals with a very low carrier density, and neither the transport nor the thermodynamic properties show any sign of the Kondo effect. The proximity of μ to E_c is indicated in $YbInCu_4$ by the Hall data and band-structure calculations[56].

Periodic Anderson model

We now discuss the thermoelectric properties of heavy fermions and valence fluctuators using the periodic Anderson model. In the first part of this subsection we derive the Fermi liquid laws and explain the universal low-temperature behavior. In the second part we discuss the anomalies that one finds in the thermopower of heavy fermions above the Kondo temperature.

Fermi liquid solution

The coherent thermal transport of heavy fermions and valence fluctuators is explained at low temperatures by the periodic Anderson model with $SU(N)$ symmetry in infinite dimensions. We consider the model with the on-site hybridization,$V_{ij} = V\delta_{ij}$, such that the charge current operators are given by Eqs. (130) and (135). This model satisfies the Jonson-Mahan theorem. Since the integral in Eq. (155) is restricted by $(-df/d\omega)$ to $|\omega| \leq k_B T$, we write in the $T \to 0$ limit $\Lambda(\omega,T) = \mathcal{N}_c(\omega)\tau(\omega,T)$ and calculate the renormalized density of states $\mathcal{N}_c(\omega)$ and the transport relaxation time $\tau(\omega,T)$ by the dynamical mean field approximation (DMFT).

The renormalized c and f Green's functions of the periodic Anderson model satisfy the Dyson equations [57, 58]

$$G_c(\mathbf{k},z) = \frac{z - \tilde{\varepsilon}_f(\omega,T) + \mu}{[z - \varepsilon_{\mathbf{k}} + \mu][z - \tilde{\varepsilon}_f(\omega,T) + \mu] - V^2}, \tag{160}$$

and

$$G_f(\mathbf{k},z) = \frac{z - \varepsilon_{\mathbf{k}} + \mu}{z - \tilde{\varepsilon}_f(\omega,T) + \mu}\, G_c(\mathbf{k},z), \tag{161}$$

where $\tilde{\varepsilon}_f(\omega,T) = E_f + \mathrm{Re}\ \Sigma_f(\omega^+,T)$ is the renormalized position and $\Sigma_f(z)$ the DMFT self-energy f electrons, which is *momentum-independent.* The retarded (advanced) Green's functions are defined for z in the upper (lower) part of the complex plane. The c electrons self-energy on the real ω-axis is

$$\Sigma_c(\omega^+) = \frac{V^2}{\omega^+ - \tilde{\varepsilon}_f(\omega,T) + \mu - \mathrm{Im}\ \Sigma_f(\omega^+,T)}, \tag{162}$$

where $\omega^{\pm} = \lim_{\delta \to 0}(\omega \pm i\delta)$.

The singularities of the of $G_c(\mathbf{k},z)$ and $G_f(\mathbf{k},z)$ are defined by the equation

$$(\omega - \varepsilon_{\mathbf{k}} + \mu)[\omega - \tilde{\varepsilon}_f(\omega) + \mu] - V^2 = 0, \tag{163}$$

which gives the QP excitations and provides for $\omega = 0$ the renormalized FS of hybridized states. Since $\Sigma_f(\omega^+,T)$ is momentum independent in the DMFT approximation, the shape of the FS is determined by the non-interacting dispersion of c states. Using the Luttinger's theorem, which requires that the k-space volume enclosed by the FS yields the total concentration of electrons $n = n_c + n_f$ in each channel, we obtain the FS from the equation $\varepsilon_{\mathbf{k}} = \mu_0$, where μ_0 is the solution of $n = \int_{-\infty}^{\mu_0} d\varepsilon\ \mathcal{N}_c^0(\varepsilon)$. Thus, the renormalized FS coincides with the non-interacting one corresponding to $n_c + n_f$ c electrons and the FS average of v^2 is given by the integral

$$v_{k_F}^2 = \int d^d\mathbf{k}\ \delta(\varepsilon_{\mathbf{k}} - \mu_0)\, v_{\mathbf{k}}^2. \tag{164}$$

The numerical calculations for the 3-dimensional periodic Anderson model with the nearest-neighbor hopping on a simple cubic lattice give $v_{k_F}^2 = (ta_l/\hbar)^2 v^2$ where $v^2 \simeq 1.4$ for $n \simeq 1/2$ and $v^2 \ll 1$ for $n \simeq 1$.

The renormalized spectral properties around the Fermi level are obtained from the Luttinger theorem. First, following [57], we expand $Re\ \Sigma_f(\omega^+, T)$ up to the terms linear in ω which gives $\omega - [\tilde{\varepsilon}_f(\omega) - \mu] \simeq (\omega - \tilde{\omega}_f) Z_f^{-1}$. Here, $\tilde{\omega}_f$ is the solution of the equation $\tilde{\omega} - [\tilde{\varepsilon}_f(\tilde{\omega}) - \mu] = 0$ and $Z_f^{-1} = [1 - \partial \Sigma_f / \partial \omega]_{\omega=0}$ is the enhancement factor. For $\mathbf{k}$ close to the FS and small ω, Eq.(163) gives the QP excitations $\omega = \Omega_{\mathbf{k}}^{\pm}$, where

$$\Omega_{\mathbf{k}}^{\pm} = \frac{1}{2}\left[(\varepsilon_{\mathbf{k}} - \mu + \tilde{\omega}_f) \pm \sqrt{(\varepsilon_{\mathbf{k}} - \mu - \tilde{\omega}_f)^2 + 4\tilde{V}^2}\right], \tag{165}$$

and $\tilde{V} = V\sqrt{Z_f}$ denotes the renormalized hybridization. To describe the excitations in intermetallic compounds with Ce (Yb) ions we put the chemical potential in the lower (upper) QP branch. The singular parts of the spectral functions can be written as

$$A_c(\varepsilon_{\mathbf{k}}, \omega) \simeq a_{\mathbf{k}}^c(\Omega_{\mathbf{k}}^{\pm})\, \delta(\omega - \Omega_{\mathbf{k}}^{\pm}), \tag{166}$$

$$A_f(\varepsilon_{\mathbf{k}}, \omega) \simeq a_{\mathbf{k}}^f(\Omega_{\mathbf{k}}^{\pm})\, \delta(\omega - \Omega_{\mathbf{k}}^{\pm}), \tag{167}$$

where the coefficients are the residues of the QP poles,

$$a_{\mathbf{k}}^c(\omega) = \left[1 + Z_f^{-1} \frac{V^2}{\omega - \tilde{\varepsilon}_f(\omega) + \mu}\right]_{\omega = \Omega_{\mathbf{k}}^{\pm}},$$

and

$$a_{\mathbf{k}}^f(\omega) = \frac{(\omega - \varepsilon_{\mathbf{k}} + \mu)^2}{V^2} a_{\mathbf{k}}^c(\omega)$$

The renormalized c and f DOS are given at low energies by the expressions [59, 60]

$$\mathcal{N}_c(\omega) \simeq \mathcal{N}_c^0(\omega + \mu - \frac{V^2}{\omega - \tilde{\varepsilon}_f(\omega) + \mu}), \tag{168}$$

$$\mathcal{N}_f(\omega) \simeq \frac{\mu - \varepsilon_{\mathbf{k_F}}}{\mu - \tilde{\varepsilon}_f(0)} \mathcal{N}_c(\omega) \tag{169}$$

where $\mathbf{k_F}$ is Fermi momentum and $\mathcal{N}_c^0(\omega)$ is the unrenormalized c-DOS, taken as a symmetric function of half-width D. Using Eq. (163) this can also be written as,

$$\mathcal{N}_f(0) = \frac{V^2}{[\mu - \tilde{\varepsilon}_f(0)]^2} \mathcal{N}_c^0(\varepsilon_{\mathbf{k_F}}). \tag{170}$$

The total DOS is obtained by multiplying the above expressions by the degeneracy N.

The physical interpretation of $\tilde{\omega}_f = [\tilde{\varepsilon}_f(0) - \mu] Z_f$ follows from the DMFT requirement that the local Green's function of the Anderson lattice coincides with the Green's function of an auxiliary Anderson impurity, such that

$$\sum_{\mathbf{k}} G_f(\mathbf{k}, z) = \frac{1}{z - E_f + \mu - \Delta(z) - \Sigma_f(z)} = G_f^{aux}(z). \tag{171}$$

The auxiliary model describes f electrons distributed over N states and hybridized with an effective c band, The effective hybridization $\Delta(z)$ follows from the condition that the self energy of the lattice and the impurity defines the same functional $\Sigma_f[G_f(z)]$. The unrenormalized impurity f level is at E_f and in the $\omega, T \to 0$ limit we approximate $\Delta(\omega) \simeq \Delta(0) = i\Delta_0$ with $\Delta_0 < 0$. The DMFT solution requires $\mathcal{N}_f(\omega) = A_f(\omega)$, where $A_f(\omega) = -Im\, G_f^{aux}(\omega^+)/\pi$ is the auxiliary spectral function. Keeping only the real part of $\Sigma_f(\omega, T)$ we write the low-frequency limit of the renormalized f-DOS as

$$\mathcal{N}_f(\omega) = \frac{1}{\pi|\Delta_0|} \frac{(\tilde{\Delta}_f)^2}{(\omega - \tilde{\omega}_f)^2 + (\tilde{\Delta}_f)^2}, \tag{172}$$

where $\tilde{\Delta}_f \simeq -\Delta_0 Z_f$. The above expression, which holds for $\omega \ll \tilde{\omega}_f$, shows that the lattice f-DOS can be approximated by the Kondo resonance centered at $\tilde{\omega}_f$ and of the width $\tilde{\Delta}_f$.

Luttinger's theorem and the DMFT self-consistency condition can now be used to relate $\tilde{\omega}_f$ or $\tilde{\Delta}_f$ to $\mathcal{N}_f(0)$. The QP branches $\Omega_{\mathbf{k}}^+$ and $\Omega_{\mathbf{k}}^-$ are separated by the hybridization gap $2\tilde{V}_f$. Close to the Brillouin zone boundary (ZB) and the zone center (ZC), we have $\Omega_{\mathbf{k}}^+ - \Omega_{\mathbf{k}}^- \simeq D[1 + 2(\tilde{V}/D)^2] \simeq D$. At the FS we have $\Omega_{\mathbf{k}_F}^- = 0$ or $\Omega_{\mathbf{k}_F}^+ = 0$, depending on the value of $n_c + n_f$. Close to the FS, the QP dispersion is very weak and $\Omega_{\mathbf{k}}^\pm$ describe two heavy QP bands of half-width $\tilde{V}_f^2/D$. Taking $n << 2$ for Ce and Eu systems, such that the chemical potential is close to the top of the lower QP branch, we find from $\Omega_{\mathbf{k}_F}^- = 0$ that $\mathbf{k}_F$ is close to the ZB, such that $\varepsilon_{\mathbf{k}_F} - \mu \simeq D$ and $\mathcal{N}_c^0(\varepsilon_{\mathbf{k}_F}) D \simeq n_c/a\pi N$. The factor $a\pi$ in the denominator, approximately accounts for the fact that the unperturbed c-DOS near the edge of the band is smaller than the average unperturbed DOS over the whole energy interval under consideration. For Yb compounds, we restrict $n_f^h << 1$ and assume that the system is more than half-filled, such that the lower QP branch is full and the upper branch is fractionally occupied. The equation $\Omega_{\mathbf{k}_F}^+ = 0$ gives $\mathbf{k}_F$ close to the zone center, such that $\varepsilon_{\mathbf{k}_F} - \mu \simeq -D$ and $\mathcal{N}_c^0(\varepsilon_{\mathbf{k}_F})(\varepsilon_{\mathbf{k}_F} - \mu) \simeq -n_c/a\pi N$. Eqs. (169) yields

$$N\mathcal{N}_f(0) Z_f^{-1} = \pm \frac{n_c}{a\pi} \frac{1}{\tilde{\omega}_f}, \tag{173}$$

where the upper (lower) sign applies to Ce and Eu (Yb) compounds in which $\tilde{\omega}_f > 0$ ($\tilde{\omega}_f < 0$) and the Kondo resonance is above (below) the chemical potential.

Since the specific heat coefficient of the periodic Anderson model in the QP approximation is given by

$$\gamma = N \frac{\pi^2 k_B^2}{3} \sum_{\mathbf{k}} \delta(\mu - \Omega_{\mathbf{k}}^\pm) = N \frac{\pi^2 k_B^2}{3} [\mathcal{N}_c(0) + \mathcal{N}_f(0) Z_f^{-1}]$$

we have the FL relationship

$$\gamma = \frac{\pi^2 k_B^2}{3} \frac{1}{T_K} \simeq \frac{\pi^2 k_B^2}{3} \frac{n_c}{a\pi \tilde{\omega}_f}, \tag{174}$$

where we assumed that the enhanced f contribution dominates. The Kondo temperature $T_K \simeq 1/\gamma$ is thus related to the center of the Kondo resonance $\tilde{\omega}_f$. The numerical coefficient a can be obtained from the NRG solution of the DMFT equations, which can provide the numerical values of γ and $\tilde{\omega}_f$.

The width of the Kondo resonance $\tilde{\Delta}_f$. can be related to γ by using Eqs. (172) and (173) which give at $\omega = 0$,

$$\frac{aN}{n_c} = |\frac{\tilde{\omega}_f}{\tilde{\Delta}_f}| + |\frac{\tilde{\Delta}_f}{\tilde{\omega}_f}|. \tag{175}$$

Thus, $|\tilde{\omega}_f/\tilde{\Delta}_f| = (aN/2n_c) \pm \sqrt{(aN/2n_c)^2 - 1} = C$. The negative sign gives $\tilde{\Delta}_f \simeq (N\pi k_B^2)/[3\gamma(1+C^2)]$, such that for $N/n_c \gg 1$ and $C \simeq n_c/aN$ we have $\tilde{\Delta}_f \simeq N\pi k_B^2/3\gamma$.

To estimate the imaginary part of $\Sigma_f(\omega)$, which is also needed for thermal transport, we use the diagrammatic analysis. In infinite dimensions this gives the FL expression [57, 52]

$$-\mathrm{Im}\Sigma_f(\omega,T) \simeq \frac{\pi}{2}[\omega^2 + (\pi k_B T)^2](N-1)\mathscr{N}_f^3(0)\Gamma_f^2, \tag{176}$$

where Γ_f is the irreducible 4-point scattering vertex for electrons with different flavors. Eq. (176) is a straightforward generalization of the result given by the 2nd order perturbation theory in which the bare interaction U_{ff} is replaced by the Γ-vertex. In the limit of large correlations, when the charge fluctuations are suppressed, the Ward identity [57, 58] gives $Z_f^{-1} = (N-1)\mathscr{N}_f(0)\Gamma_f$, such that $\gamma = (\pi^2 k_B^2/3)N(N-1)\mathscr{N}_f^2(0)\Gamma_f$ and

$$-(N-1)\mathscr{N}_f(0)\mathrm{Im}\left[\frac{\partial^2 \Sigma_f(\omega)}{\partial \omega^2}\right]_{\omega=0^+} \simeq \frac{9\gamma^2}{\pi^3 k_B^4 N^2} \simeq \frac{1}{\pi \tilde{\Delta}_f^2}. \tag{177}$$

The transport relaxation time can now be obtained by substituting $\Sigma_c(\omega^+, T)$ in Eq. (159) which yields

$$\tau(\omega,T) \simeq \frac{[\omega - \tilde{\varepsilon}_f(\omega,T) + \mu]^2 + [\mathrm{Im}\,\Sigma_f(\omega^+,T)]^2}{V^2[\delta - \mathrm{Im}\,\Sigma_f(\omega^+,T)]}. \tag{178}$$

The singularity at $T = 0$ is avoided by taking the limit $\omega \to 0$ before $\delta \to 0$.

The transport coefficients of the periodic $SU(N)$ Anderson model in the FL regime are obtained by making the lowest order Sommerfeld expansion of Eq. (155), which gives

$$L_{mn} = \left[\omega^{m+n-2}\Lambda(\omega,T)\right]_{\omega=0} + \frac{\pi^2 k_B^2 T^2}{6}\left\{\frac{\partial^2}{\partial \omega^2}\left[\omega^{m+n-2}\Lambda(\omega)\right]\right\}_{T,\omega=0}.$$

The correlation function $\Lambda(\omega)$ is obtained by substituting for $\mathscr{N}_c(\omega)$ and $\tau(\omega,T)$ the expressions derived above. Neglecting the energy dependence of $\mathscr{N}_c^0(\omega)$ for $\omega \simeq 0$ we obtain for the electrical resistance of N parallel channels the result,

$$\rho(T) \simeq \frac{(1+C^2)^2}{(N-1)N}\frac{3\mathscr{V}a^2D^2}{5\hbar e^2 v_{k_F}^2 n_c^2}\left(\frac{3\gamma T}{\pi k_B}\right)^2, \tag{179}$$

where we approximated $N\mathcal{N}_c(0) \simeq n_c/\pi a D$ in Eq.(170) or (169) and used for $Im\ \Sigma_f(\omega,T)$ the expression Eq. (176). When $N/n_c \gg 1$ and $C \ll 1$, the FL law reads[52]

$$\rho(T) \simeq A\frac{\gamma^2 T^2}{(N-1)Nn_c^2}\left(\frac{3}{\pi k_B}\right)^2 = \frac{A\pi^2}{(N-1)N}\left(\frac{T}{T_K}\right)^2, \tag{180}$$

where $A = 3\mathcal{V}D^2/(5\hbar e^2 v_{k_F}^2)$.

The Seebeck coefficient is given by

$$\frac{\alpha(T)}{T} = -\frac{\pi^2}{3}\frac{k_B^2}{|e|}\left\{\frac{1}{\mathcal{N}_c(\omega)}\frac{\partial\mathcal{N}_c(\omega)}{\partial\omega} + \frac{1}{\tau(\omega)}\frac{\partial\tau(\omega)}{\partial\omega}\right\}_{\omega,T=0} \tag{181}$$

and we shall neglect the first term assuming $d[\ln\mathcal{N}_c(\omega)]/d\omega \ll d[\ln\tau(\omega)]d\omega$ for a slowly varying $\mathcal{N}_c^0(\omega)$. Equations (169), (173), (176), and (178) give $d[\ln\tau(\omega)]/d\omega|_{\omega=0} \simeq \mp 2a\pi N\mathcal{N}_f(0)/n_c Z_f$, such that the low-temperature Seebeck coefficient becomes [61, 62]

$$\alpha(T) = \pm\frac{2a\pi}{|e|}\frac{\gamma T}{n_c} = \pm\frac{2\pi^2}{3}\frac{k_B}{|e|}\frac{T}{T_K}. \tag{182}$$

The upper (lower) sign applies to Ce and Eu (Yb) systems and $T_K/k_B = a\pi\tilde{\omega}_f/n_c$ defines the lattice Kondo scale.

The Sommerfeld expansion also gives $L_{22}/L_{11} = \pi^2 k_B^2 T^2/3$ and $L_{12}/L_{11} = \mp 2a\pi\gamma T^2/n_c$, so that the thermal conductivity becomes

$$\kappa(T) = \mathcal{L}_0\, T\sigma(T)\left[1 - \frac{4\pi^2}{3}\left(\frac{T}{T_K}\right)^2\right], \tag{183}$$

The square bracket provides corrections to the WF law. In normal metals, T_K should be replaced by $E_F/k_B \gg T$ and the WF law holds. In correlated systems, T_K can be small and the deviations from the WF law can be large (although Eqn. (183) only holds for $T << T_K$).

The renormalization of the average velocity in Eq.(180) offers a possible explanation of the sharp and asymmetric maximum of the A coefficient which is observed in resistivity measurement of Ce intermetallics under pressure [63]. Consider, for example, the PAM with 3 doublets split by the CF. At low temperatures the scattering of c-electrons on two lowest doublet gives rise to the Kondo resonances which accommodate most of the f-charge, i.e., there are $n \simeq n_c + 1/2$ electrons in each of the resonant channels. The 4 excited f states give rise to simple scattering with constant phase shift. These non-resonant channels have $n \simeq n_c$ electrons; they contribute to the residual resistivity but not to the temperature dependent part of the resistivity.

The heavy mass is found in two resonant channels with $n = n_c + 1/2$ electrons, provided $\mathbf{k}_F$ is close to the edge of the BZ. But this makes $v_{K_F}^2$ quite small and gives rise to the T-dependent resistivity, $\rho(T) \simeq \rho_0 + \tilde{A}T^2$. In Ce compounds, heavy mass is seen in systems with large $\tilde{A}$. In these systems, pressure gives rise to a further increase of $\tilde{A}(P)$, which might be due to the reduction of $v_{k_F}^2$, because pressure increases hybridization and shifts the QP bands further apart. Since $\mathbf{k}_F$ is fixed by the Luttinger theorem, pressure reduces the dispersion and $v_{k_F}^2$. Of course, the reduction of n_f enhances T_K but for μ close to the band edge, the change in n_f cannot be large and T_K might not increase sufficiently to compensate the reduction of $v_{k_F}^2$. This might explain the initial increase of $\tilde{A}$ with pressure.

Above some critical pressure, the system becomes a valence fluctuator, i.e., the degeneracy of the f-state increases and the CF excitations vanish. All the f-channels are now equivalent and single f-electron is distributed over 6 channels. There are $n_c + 1/6$ electrons in each channel and $\mathbf{k}_F$ is shifted away from the edge of the BZ. The FS average of $v_{K_F}^2$ and the characteristic energy are both enhanced, which reduces the coefficient $\tilde{A}$, as seen in Wilhelm's data [47]. A further increase of pressure increases the characteristic temperature, without changing $v_{K_F}^2$, i.e., above the critical pressure A decreases. Thus, $\tilde{A}(P)$ exhibits an asymmetric maximum.

The high-temperature 'poor man's' solution

At elevated temperature, the FL laws breaks down and to obtain $\alpha(T)$ we need the full solution of the periodic Anderson model with the CF splittings and/or large degeneracy. Such a solution is not available at present and to estimate $\alpha(T)$ we use a 'poor man's' approach. We assume that the conduction electrons scatter incoherently on the $4f$ ions and calculate the transport relaxation time in the T-matrix approximation. We write $\Sigma_c(k,\omega^+) \simeq n_i T_{kk}(\omega^+)$, where $T_{kk}(\omega^+)$ is the single-ion scattering matrix on the real axis, and for the stoichiometric compounds set the concentration of $4f$ ions to $n_i = 1$. Since transport integrals are restricted to the Fermi window, we average $T_{kk}(\omega^+)$ over the FS and calculate $\tau(\omega)$ using Eq. (159). However, for a complete description, the crystal field (CF) splitting has to be taken into account.

In the case of a single scattering channel (no CF splitting) the vertex corrections to the T-matrix vanish by symmetry and the conduction electron's self energy in Eq. (159) is given by $\Sigma_c(\omega^+) = V^2 G_f(\omega^+)$, where $G_f(\omega^+)$ is the retarded Green's function of the effective N-fold degenerate single impurity Anderson model. When the degeneracy is lifted by the CF splitting the vertex corrections do not vanish but we neglect them anyway and use

$$\Sigma_c(\omega^+) = \sum_{\Gamma} V_{k_F\Gamma} G_\Gamma(\omega^+) V_{\Gamma k_F}, \tag{184}$$

where $V_{\Gamma k_F} = \langle \Gamma | V | \mathbf{k} \rangle$ is the FS average of the matrix element for the scattering between the c-electrons (in the $\mathbf{k}$-state) and the f-state (belonging to the irrep Γ) and $G_\Gamma(\omega^+)$ is the corresponding Green's function of the single impurity Anderson model with CF splitting.

As regards the validity of the 'poor man's' approach we point out that the DMFT+NRG solution[64] of the spin-1/2 Anderson lattice shows that the electrical resistance $\rho(T)$ increases rapidly and is very large at temperature T_K where $\alpha(T)$ has a maximum. For $T \geq T_K/2$, the thermopower of the lattice model[64] is very similar to the exact results[65] obtained for the spin-1/2 Anderson impurity. This indicates that the 'poor man's' approach can be used to describe the stoichiometric compounds at temperatures above $T_K/2$, provided the single impurity scattering is solved by methods which can deal with large Coulomb correlation and the CF splitting.

The experimental results on the heavy Fermion and valence fluctuators provide additional support for the 'poor man's' approach. The data show that the residual resistance of ternary and quaternary compounds like $EuCu_2(Ge_{1-x}Si_x)_2$, $CePt_{1-x}Ni_x$ and $YbIn_{1-x}Ag_xCu_4$ grows rapidly with x and for $0.3 \leq x \leq 0.8$ the mean free path is reduced by disorder to about a single lattice spacing. In these random alloys the electron propagation is incoherent even at $T = 0$ and the single impurity model should apply down to the lowest temperatures. At low doping and in stoichiometric compounds, the impurity description breaks down at low temperatures where ρ_0 is small. However, in these systems $\rho(T)$ and $\alpha(T)$ grow rapidly with increasing temperature (RT) and attain large maxima at T_K^ρ and T_K, respectively. In the case of degenerate f-states or small CF splitting, the data show[66, 67, 68] $T_K^\rho < T_K < RT$ and for $T \geq T_K^\rho/2$ there is not much difference between the stoichiometric compounds and doped systems, i.e., the impurity model applies. For large CF splitting, $\rho(T)$ has two maxima: a low-temperature one at T_K^ρ and a high-temperature one at T_ρ. The thermopower also exhibits two maximal: a low-temperature one at $T_K > T_K^\rho$ and a high-temperature one at T_S. For $T \geq T_K/2$, where the mean free path is short, the thermopower of periodic systems exhibits the same qualitative features as in random alloys, and can be explained in terms of impurity scattering. The experimental data show that the functional form of $\alpha(T)$ is strongly affected by pressure or chemical pressure. The fact that all the qualitative features of the pressure-induced variations of $\alpha(T)$ can be successfully explained by impurity scattering justifies, a posteriori, the 'poor man's' approach.

The thermoelectric properties of the single impurity Anderson model with the CF split f-states are obtained by the NCA, which is explained in detail in Refs. [69] and [70]. In the limit of large asymmetry and infinite f-f correlation, the Kondo temperature T_K, obtained from the low-energy peak of the spectral function[71], agrees with the exact result[72], $T_K = 3\gamma/\pi k_B$. $\gamma = (\pi^2 k_B^2/3)N\mathcal{N}_f(\mu)Z_f^{-1}$ is the impurity contribution to the specific heat coefficient, $\mathcal{N}_f(\mu) = (1/\pi\Gamma)\sin^2(\pi n_f/N)$ is the f-DOS at the Fermi level, and Z_f is the renormalization factor. For a structureless c-DOS, the NCA shows that T_K depends sensitively on the degeneracy of the f-state. A small T_K is found for small N and large Δ_{CF}, while a large T_K is found for large N and small Δ_{CF}. For a given N and Δ_{CF}, the Kondo scale is a monotonic function of n_f. It has a minimum at $n_f = 1$ and increases rapidly as n_f is reduced. In what follows, we discuss the behavior of $\alpha(T)$ assuming that pressure or doping increase the coupling constant g and reduce n_f but do not change Δ_{CF}.

We consider first a N-fold degenerate f-state and show typical NCA results[69, 70, 62] in Fig. 1, where thermopower is plotted as a function of temperature for several values of E_f. The calculations are performed for a half-filled, semielliptic c-band, no CF splitting (N=8), and a constant hybridization Γ. A decrease of E_f gives rise to an increase of g, which mimics the effect of pressure in Eu intermetallics. The thermopower is characterized by the Kondo maximum α_S at temperature $T_S \simeq T_K$. The high-temperature behavior depends strongly on the value of n_f. [Since n_f can be temperature-dependent we characterize the system by $n_f(T_K)$]. For $n_f \simeq 1$, the thermopower has a large high-temperature slope, changes sign at $T_0 > T_K$ and assumes large negative values above

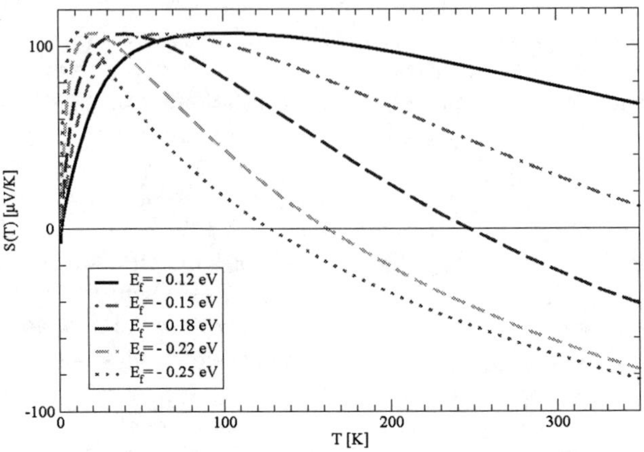

FIGURE 1. Thermopower of the single impurity Anderson model of a 8-fold degenerate *f*-state calculated by the NCA for fixed hybridization $\Gamma = 0.015$ eV is plotted as a function of temperature for several values of E_f, as indicated in the figure. The values of $n_f(T_K)$ are 0.76, 0,81, 0,.86 0.91, and 0.93 for $-E_f$ =0.12, 0.15, 0.18, 0.22, and 0.25, respectively.

T_0. The T_S and T_0 increase, and the high-temperature slope of $\alpha(T)$ decreases with decreasing n_f. For $n_f < 0.7$, we still find a shallow maximum of $\alpha(T)$ below RT but the high-temperature slope is very small and the sign-change does not occur. A similar behavior is obtained if the coupling constant g is reduced by increasing Γ. For smaller N, we find the same qualitative features but α_S and T_S are reduced; for $N = 2$ and $n_f \simeq 1$, the asymmetry of the model is much reduced and the Kondo maximum is almost completely suppressed.

The CF splitting leads to additional features which we explain by the example of an *f*-ion with two CF states separated by Δ_{CF}. The respective degeneracies of the ground and the excited state are M and M', where $M + M' = N$. The system now has two characteristic low-energy scales: the Kondo temperature T_K and a larger scale $T_K^N \gg T_K$ which comes into play[73] when the excited CF states become significantly populated at temperature $T_\Delta \simeq \Delta/2$. For $T \geq T_\Delta$, the thermopower can be approximated by the function $\alpha_N(T)$ which describes an effective N-fold degenerate *f*-state with Kondo temperature T_K^N and exhibits all the features discussed in the previous paragraph. For $T < T_\Delta$, the excited CF states are unoccupied and the properties are determined by the lowest CF state which is M-fold degenerate (typically, $M \ll N$). Thus, the thermoelectric response of a CF split *f*-level is described at low temperatures by an effective M-fold degenerate Anderson model with Kondo scale T_K. All other parameters being the same, the main difference between this effective model and a simple M-fold degenerate model with Kondo scale $T_K^M = \lim_{\Delta\to\infty} T_K$ is that $T_K \gg T_K^M$. The enhancement of T_K is due to the virtual transitions from the ground to the excited CF states. The function $\alpha_M(T)$ which approximates $\alpha(T)$ at low temperatures exhibits all the usual Kondo features. For $n_f \simeq 1$, it has a Kondo maximum at T_K and changes sign at $T_0 > T_K$; in the case of a doublet ground state the Kondo maximum is very small and $\alpha(T)$ is negative down to the lowest accessible temperatures. For $0.7 \leq n_f \leq 0.95$, the Kondo maximum is enhanced

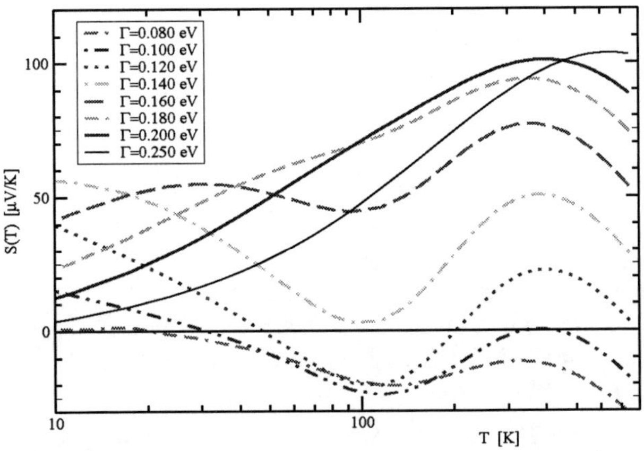

FIGURE 2. Thermopower of an f-ion with the ground state doublet and excited quartet calculated by the NCA for the CF splitting $\Delta = 0.07$ eV is plotted as a function of temperature for several values of the hybridization strength Γ, as indicated in the figure. The two bottom curves describe the type (a) Kondo system, the third curve from the bottom is type (b), the two middle curves are type (c) and the third one from the top describes the type (d) Kondo systems. The two upper curves are type (e) and describes valence fluctuators.

with respect to $n_f \simeq 1$, the high-temperature slope of $\alpha(T)$ is reduced, and the sign-change shifted to $T_0 \gg T_K$. For $n_f \leq 0.7$ the maximum of $\alpha(T)$ is further enhanced but the sign-change is absent. Of course, for $T \geq T_\Delta$ the excited CF states come into play and $\alpha_M(T)$ ceases to be physically relevant.

These effects are illustrated for a ground state doublet and an excited quartet in Fig. 2, where $\alpha(T)$ is plotted as a function of temperature for various values of $\Gamma(p)$. The CF spitting Δ_{CF} is the same for all the curves. An increase of pressure increases $\Gamma(p)$ and $g(p)$, and reduces n_f, which has a drastic effect on the functional form of $\alpha(T)$. We assume at ambient pressure $n_f \simeq 1$ and choose $\Gamma(0)$ and $g(0)$ such that $\alpha(T) < 0$ at RT. The corresponding NCA spectral function has well defined CF excitations[69, 70] which show that the low-energy scales satisfy $T_K \ll T_K^N \ll T_\Delta$. >From the previous discussion we expect $\alpha(T)$ with a very weak Kondo maximum at T_K, a sign-change at $T_0 > T_K$, and large negative values for $T \gg T_0$. Nothing particular happens at $T \simeq T_K^N$, where the excited CF states are still unoccupied. At T_Δ, the excited CF states become thermally populated and the functional form of $\alpha(T)$ changes from $\alpha_M(T)$ to $\alpha_N(T)$. Thus, a system with a low Kondo scale and large CF splitting has $\alpha(T)$ with a negative maximum around T_Δ and a deep negative minimum at low temperatures. The thermopower of that shape is illustrated by $\Gamma = 0.08$ and $\Gamma = 0.10$ curves in Fig. 2, classified[12, 70] as type (a).

If $\Gamma(p)$ and $g(p)$ increase, such that $0.8 \leq n_f \leq 0.95$, the NCA shows that the low-energy CF excitations are still resolved and $T_K(p) < T_K^N(p) < T_\Delta$. In this parameter range the Kondo maximum of $\alpha(T)$ is enhanced and shifted to higher temperatures, such that $\alpha_M(p,T) > \alpha_M(T)$ for $T \geq T_K(p)$. The values of T_K, $T_0(p)$ and $T_K^N(p)$ are much closer to T_Δ than at ambient pressure, and $T_0^N(p)$ is now above T_Δ. Thus, at the crossover we have $\alpha_M(p,T) < 0$ and $\alpha_N(p,T) > 0$. Since $\alpha_N(p,T_\Delta)$ is enhanced with respect to the $p = 0$ values, pressure enhances the Kondo maximum and brings it closer to the CF maximum. The temperature interval in which $\alpha(T) < 0$ shrinks with pressure, while T_S does not change. These features are demonstrated by the $\Gamma = 0.12$ curve in Fig. 2, which is classified as type (b).

In the pressure range such that $0.75 \leq n_f \leq 0.8$ the sign-change of $\alpha_M(T)$ is pushed to even higher temperatures and at large enough pressure (Γ) we have eventually $T_0(p) \geq T_\Delta$. The doublet-sextet crossover starts from $\alpha_M(T_\Delta) \geq 0$ and $\alpha(T)$ still exhibits two well resolved peaks but is always positive. These features are demonstrated by the $\Gamma = 0.14$ and $\Gamma = 0.16$ curves in Fig. 2, which are classified as type (c). A further increase of pressure gives $0.7 \leq n_f \leq 0.75$, which brings $T_K(p)$ so close to T_Δ that the Kondo and the CF peak cannot be resolved any more. The $\alpha(T)$ exhibits a single peak with a shoulder on the low-temperature side, as shown by the $\Gamma = 0.18$ curve in Fig. 2, which is classified as type (d). Note, as long as the low-energy CF excitations are well defined, the thermopower has a peak at temperature $T_S \simeq T_\Delta$ and the magnitude of this peak increases with pressure.

If $\Gamma(p)$ and $g(p)$ become very large and n_f drops below 0.7, the spectral function does not show the CF excitations any more. The thermopower acquires a single maximum at T_S which is unrelated to Δ_{CF}. The difference with respect to Kondo systems is that pressure shifts this thermopower peak to higher temperatures without changing its magnitude. Such behavior is typical of valence fluctuators and is demonstrated by the $\Gamma = 0.20$ and $\Gamma = 0.25$ curves in Fig. 2, which are classified as type (e). The curves in Fig. 1, which describe the system without the CF splitting such as Eu intermetallics, are also of this type. The Yb systems are characterized by an *f*-hole and the qualitative features of $\alpha(T)$ are obtained by reflecting ('mirror imaging') the curves in Figs. 1 or 2 on the temperature axis.

Falicov-Kimball model

The thermoelectric properties, the anomalous magnetic response of *f*-electrons and the metal-insulator transition of the conduction states that accompanies the valence-change transition in $YbInCu_4$-like intermetallic compounds are very well described by the spin-degenerate Falicov-Kimball model. The model considers a lattice of localized *f*-sites, which can be either occupied or empty, and conduction states which are delocalized via a nearest-neighbor hopping. The two types of electrons interact via a short-range Coulomb interaction and share a common chemical potential, which controls the total number of electrons $n = n_c + n_f$. The occupation of the *f*-states, which can be split into several CF levels, is restricted to $n_f < 1$. For a given total number of electrons, thermal fluctuations can change the average *f*-occupation by transferring electrons or holes from the conduction band to the *f*-states and *vice versa*. The interaction has to be large enough to open a gap in the conduction band. We also assume that at low temperatures μ is within the lower (or upper) Hubbard band, so that the ground state is metallic. Since the model neglects quantum fluctuations, the ground state has no *f*-holes and the conduction electrons are essentially free.

The Hamiltonian of the spin-one-half Falicov-Kimball model can be written as[74]

$$H = -\frac{t^*}{2\sqrt{d}} \sum_{\langle i,j\rangle\sigma} c_{i\sigma}^\dagger c_{j\sigma} + E_f \sum_i w_i + U \sum_{i\sigma} w_i c_{i\sigma}^\dagger c_{i\sigma}, \tag{185}$$

where $c_{i\sigma}^\dagger$ ($c_{i\sigma}$) is the electron creation (annihilation) operator for an electron at site i with spin σ, E_f is the energy level of the localized electrons, w_i is a variable that equals zero or one and corresponds to the localized electron number, and U is the interaction strength. The hopping integral is scaled with the spatial dimension d so as to have a finite result in the limit [75] $d \to \infty$; we measure all energies in units of $t^* = 1$. We work on a hypercubic lattice where the noninteracting density of states is a Gaussian $\rho(\varepsilon) = \exp(-\varepsilon^2)/\sqrt{\pi}$.

The Falicov-Kimball model can be solved exactly by employing dynamical mean field theory [76, 77, 78]. Because the self energy $\Sigma(z)$ is local, the local Green's function satisfies

$$G(z) = \int d\varepsilon \rho(\varepsilon) \frac{1}{z + \mu - \Sigma(z) - \varepsilon}, \tag{186}$$

with z anywhere in the complex plane (we suppress the spin index here). The self energy, local Green's function, and effective medium G_0 are related by

$$G_0^{-1}(z) - G^{-1}(z) = \Sigma(z), \tag{187}$$

and the Green's function also satisfies

$$G(z) = (1 - w_1) G_0(z) + w_1 \frac{1}{G_0^{-1}(z) - U}. \tag{188}$$

Here w_1 is the average concentration of localized electrons,

$$w_1 = 2\exp[-\beta(E_f - \mu)] Z_\uparrow(\mu - U) Z_\downarrow(\mu - U)/Z, \tag{189}$$

with $Z = Z_\uparrow(\mu) Z_\downarrow(\mu) + 2\exp[-\beta(E_f - \mu)] Z_\uparrow(\mu - U) Z_\downarrow(\mu - U)$ and

$$Z_\sigma(\mu) = 2e^{\beta\mu/2} \prod_n \frac{i\omega_n + \mu - \lambda_\sigma(i\omega_n)}{i\omega_n}. \tag{190}$$

The factor of two arises from the spin degeneracy of the f-electrons and the constraint that no more than one f-electron is allowed on any site. The symbol $\lambda_\sigma(z)$ is defined from the effective medium via $\lambda_\sigma(i\omega_n) = i\omega_n + \mu - G_{0\sigma}^{-1}(i\omega_n)$, $\omega_n = \pi T(2n+1)$ is the fermionic Matsubara frequency, and $\beta = 1/T$. The algorithm for determining the Green's function begins with the self energy set equal to zero. Then Eq. (186) is used to find the local Green's function. The effective medium is found from Eq. (187) and the localized electron filling from Eq. (189). The new local Green's function is then found from Eq. (188) and the new self energy from Eq. (187). This algorithm is repeated until it converges. The transport coefficients are obtained in the limit of infinite dimensions by substituting the exact conduction electron Green's function into Eq. (156) and evaluating the integrals for transport coefficients numerically[79, 78].

The properties of the model exhibit all the features seen in the high-temperature phase of the intermetllic with the valence-change transition. The calculations show that the degenerate f-states become fractionally occupied at finite temperature and that the additional paramagnetic entropy of these excited states competes for the free energy with the excitation energy, the kinetic energy of the conduction electrons, and the interaction energy[78]. This gives rise to a valence transition at a temperature T_V, such that a substantial number of electrons (in Eu compounds) or holes (in Yb compounds) are transferred from the conduction band to the $4f$ ions. The onset of the $4f$ paramagnetism is accompanied by the reconstruction of the interacting density of conduction states and the shift of μ into the gap. Below T_V the entropy of the system is given by the entropy of the conduction states. Above T_V the entropy is dominated by the contribution of the localized, paramagnetic states. At intermediate temperatures the behavior can be quite complex[79, 78], because both the degeneracy of the f-states and the number of charge carriers change at T_V. The large reduction of the $\alpha/\mathscr{S}$ ratio at T_V is an indication of the reduction of Fermi volume.

As regards the thermal transport, the calculations show[79, 78] that the electrical resistance of the paramagnetic phase is large and has a maximum at a temperature $T^* \gg T_V$, which is of the order of the gap, or the pseudo-gap in the density of states. The thermopower obtained by the DMFT is weakly temperature dependent and its sign depends on the band filling. The maximum of $\alpha(T)$ is also at T^*. The dc-conductivity, thermopower, and electronic contribution to the thermal conductivity, obtained by the DMFT solution for $U = 2$ and five different w_1 values are shown in Fig. 3 (for details see [78]). For these parameters, there is always a region of exponentially small DOS near the chemical potential at low temperature [but in this region $\tau(\omega)$ decreases only as a power law]. As the localized electron concentration w_1 moves away from 0.5, the high-temperature thermopower increases in magnitude due to the asymmetry in the

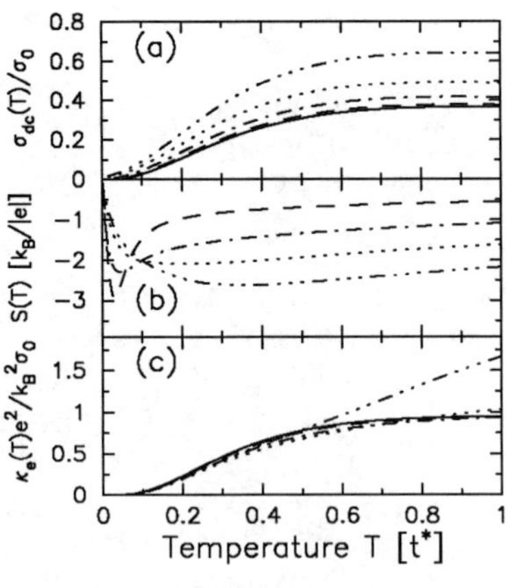

FIGURE 3. (a) DC conductivity, (b) thermopower, and (c) electronic contribution to the thermal conductivity, for the spinless FK model with $\rho_e = 1 - w_1$ and $U = 2$. Five fillings are shown: (i) $w_1 = 0.5$ (solid line); (ii) $w_1 = 0.4$ (dashed line); (iii) $w_1 = 0.3$ (chain-dotted); (iv) $w_1 = 0.2$ (dotted); and (v) $w_1 = 0.1$ (chain-triple-dotted). From Freericks and Zlatic, 2003 [78]

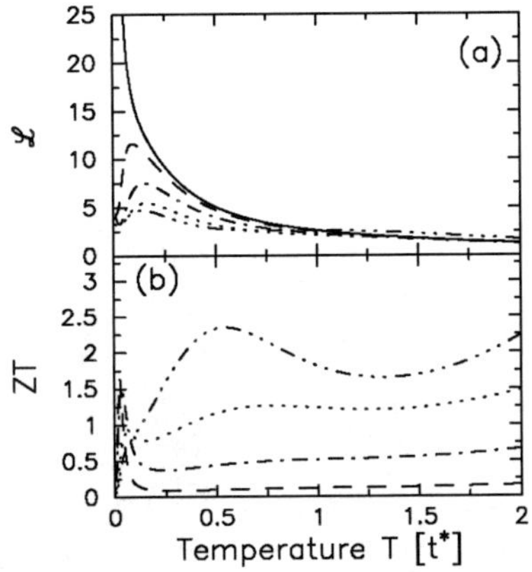

FIGURE 4. (a) Lorenz number $\mathscr{L}k_B^2/e^2 = \kappa_{el}/\sigma_{dc}T$ and (b) electronic thermoelectric figure of merit $ZT = T\sigma_{dc}S^2/\kappa_{el}$ for the spinless FK model with $\rho_e = 1 - w_1$ and $U = 2$. Five fillings are shown: (i) $w_1 = 0.5$ (solid line); (ii) $w_1 = 0.4$ (dashed line); (iii) $w_1 = 0.3$ (chain-dotted); (iv) $w_1 = 0.2$ (dotted); and (v) $w_1 = 0.1$ (chain-triple-dotted). From Freericks and Zlatic, 2003 [78]

DOS (it must vanish at 0.5 due to particle-hole symmetry) and the low-temperature thermopower shows a sharp peak for fillings close to half filling (the sign is hole-like, because the DOS from the lower Hubbard band dominates the transport coefficients at low temperature); the dc conductivity and thermal conductivity both vanish at low T due to the "gap" as well. The thermoelectric figure-of-merit $ZT = T\sigma_{dc}S^2/\kappa_{el}$ is plotted in Fig. 4—we find it is larger than one at high T for $w_1 < 0.22$, and for fillings close to half filling, there is a low-temperature spike in ZT that can become larger than one over a narrow temperature range. The spike at low T is due to the large peak in S and the small thermal conductivity; but the phonon contribution to the thermal conductivity can sharply reduce ZT if the phonon thermal conductivity is much larger than the electronic thermal conductivity (this all electronic calculation provides only an upper bound to ZT). The Lorenz number is also plotted in Fig. 4. It gets huge at half filling, but becomes more metallic ($\approx \pi^2/3$) as the filling moves further away from half filling. It is not a constant even at low temperature because the system is not a Fermi liquid.

The above discussion shows that the DMFT calculations for a parameter set which yields the valence change transition at $T_V = 50$ K and opens a pseudo-gap of the order of $T^* \simeq 500$ K, explains the main features of the magnetic susceptibility, the XPS data, and the optical conductivity of the stoichiometric compound $YbInCu_4$ at temperatures above T_V. The calculated thermopower[79] is of the order of a few $\mu V/K$, and its sign is either positive or negative, depending on the band filling and the shape of the conduction band. The proximity of μ to the pseudo-gap would lead (in a non-interacting system) to a shallow minimum of $\alpha(T)$ at a temperature of the order of T^*, but in an interacting system, the valence-change transition destabilizes the semiconducting phase, and gives rise to a discontinuity of $\alpha(T)$ at T_V. The low-temperature FL state has a large characteristic temperature and $|\alpha(T)|$ is a linearly increasing function of temperature. Thus, a cusp or even a discontinuity appears in $\alpha(T)$ at T_V, in good agreement with the experimental data.

ACKNOWLEDGMENTS

I am pleased to acknowledge many useful discussions with R. Monnier and J.K. Freericks. They have contributed substantially to the research activity I have presented in these lecture notes.

REFERENCES

1. H. B. Callen, *Phys. Rev.* **73**, 1349 (1949).
2. J. M. Luttinger, *Phys. Rev.* **135** (1964).
3. C. A. Domenicali, *Rev.Mod.Phys.* **26**, 237 (1954).
4. G. D. Mahan, *Many-Particle Physics*, Plenum, New York, 1981.
5. L. D. L. E. M. Lifshitz, and L. P. Pitaevskii, *Electrodynamics of Continuous Media*, Elsevier Butterworth-Heinemann, Oxford, 1984.
6. A. F. Ioffe, *Semiconductor Thermoelements and Thermoelectric Cooling*, Infosearch, London, 1954.
7. M. Jonson, and G. D. Mahan, *Phys. Rev. B* **21**, 4223 (1980).
8. M. Jonson, and G. D. Mahan, *Phys. Rev. B* **42**, 9350 (1990).
9. G. D. Mahan, *G. D. Mahan*, Academic Press, San Diego, 1998, vol. 51 of *Solid State Physics*, p. 81.
10. J. K. Freericks, V. Zlatić, and A. M. Shvaika, *Phys. Rev. B* **75**, 0135133 (2007).
11. J. Sakurai, H. Takagi, S. Taniguchi, T. Kuwal, Y. Isikawa, and J.-L. Tholance, *J. Phys. Soc. Japan Suppl. B* **65**, 49 (1996).
12. P. Link, D. Jaccard, and P. Lejay, *Physica B* **225** (1996).
13. V. Zlatić, I. Milat, B. Coqblin, G. Czycholl, and C. Grenzbach, *Phys. Rev.* (2003).
14. G. Sparn, W. Lieka, U. Gottwick, F. Steglich, and N. Grewe, *J. Magn. Magn. Mater.* **47** (1985).
15. D. Jaccard, J. M. Mignot, B. Bellarbi, A. Benoit, H. F. Braun, and J. Sierro, *J. Magn. Magn. Mater.* (1985).
16. D. Huo, T. Kuwai, T. Mizushima, Y. Isikawa, and J. Sakurai, *Physica B* (2002).
17. P. B. van Aken, H. J. van Daal, and K. H. J. Buschow, *Phys. Lett. A* **49**, 201 (1974).
18. R. Cibin, D. Jaccard, and J. Sierro, *J. Magn. Magn. Mater.* **108**, 107 (1992).
19. D. Huo, K. Mori, T. Kuwai, S. Fukuda, Y. Isikawa, and J. Sakurai, *Physica B* (2000).
20. J. Sakurai, D. Huo, D. Kato, T. Kuwai, Y. Isikawa, and K. Mori, *Physica B* (2000).
21. H. Wilhelm, and D. Jaccard, *Phys. Rev.* **69** (2004).
22. K. Behnia, D. Jaccard, and J. Flouquet, *J. Phys.: Condens. Matter* (2004).
23. F. Steglich, *Festkoeorperprobleme* **17**, 319 (1977).
24. Y. Onuki, and T. Komatsubara, *J. Magn. Magn. Mater.* **63** (1987).
25. J. Sakurai, J. C. G. Sal, and J. R. Fernandez, *J. Magn. Magn. Mater* **140** (1995).
26. A. Amato, D. Jaccard, J. Sierro, P.Haen, P. Lejay, and J. Flouquet, *J. Low. temp. Phys.* **77**, 195 (1989).
27. J. Sakurai, T. Ohyama, and Y. Komura, *J. Magn. Magn. Mater.* **47** (1985).
28. E. Bauer, *Adv. Phys* **40**, 417 (1991).
29. D. Jaccard, J. M. Mignot, A. B. B. Bellarbi, H. F. Braun, and J. Sierro, *J. Magn. Magn. Mater.* **47** (1985).
30. C. Fierz, D. Jaccard, and J. Sierro, *J. Appl. Phys.* **63**, 3899 (1988).
31. D. Jaccard, K. Behnia, and J. Sierro, *Phys. Lett. A* **163**, 475 (1992).
32. Y. Bando, J. Sakurai, and E. V. Sampathkumaran, *Physica B* **186** (1993).
33. E. Gratz, E. Bauer, R. Hauser, N. Pillmayr, G. Hilscher, H. Mueller, and B. Barbara, *J. Magn. Magn. Mater.* (1988).
34. J. Sakurai, H. Kamimura, and Y. Komura, *J. Magn. Magn. Mater.* (1988).
35. D. Huo, K. Mori, T. Kuwai, H. Kondo, Y. Isikawa, and J. Sakurai, *J. Phys. Soc. Japan* **68**, 3377 (1999).
36. M. Očko, C. Geibel, and F. Steglich, *Phys. Rev.* **64**, 195107 (2001).

37. P. Link, D. Jaccard, and P. Lejay, *Physica B* **225** (1996).
38. P. Link, D. Jaccard, , and P. Lejay, *Physica B* **223** (1996).
39. K. Alami-Yadri, D. Jaccard, and D. Andreica, *J. Low Temp. Phys.* **114**, 135 (1999).
40. G. Nakamoto, T. Nobata, S. Ueda, Y. Nakajima, T. Fujioka, and M. Kurisu, *Physica B* **259** (1999).
41. R. Casanova, D. Jaccard, C. Marcenat, N. Hamdaoui, and M. J. Besnus, *J. Magn. Magn. Mater.* **90** (1990).
42. O. Trovarelli, C. Geibel, B. Buschinger, R. Borth, S. Mederle, M. Grosche, G. Sparn, and F. Steglich, *Phys. Rev.* **60**, 1136 (1999).
43. D. Andreica, K. Alami-Yadri, D. Jaccard, A. Amato, and A. Schenk, *Physica B* **259** (1999).
44. H. J. van Daal, P. B. van Aken, and K. H. J. Buschow, *Phys. Lett. A* **49**, 246 (1974).
45. I. Aviani, M. Miljak, V. Zlatic, K.-D. Schotte, C. Geibel, and F. Steglich, *Phys. Rev.* **64**, 184438 (2001).
46. H. Wilhelm, and D. Jaccard, *Phys. Rev.* **66**, 064428 (2002).
47. H. Wilhelm, K. Alami-Yadri, B. Revaz, and D. Jaccard, *Phys. Rev. B* **69**, 3651 (1999).
48. A. Demuer, A. T. Holmes, and D. Jaccard, *J. Phys.: Condens. Matter* **14** (2002).
49. H. Wilhelm, D. Jaccard, V. Zlatić, R. Monnier, B. Delley, and B. Coqblin, *J. Phys.: Condens. Matter* **17**, 823 (2005).
50. J. Sakurai, and Y. Isikawa, *J. Phys. Soc. Japan* **74**, 1926 (2005).
51. K. Kadowaki, and S. B. Woods, *Solid State Commun.* **71**, 1149 (1987).
52. H. Kontani, *J. Phys. Soc. Japan* **73**, 515 (2004).
53. N. Tsujii, H. Kontani, and K. Yoshimura, *Phys. Rev. Lett.* **94**, 057201 (2005).
54. A. L. Cornelius, J. M. Lawrence, , J. Sarrao, Z. Fisk, M. F. Hundley, G. H. Kwei, J. D. Thompson, C. H. Booth, and F. Bridges, *Phys. Rev. B* **56**, 7993 (1997).
55. M. Očko, J. L. Sarrao, I. Aviani, D. Drobac, I. Ž ivković, and M. Prester, *Phys. Rev. B* **51**, 975102 (2003).
56. J. L. Sarrao, *Physica B* **259 & 261**, 129 (1999).
57. K. Yamada, and K. Yosida, *Prog. Theor. Phys.* **76**, 681 (1986).
58. K. Yamada, *Electron Correlation in Metals*, Cambridge University Press, Cambridge, 2004.
59. N. Vidhyadhiraja, and D. Logan, *European Phys. Jour. B* **39**, 313 (2004).
60. D. E. Logan, and N. S. Vidhyadhiraja, *J. Phys: Cond. Matter* **17**, 2935 (2005).
61. K. Miyake, and H. Kohno, *J. Phys. Soc. Japan* **74**, 254 (2005).
62. V. Zlatić, R. Monnier, J. Freericks, and K. W. Becker, *Phys. Rev. B* **76**, 085122 (2007).
63. H. Wilhelm, D.Jaccard, V. Zlatic, R. Monnier, B. Delley, and B. Coqblin, *J. Phys: Cond. Matter* **17**, S830 (2005).
64. C. Grenzebach, F. B. Anders, G. Czycholl, and T. Pruschke, *Phys. Rev. B* **74**, 195119 (2006).
65. T. Costi, A. Hewson, and V. Zlatić, *J. Phys: Cond. Matter* **6**, 2519 (1994).
66. M. Kim, H. T. Y. Nakai, M. Sera, T. T. F. Iga, and S. Kunii, *J. Phys. Soc. Japan* **75**, 064704 (2006).
67. S. Kobayashi, Y. Yoshino, S. Tsuji, M. Sera, and F. Iga, *J. Phys. Soc. Japan* **72**, 25 (2002).
68. Z. Hossain, C. Geibel, N. Senthilkumaran, M. Deppe, M. Baenitz, F. Schiller, and S. L. Molodtsov, *Phys. Rev. B* **69**, 014422 (2004).
69. N. E. Bickers, D. L. Cox, and J. Wilkins, *Phys. Rev. B* **36**, 2036 (1987).
70. V. Zlatić, and R. Monnier, *Phys. Rev. B* **71**, 165109 (2005).
71. B. Horvatić, and V. Zlatić, *Phys. Rev. B* **30**, 6717–6731 (1984).
72. A. Hewson, *The Kondo Problem to Heavy Fermions*, Cambridge University Press, Cambridge, 1993.
73. K. Hanzawa, K. Yamada, , and K. Yosida, *J. Magn. Magn. Mater.* **47 & 48**, 357 (1985).
74. L. M. Falicov, and J. C. Kimball, *Phys. Rev. Lett.* **22**, 997 (1969).
75. W. Metzner, and D. Vollhardt, *Phys. Rev. Lett.* **62**, 324 (1989).
76. U. Brandt, and C. Mielsch, *Z. Phys. B* **75**, 365 (1989).
77. U. Brandt, and C. Mielsch, *Z. Phys. B* **79**, 295 (1990).
78. J. K. Freericks, and V. Zlatić, *Rev. Mod. Phys.* **75**, 1333 (2003).
79. J. K. Freericks, and V. Zlatić, *Phys. Rev. B* **64**, 245118 (2001).

PART II

PARTICIPANT CONTRIBUTIONS

THE STATISTICAL EQUIVALENTS OF THE $t-U$ AND $t-t'-U$ MODELS

G. E. Akpojotor

Physics Department, Delta State University, Abraka, Nigeria

Abstract. The Hubbard (t-U) model is the simplest model for studying strongly correlated systems.The t-part of this model is the kinetic part and the U-part is the interacting potential. In order to enhance the itinerancy of the kinetic part, some workers often include the hopping of the electrons to the next nearest neighbour (NNN) leading to the $t-t'-U$ model. In this analytical study, we have developed from lattice diagrams, statistical distribution operators which simply by adding and subtracting one or two appropriately to sites in the basis states yield the same results as the kinetic part of the Hubbard Hamiltonian. By introducing the interacting Coulomb potential, we obtained the same ground state energies and wavefunctions for lattices in one dimension (1D), 2D and 3D as with the variational t-U and $t-t'-U$ models. The results and the implication of this equivalence are discussed with respect to magnetism and superconductivity

PACS: 71.10.Fd, 05.50.+q, 31.15.xt

INTRODUCTION

Strong electron correlated systems exhibits some of the most intriguing phenomena in condensed matter physics [1, 2]. One basic feature common to all these materials is that they are made of electrons that are neither fully itinerant nor fully localized on their atomic sites [3]. Consequently, neither band theory [4] which is successful for many itinerant systems nor localization theory [5] which is used to study systems with localized carriers has been successful for these materials [6]. The Hubbard model proposed in 1963 [7] is universally considered the simplest minimal description of the strongly correlated systems as it has a kinetic part that is hoped to account for the itinerancy and a Coulombic interaction to represent the localization of the electrons. Inspite of this simplicity, the properties of the Hubbard model have been well determined only in the one dimension (d = 1) limit where an exact Bethe ansatz solution was obtained in 1968 [8] and in the infinite dimension ($d=\infty$) limit where an exact solution has been obtained by Dynamical Mean Field theory (DMFT) in 1989[8,10]. Thus there is no general consensus for the properties at finite dimensions which is the realm for the real materials. It is pertinent to remark that the model has been investigated with all available theoretical tool that is hoped to be capable of extracting relevant information from the model, as can be found in its rich and vast literature. The obvious implication is that the lack of consensus understanding of the Hubbard model in the finite dimension may be due to the limitations within the various theoretical methods or we have not develop the appropriate tool. So even as the various experts continue to seek means to remove the limitations in their methods or develop new ones till we get to our target, it is necessary to give an insight into the itinerant and localized behaviour of the Hubbard systems for d

CP1014, *Lectures on the Physics of Strongly Correlated Systems XII: Twelfth Training Course,*
edited by A. Avella and F. Mancini

= 1, 2, 3 using a pedagogical tool. This is the purpose of this paper and it becomes even more imperative in a training course on strongly correlated systems for young scientists as in the present forum.
The pedagogical method to be used is a highly simplified correlated variational approach (HSCVA). In this analytical method, we develop very simple formulars from periodic lattice diagrams (d = 1, 2, 3) to statistically operate the hopping of the electrons and then show that these formulars are equivalent to the activation of these lattices with the kinetic parts of the Hubbard ($t-t'-U$) Hamiltonian. The Coulombic interaction is then added to the statistical operators and the resulting Hamiltonian is equivalent the $t-t'-U$ Hamiltonian. This equivalence enables us to obtain a highly simplified formulation of the variational ground state energies and wavefunctions for lattices in d = 1, 2, 3 as direct products of the competition between the kinetic part and the Coulombic part of the Hubbard model. Thus we could gain an insight into the electronic behaviour at various regimes of this competition for all three dimensions.
The plan of this paper is the following. In section II, we will develop the statistical equivalents of the t-U and $t-t'-U$ models and then use them to obtain the ground state energies and wavefunctions for some arbitrary-sized lattices in d = 1, 2, 3, using the HSCVA. This will be followed by the discussion of the results in section III. Before concluding, we will discuss how to extend our simple approach to include other interaction terms that will enable us also peep further into strongly correlated systems in these extended Hubbard model. Further, we will also highlight how to extend the HSCVA to the few particles problem [11]

OBTAINING THE GROUND STATE ENERGY AND WAVEFUNCTIOM

Formulating the Statistical Operators

To enhance the understanding of the study, it will be appropriate to study the two electrons, 4 sites (N = 4) problem which is a 1D system before moving to larger lattices and higher dimensions. The electronic state is $|i\sigma, j\bar{\sigma}>$, where i and j represent the site positions of the particles [12] with i = j for onsite states and $i \neq j$ for intersite states, while the σ and $\bar{\sigma}$ represent the spins defined by $\sigma(\bar{\sigma}) = \uparrow(\downarrow), \downarrow(\uparrow)$.
If we consider the separation, L, between the sites to be in terms of single lattice spacing in Figure 1, that is, $L_a = |i-j|a$, the electronic singlet states and their separations are respectively $L_a = 0$ for onsite, $L_a = a$ for nearest neighbour (NN) sites and $L_a = 2a$ for next nearest neighbour (NNN) sites:
$|1\uparrow,1\downarrow>, |2\uparrow,2\downarrow>, |3\uparrow,3\downarrow>, |4\uparrow,4\downarrow>; |1\uparrow,2\downarrow>, |1\downarrow,2\uparrow>, |2\uparrow,3\downarrow>, |2\downarrow,3\uparrow>, |3\uparrow,4\downarrow>, |3\downarrow,4\uparrow>, |1\uparrow,4\downarrow>, |1\downarrow,4\uparrow>; |1\uparrow,3\downarrow>, |1\downarrow,3\uparrow>, |2\uparrow,4\downarrow>, |2\downarrow,4\uparrow>$.
Thus the total classes of possible separations, S, for the 16 states is 3 which can be related to the various separations in a compact form (with the a = 1) as

$$L_c = 0, 1, 2, ..., S-1 \quad (1)$$

where the S has been obtained from our study of lattices in all three dimensions as S =1/2 (N + 2) for even N sites, S = 1/2 (N + 1) for odd N sites, S = 1/8 (N + 4)(N + 2) for even N x N sites, S = 1/8 (N + 3)(N + 1) for odd N x N sites, S = 1/48 (N + 6)(N + 4)(N + 2) for even N x N x N sites and S = 1/48 (N + 5)(N + 3)(N + 1) for odd N x N x N sites.

Now, if we introduce a statistical distribution operator called the hopperer, h_t, which can

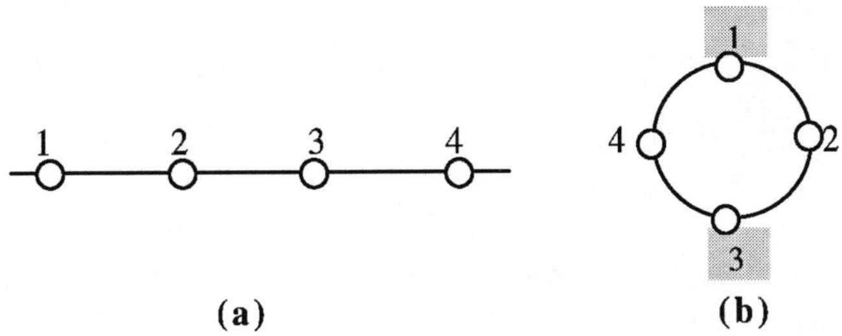

FIGURE 1. The 1D lattice with four sites ($N = 4$): (a) linear lattice so that there are edge effects hence have no periodic boundary conditions, and (b) ring lattice so that there is no edge effect hence have periodic boundaries.

hop any of the particle to its NN sites simply by subtracting one and adding one, with the condition that the Pauli exclusion principle must be obeyed, observe from Figure 1 that the new states will be for state $|2\uparrow,2\downarrow>$ say; $|1\uparrow,2\downarrow>, |3\uparrow,2\downarrow>, |2\uparrow,1\downarrow>, |2\uparrow,3\downarrow>$ It follows then that for any of the particle states of a given lattice denoted by the basis state,$|i\sigma, j\bar{\sigma}>$ say,

$$h_t|i\sigma, j\bar{\sigma}> = -t(\pm 1|i\sigma, j\bar{\sigma}> \\ = -t[|(i+1)\sigma, j\bar{\sigma}>, |(i-1)\sigma, j\bar{\sigma}>, |i\sigma, (j+1)\bar{\sigma}>, |i\sigma, (j-1)\bar{\sigma}> \quad (2)$$

where -t is the NN hopping amplitude.

Similarly, it can be shown for 2D that for the basis state,$|ii\sigma, jk\bar{\sigma}>$ say,

$$h_t|ii\sigma, jk\bar{\sigma}> = -t[|(i\pm 1)i\sigma, jk\bar{\sigma}>, |i(i\pm 1)\sigma, jk\bar{\sigma}>, \\ |ii\sigma, (j\pm 1)k\bar{\sigma}>, |ii\sigma, j(k\pm 1)\bar{\sigma}>, \quad (3)$$

And for 3D in which a basis state is denoted by $|iii\sigma, jkl\bar{\sigma}| >$ say, that

$$h_t|iii\sigma, jkl\bar{\sigma}> = -t[|(i\pm 1)ii\sigma, jkl\bar{\sigma}>, |i(i\pm 1)i\sigma, jkl\bar{\sigma}>, \\ |ii(i\pm 1)\sigma, jkl\bar{\sigma}>, |iii\sigma, (j\pm 1)kl\bar{\sigma}>, |iii\sigma, j(k\pm 1)l\bar{\sigma}>, |iii\sigma, jk(l\pm 1)\bar{\sigma}> (4)$$

Following the same procedure, we can define a statistical operator that hops the electrons to the NNN in 1D lattices for $|i\sigma, j\bar{\sigma}>$ say, as

$$h'_t|i\sigma, j\bar{\sigma}> = t'[|(i\pm 2)\sigma, j\bar{\sigma}>, |i\sigma, (j\pm 2)\bar{\sigma}>, \quad (5)$$

where t' indicates NNN hopping amplitude.

The cases of 2D and 3D lattices, however, cannot be achieved merely by adding and subtracting 2 since we have more than one direction. Instead, we add 1 to each of any two directions, subtract 1 from each of any two directions and then add and subtract 1 alternatively from each of any two directions. Thus statistically, the NNN distribution of any two electrons in any 2D lattice will be for any basis state, $|ii\sigma, jk\bar{\sigma}>$ say,

$$h'_t|ii\sigma, jk\bar{\sigma}> = t'[|(i\pm1)(i\pm1)\sigma, jk\bar{\sigma}>, |(i\pm1)(i\mp1)\sigma, jk\bar{\sigma}>, \\ |ii\sigma, (j\pm1)(k\pm1)\bar{\sigma}>, |ii\sigma, (j\pm1)(k\mp1)\bar{\sigma}>], \tag{6}$$

And for 3D lattices, it will be for the basis state, $|iii\sigma, jkl\bar{\sigma}>$ say,

$$h'_t|iii\sigma, jkl\bar{\sigma}> = t'[|(i\pm1)(i\pm1)i\sigma, jkl\bar{\sigma}>, |(i\pm1)(i\mp1)i\sigma, jkl\bar{\sigma}>, \\ |(i\pm1)i(i\pm1)\sigma, jkl\bar{\sigma}>, |(i\pm1)i(i\mp1)\sigma, jkl\bar{\sigma}>, |i(i\pm1)(i\pm1)\sigma, jkl\bar{\sigma}>, \\ |i(i\pm1)(i\mp1)\sigma, jkl\bar{\sigma}>, |iii\sigma, (j\pm1)(k\pm1)l\bar{\sigma}>, |iii\sigma, (j\pm1)(k\mp1)l\bar{\sigma}>, \\ |iii\sigma, (j\pm1)k(l\pm1)\bar{\sigma}>, |iii\sigma, (j\pm1)k(l\mp1)\bar{\sigma}>, |iii\sigma, j(l\pm1)(k\pm1)\bar{\sigma}>, \\ |iii\sigma, j(l\pm1)(k\mp1)\bar{\sigma}>], \tag{7}$$

The Hubbard model

The Hubbard $(t - t' - U)$ model is given by

$$H_H = H_t + H_{t'} + H_U, \tag{8}$$

with

$$H_t = -t\left[\sum_{<i,j>,\sigma} C^+_{i\sigma}C_{j\sigma} + H.C\right], \quad H_{t'} = t'\left[\sum_{<<i,j>>,\sigma} C^+_{i\sigma}C_{j\sigma} + H.C\right]$$

and

$$H_U = U\Sigma_i n_i \uparrow n_i \downarrow,$$

where $C^+_{i\sigma}(C_{j\sigma})$ and n_i are respectively the creation (annihilation) and number operators in the Wannier state on the ith(jth) site, H.C. is the Hermitian conjugate and U is the onsite Coulomb potential.

In Eq.(8), the kinetic parts also known as the kinetic-energy operators, H_t and $H_{t'}$ [1], of the Hubbard $(t - t' - U)$ Hamiltonian actually deals with the distribution of the electrons in a given system controlled only by the Pauli exclusion principle when decoupled from the interaction part. The H_t is expanded for the N = 4 problem and then operated on its basis states. The emerging new states were the same with those from our statistical operator, that is for any of the basis states considered so far, $|...\sigma, ...\bar{\sigma}>$ say, $H_t|...\sigma, ...\bar{\sigma}> = h_t|...\sigma, ...\bar{\sigma}>$. This equivalence is valid in all the three dimensions. Similarly we can show that $H_{t'}|...\sigma, ...\bar{\sigma}> = h_{t'}|...\sigma, ...\bar{\sigma}>$.

The implication here is that we can achieve the hopping of the Hubbard electrons both for the NN and NNN simply by adding and subtracting one or two appropriately to the positions of sites in the basis states. Thereafter we can introduce the Coulombic interaction to model the localization of the electrons and thereby kick starting the competition between itinerancy and localization. For consistency, we couple the Coulombic interaction part, H_U of the Hubbard Hamiltonian to our statistical operators to obtain the full Hamiltonian of the statistical model expressed in compact form as

$$H_S = h_{ij} + U\Sigma_i n_i \uparrow n_i \downarrow, \tag{9}$$

where as usually defined, $t_{ij} = -t$ for NN hopping and $t_{ij} = t'$ for NNN hopping. Eq. (13) is the equivalent of the Hubbard $(t - t' - U)$ Hamiltonian compactly written as

$$H_H = t_{ij}\left[\sum_{<i,j>,\sigma} C^{+}_{i\sigma} C_{j\sigma} + H.C\right] + U\Sigma_i n_i \uparrow n_i \downarrow, \tag{10}$$

The ground state energy and wavefunction

The variational ground state energy is given by

$$E_g = \frac{\langle\psi|H|\psi\rangle}{\langle\psi|\psi\rangle} \tag{11}$$

where the H is the model Hamiltonian.
The ket in the Hilbert space is the trial wave function patterned after an exact case[13] and is defined as

$$|\psi\rangle = \sum_{L_c=0}^{S-1} X_{Lc}|\psi_{Lc}\rangle \tag{12}$$

where the X is the variational parameter, the L_c and S retain their earlier meanings while the $|\psi_{Lc}\rangle$ are the electronic states for the various separations. Chen and Mei[13] using the correlated variational approach(CVA) obtained two separate expression for the ground state energy for 1D and 2D and said their method can be extended to 3D in principle. Our formulation is to extend the method to 3D and also simplify the CVA. So adopting the approach in Ref. [13] mutatis mutandis, we obtain an expression for the variational ground state energy for NN and NNN hopping as

$$E_g = \frac{-t\left[\Sigma_{Lc=0}^{S-1} T_L n_{Lc} X_{Lc} X_L\right] + t'\left[\Sigma_{Lc=0}^{S-1} T_L{}' n_{Lc} X_{Lc} X_L{}'\right] + U n_0 X_0^2}{\Sigma_{Lc=0}^{S-1} X_0^2}, \tag{13}$$

where L and L' are the separations of the new states produced by using H_t and $H_t{}'$ to operate on a state with separation L_c for NN and NNN hopping respectively, while T_L and $T_L{}'$ are the number of such new states for the various L and L' respectively. The n_{Lc} in Eq.(13) is the total number of states in the various separations, L_c, of the non-interacting system. Observe that when t' = 0, the expression becomes only for NN hopping and will lead to the same equations for 1D and 2D as those of Ref. [13]. Further, observe that the operation of the Coulombic part of the Hamiltonian yields only a product of the U term and the total number of onsite states, n_0. So if U = 0, we will have only the kinetic part which is non-interacting.

Eq. (13) is applicable in all the three dimensions. Like the two separate aforementioned equations in Ref. [13], Eq. (13) has to be minimized with respect to the variational parameters to obtain an S x S matrix form of the interaction. It is this matrix form that is solved to obtain the smallest eigenvalue which becomes the ground state energy for that system while its corresponding eigenvectors are the variational parameters which are substituted into Eq. (12) to obtain the ground state wavefunction. The minimization process, however, could become tedious especially as the lattice size increases as well as we move to higher dimensions. Thus as a further simplification, we develope an expression to directly obtain the matrix form of the interaction and that is also applicable in all the three dimensions:

$$\left[E\delta_{L_cL} - \left(\frac{U}{t}\right)\delta_{0L_c}\delta_{0L} + (2T_L)_{L_cL} - \left(\frac{t'}{t}\right)(2T_{L}')_{L_cL'}\right][X_{Lc}] = 0 \tag{14}$$

where $E = E_g/t$ is the total ground state energy spectrum and its smallest value is the ground state energy of the systems which will be denoted by $E(G)$.

DISCUSSION OF RESULTS AND CONCLUSION

The ground state energies and ground state wavefunctions of several arbitrary-size lattices in all the three dimensions are obtained. Starting with the limiting case of the Hubbard systems which is when the interaction is switched off (i.e. U = 0), the expected behaviour of the ground state energy and wavefuntion of the non-interacting system of electrons are observed. First, the ground state energy is expected to be the same for systems of the same dimensions, irrespective of their sizes. This is observed in our results as the values of the ground state energies for lattices in the various dimensions are the same, that is, all the ID lattices have $E(G)$ = -4, all the 2D lattices have $E(G)$ = -8 and all the 3D lattices have $E(G)$ = -12. Further, it is obvious that even in the absence of any form of interaction between the two electrons, they require the lowest energy to hop in 3D and the highest energy in 1D. Secondly, the wavefunction of the non-interacting system should exhibit that the electrons have equal probability of being found at any given site because they are free to hop to any available sites in the system, irrespective of the lattice separation distance. Thus the probability of double occupancy is the same as single occupancy. This is also the case with our results as the variational parameters of each of the lattices (i.e.X_{Lc}) are of equal weight when U = 0. Further, the values of the variational parameters is the same for lattices having the same total number of possible separations, irrespective of their dimensions. For example the 1D N = 10 and 2D N x N = 5 x 5 have S = 6 and they have $X_1 = X_2 = X_3 = X_4 = X_5 = 0.4082$; the 2D N x N = 7 x 7 and the 3D N x N x N = 5 x 5 x 5 have S = 10 and they have $X_0 = X_1 = X_2 = \ldots = X_3 = X_9 = 0.3162$. This consequence of the normalization condition, is a successful check of the HSCVA.

Switching on the competition between itinerancy and localization by including the Coulombic interaction, we can observe the behaviour of the electrons and consequently the propeties of the system as the dimensionless interaction strenght, U/4t, is varied. As shown in Figures 2(a) - 1(d), the ground state energy increases as U/4t is increased in all

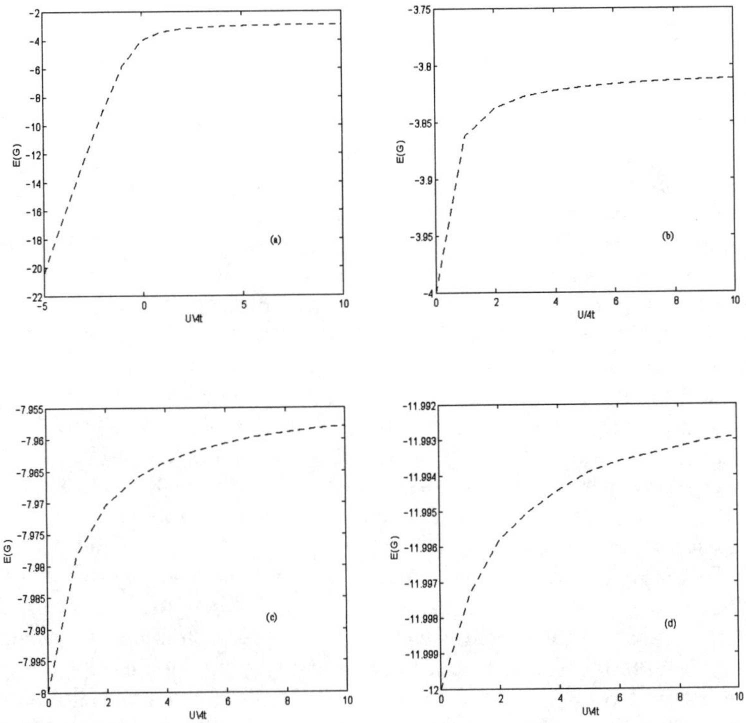

FIGURE 2. The ground state energy, E(G), as a function of the interaction strength, U/4t for : (a) N = 4, (b) N = 10, (c) N x N = 10 x 10 and (d) N x N x N = 10 x 10 x 10.

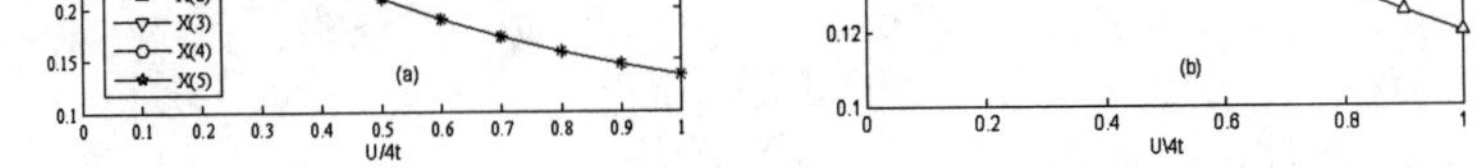

FIGURE 3. The variational parameters, X, as a function of the interaction strength, U/4t, for: (a) N = 10 with all separations (i.e Lc = 0, 1, . . . , S-1) and (b) for N x N = 10 x 10 with only the smallest and largest separations (i.e. X(0)and X(20)).

the three dimensions. The negative U/4t induces almost a rapid response on the ground state energy while a positive U/4t induces a gradual response on it. This agrees with the general belief that the negative U/4t regime will enhance the interaction of the electrons while the positive U/4t will suppress it. This observation becomes clearer in the ground state wavefunction. Observe in Figures 3(a) and 3(b) in which we considered only the positive U/4t for simplicity, that the electrons prefer to stay apart for large positive U/4t. This is monitored in the ground state wavefunction by the parameters X_{LS-2} and X_{LS-1} which in Figure 3 increase as U/4t is increased. It is easy to see that the tendency to double occupancy is less favoured in this regime. However, for smaller values of U/4t, the electrons prefer to come close together and the tendency to double occupancy is more favoured. This is monitored by $X_0(=X(0))$ and $X_1(=X(1))$ in Figure 3. Similar graphs were obtained in Ref. [13] for the pair correlation function vs U/4t. This is one of the interesting features of the HSCVA: the ground state energy and wavefunction are enough to see where the physics of the system lies. As we have observed from the discussion of the results, the negative U will be suitable for studying systems with attractive carriers as in high temperature superconductivity [14] while the positive U will be suitable for systems of non-attractive carriers as in magnetism [15].

In the discussion so far, it is obvious that the parameter regime of the competition is obtained by varying U. One way to theoretically varies the hopping in the literature is to allow NNN hopping leading to the $t-t'-U$ model. Such attempts in the literature [16, 17] are to show that the stability of ferromagnetism is intimately linked with the structure of the underlying lattice and kinetic energy (i.e. hopping) of the electrons. This effect can be explored by enhancing only the kinetic energy since the lattice structure only enters the interaction via the kinetic energy. The ground state energy and wavefunction for arbitrary lattices are also obtained for the $t-t'-U$ Hamiltonian using the HSCVA. It is observed that by varying the hopping interaction, t'/t, the interaction is enhanced at all $t' \neq 0$ in all three dimensions but do not produce ferromagmagnetism. This agrees with the view that the Hubbard model being a singlet system can produce ferromagnetism only in special cases [18]. If we have to account for ferrogmagnetism, then we need a model that allows triplet pairing and this requires also including other interacting terms. This can be achieved with the HSCVA by including triplet states to the basis states in Eq. (12) and then studying the t-U-V-J model where V is the NNN Coulombic term and J is the NNN exchange interactions. Amadon and Hirsch have already studied this model for 1D systems [19]. However, they spotted a controversial observation that in their exact calculation, the V term suppresses the tendency to ferromagnetism particularly for small U for all band fillings while mean field theory (MFT) predicts no effect of V. It will be interesting, therefore, to use the HSCVA to study this model especially as it is applicable to d = 1, 2, 3.

Another interesting extension of the HSCVA is to generalize it for few particles' problems which have become more attractive in recent times because of the possible application in quantum dots, mesoscopic systems and cold gases (see Ref [11] and references therein). We have observed that the rules to set up the various separations for the general few particle problem is the same as the case of two electrons. Since one of the main ingredients of the HSCVA is the ability to obtain these separations, its generalization to higher numbers of particles is possible in principle.

Acknowledgement

I thank J. O. Idiodi, E. A. Enaibe and E. Enukpere for very useful discussion. I also appreciate the inspirational discussion with M. Kulic, K. Matho and A. Liebsch. I acknowledge that the final preparation of this paper was done with facilities at the Max Planck Institute for Physics of Complex Systems, Dresden, Germany. Finally, I express my gratitude to the organizers of the XII Training Course in the Physics of Strongly Correlated Systems for partial support and hospitality in Vietri sul Mare (Salerno), Italy. In particular, I am grateful to Adolfo Avella for his help and efforts to make the trip possible for me and my colleague, Simon Ehika.

References

1. M. Imada et al., Rev. Mod Phys. 70, 1039 (1998)
2. A. Georges, in Lectures on the Physics of Highly Correlated Electron Systems VIII (ed. by A. Avella and F. Mancini), AIP Con. Proc. 715, 3 (2004)
3. G. Kotliar and D. Vollhardt, Phys. Today 57, 53 (2004)
4. J. Simpleton, Band Theory and Electronic Properties of Solids, Oxford University Press (2001)
5. G. G. Naumis, Physica A 372, 243 (2006)
6. A.M. Tsvelik (editor), New Theoretical Approaches to Strongly Correlated Systems, Kluwar Academic Publishers (2001)
7. J. Hubbard, Proc. Roy. Soc. London, Series A276, 238 (1963)
8. E. H. Lieb and F. Y. Wu, Phys. Rev. Lett. 20, 1445 (1968)
9. W. Metznar and D. Vollhardt, Phys. Rev. Lett. 62, 324 (1989)
10. A. Georges et al., Rev. Mod. Phys. 68, 13 (1996)
11. E. Kovacs and Z. Gulacsi, J. Phys. A: Math. Gen. 38, 10273 (2005)
12. M. Fabrizio, Phys. Rev. B. 76, 165110 (2007)
13. L. Chen and C. Mei, Phys. Rev. B39, 9006 (1989)
14. A. Montorsi and D. K. Campbell Phys. Rev. B53, 5153 (1996)
15. N. Bulut et. Al., Phys. Rev. B50, 9623 (1994)
16. K. Penc et al., Phys. Rev. B 54, 4056 (1996)
17. M. Ulmke, Eur. Phys. J. B1: 301 (2000)
18. J. E. Hirsch, Phys. Rev. B 56, 11022 (1997)
19. J. C. Amadon and J. E. Hirsch, Phys. Rev B54, 6364 (1996)

Gigantic excited state optical nonlinearities in one-dimensional Mott-Hubbard Insulators

Haranath Ghosh

Max Planck Institute for complex systems, Nöthnitzer strasse 38, Dresden-01187, Germany

Abstract. Ground as well as excited state non-linear optical properties of one-dimensional Mott-Hubbard insulators are discussed in detail. One dimensional strongly correlated materials are predicted to have several orders-of-magnitude larger *excited state* optical non-linearities in comparison to that from the ground state. One dimensional strongly correlated materials like Sr_2CuO_3, halogen bridged *Ni* compounds etc. have excited one- and two-photon allowed states almost energetically degenerate which leads to order(s)-of-magnitude larger dipole coupling between them in comparison to that between the ground and optical state. This causes several orders-of-magnitude enhancement in the optical non-linearities obtained from the first two-photon state in the wavelength region suitable for terahertz communications. Our results and conclusions are based on exact numerical calculations of extended Hubbard model suitable strongly correlated systems. We also discuss ground state non-linear properties like third-harmonic generation, two-photon absorption, electro-absorption from the theoretical model and compare with experiments.

Keywords: Strongly correlated electron systems, Optical properties: Theory, Models and numerical simulation, Nonlinear optics
PACS: 71.27.+a, 78.20.Bh, 42.65.-k

INTRODUCTION

Electron-electron correlation in condensed matter systems exhibited a large number of wonders like superconductivity, magnetism, giant magneto resistance and many others [1] ; enhanced optical non-linearity is also proved to be the manifestations of strong coulomb repulsion among conducting electrons together with quantum confinement [2, 3, 4, 5]. Research on nonlinear optics of strongly correlated one dimensional systems originated from [2, 3], both of which principly showed the occurrence of nearly degenerate one- and two-photon states due to strong coulomb correlation. These nearly degenerate one- and two-photon states are also strongly dipole coupled. Spin-charge separation on Sr_2CuO_3 system [4, 6, 7, 8], joint influence of charge transfer gap and on-site [5] repulsion was emphasized. Potential of Sr_2CuO_3 single crystals as a non-linear optical material for all optical switching devices was pointed out [9]. Similarly for Halogen-bridged Nickel chain compounds ultrafast optical switching to a metallic state by photoinduced Mott- transition was reported [10]. Further important developments like evidence of two types of charge transfer excitations in Sr_2CuO_3 was found from electron energy-loss spectroscopy analysis [11]. Higher dimensional counter parts of these materials (*e.g*, $Sr_2CuO_2Cl_2$) are found to have an order of magnitude lower optical non-linearity [12, 14]. Even then such stimulating science and application research in one-dimension has also lead to the study of non-linear optical properties in higher dimensional Mott-Hubbard systems [13] too.

CP1014, *Lectures on the Physics of Strongly Correlated Systems XII: Twelfth Training Course*, edited by A. Avella and F. Mancini

Subsequently, large third-order optical non-linearity in one dimensional $Cu-O$ chain systems like Sr_2CuO_3, Ca_2CuO_3 were investigated by third harmonic generation (THG) spectroscopy [15], some of these results were reconfirmed by A. Maeda *et al.*, [16] by Z-scan technique and by M. Ono *et al.*, [17] alongwith detailed studies on Ni-Halogen chain compounds. There are some discrepancies in these studies too which we discuss later. One dominant one photon peak at $\omega_A = 0.61$ eV, a two photon resonant peak $\omega_B = 0.87$ eV were observed together with an *oscillatory structure* at $\omega_C \sim 2 \times \omega_B$ and $\omega_A \simeq 2\omega_B/3$, in case of Sr_2CuO_3. Thus the one-photon state is at energy $0.61 \times 3 = 1.83$ eV whereas the two-photon state at $0.87 \times 2 = 1.74$ eV —- the two-photon state is below the one-photon state. This observation is contrary to the other earlier studies, the two-photon state is above the one-photon state [2, 3]. We shall show that this energy ordering of low energy excited states is extremely significant for photoinduced absorption (PA), excited state optical nonlinearities etc. Furthermore, determination of locations of one- and two-photon states obtained in [3] by three-level system fitting procedure seems to be doubtfull because we shall show that the concerned two-photon state corresponds to the dip of the Im $\chi^{(3)}(-\omega;0,0,\omega)$. On the contrary, locations of two-photon states in [3, 17] appears to be away from the dip of Im $\chi^{(3)}(-\omega;0,0,\omega)$. This would lead to the fact that the energy difference between the one- and two-photon states would be even smaller than predicted and might be reversed ordered as well similar to the observation in [15].

In this paper, one of our prime interests are theoretical prediction of huge *excited state* optical non-linearities compared to that from the ground state in these classes of materials. Here first, we theoretically demonstrate the observed third harmonic generation (THG) spectrum and that the *oscillatory structure* at ω_C is another higher energy two-photon state which can be verified experimentally. We also present two photon absorption (TPA), Electro-absorption (EA) theoretically and compare with available experimental data. These results of ground state non-linear as well as linear properties compare well with experiments. Then we consider the case of third-order nonlinear optical susceptibilities from an excited state rather than the ground state to predict a general enhancement [18] in 1-d Mott-Hubbard insulators. We predict from our theoretical calculation that these classes of materials would have *several orders-of-magnitude larger* excited state optical non-linearities (*e.g,* excited state- THG, -TPA, -EA) compared to that from the ground state. Furthermore, choice of large optical non-linear medium depends not only on the ground state but also the excited states matter [20]. Nonlinear optical materials with large third-order nonlinear susceptibilities, fast response time, low loss and operability at room temperature are indispensable for next generation all-optical switching, computing devices, because the magnitude of this quantity strongly influences on the instrument performance. One of the pioneering experimental demonstration of excited state enhancement of optical nonlinearities was by Rodenberger *et al.*, in linear conjugated molecules [18] showing around two orders of magnitude enhancement in non-linearity. 50 to 160 fold increase in third order non-linearity was also observed in bacteriochlorophylls [19]. However, there are no studies yet on excited state non-linearities on one dimensional Mott-Hubbard insulators (as mentioned earlier) and given earlier experimental demonstrations in π-conjugated organic molecules [18, 19] we hope experimental verification of our theoretical predictions will also be possible.

THEORETICAL MODEL CALCULATION

One of the most widely used theoretical model for studying strongly correlated electron system is known as Hubbard model. One dimensional Mott-Hubbard insulators that we are concerned about are Sr_2CuO_3, Ca_2CuO_3 and $[Ni(chxn)_2X]Y_2$ where $X \equiv$ Halogen (like Cl, Br etc.), chxn $\equiv$ cylohexanediamine) and Y $\equiv$ counter ion like Cl, Br, NO_3. One dimensional $Cu-O$ chains are composed of CuO_4 quadrilateral structures with shared corner oxygens, in which the Cu ion is divalent (spin quantum number S = 1/2) and one unpaired electron exists in the overlap of the p_x,p_y orbitals of O and the $d_{x^2-y^2}$ orbitals of Cu. Because of large on-site Coulomb repulsion energy U_{cu} (described below) acting on Cu ions, the Mott-Hubbard gap is opened in the Cu $3d_{x^2-y^2}$ band. Similarly, in case of halogen bridged Ni-Hallides, Ni^{3+} ions and X^- are arranged alternately along the chain axis, forming purely 1D electronic state composed of the p_z orbitals of X and the d_{z^2} orbitals of the Ni.Thus both the cuprates as well as the hallogen bridged Ni compounds described above are one dimensional Mott-Hubbard insulators (involving two different kinds of bands Cu(Ni)-$3d_{x^2-y^2}(3d_{z^2})$ and O(X)-$2p_{x,y}(2p_z)$. Therefore, we use two band extended Hubbard model corresponding to Cu -$3d_{x^2-y^2}$ (Ni - $3d_{z^2}$) and O- $p_{x,y}(X-p_z)$.

$$H = \sum_{<ij>,\sigma} t_{ij}(c^\dagger_{i\sigma}c_{j\sigma}+h.c)+\sum_{i\sigma}(-1)^i\varepsilon n_i+\sum_i U_i n_{i\uparrow}n_{i\downarrow}+\sum_i V n_i n_{i+1} \qquad (1)$$

The first term of the above equation represents hopping probability between Cu(Ni) and O(X) sites, the 2nd term the on-site energies of the Cu(Ni) and $O(X)$ and thereby defining the charge transfer energy $2\varepsilon = \varepsilon_{O(X)} - \varepsilon_{Cu(Ni)}$. The third term represents the most dominant term in the Hamiltonian known as onsite coulomb repulsion, the energy felt by two electrons occupying the same Cu (Ni) -3d or O (X)- p orbital with opposite spin. The fourth term is known as intersite coulomb repulsion (V), repulsive energy felt by two electrons occupying the neighboring Cu (Ni) and O (X) sites. We solve this Hamiltonian exactly numerically for a ring of 12 sites containing 6 Cu (Ni) and 6 O (X) sites. Details of the model and method of calculation is beyond the scope of this article, these are described in detail in the references [21]. Here we describe our main results presented in the Figures below.

RESULTS AND DISCUSSIONS

Theoretical prediction of third order susceptibility of one dimensional Mott-Hubbard insulators relevant for two-photon absorption (TPA) as well as intensity dependent refractive index (n_2) are shown in upper left panel of the Figure 1. The left upper panel **(a)** represents imaginary part of the third order susceptibility whereas **(b)** the real part. Both the real and imaginary parts show peaked structure around 0.88 eV (optical fiber comunication length). The dashed curve in **(b)** corresponds to higher value of linewidth parameter $\Gamma_a = 0.4$ eV compared to the solid line case, $\Gamma_a = 0.3$ eV. Theoretical predictions of THG **(a)**, TPA **(b)**, EA **(c)** are shown in the upper middle panel. This figure conclusively proves that the third peak at ω_C is a two-photon resonant one which could be seen in the TPA spectra **(b)**. The same should be observed in other one dimensional

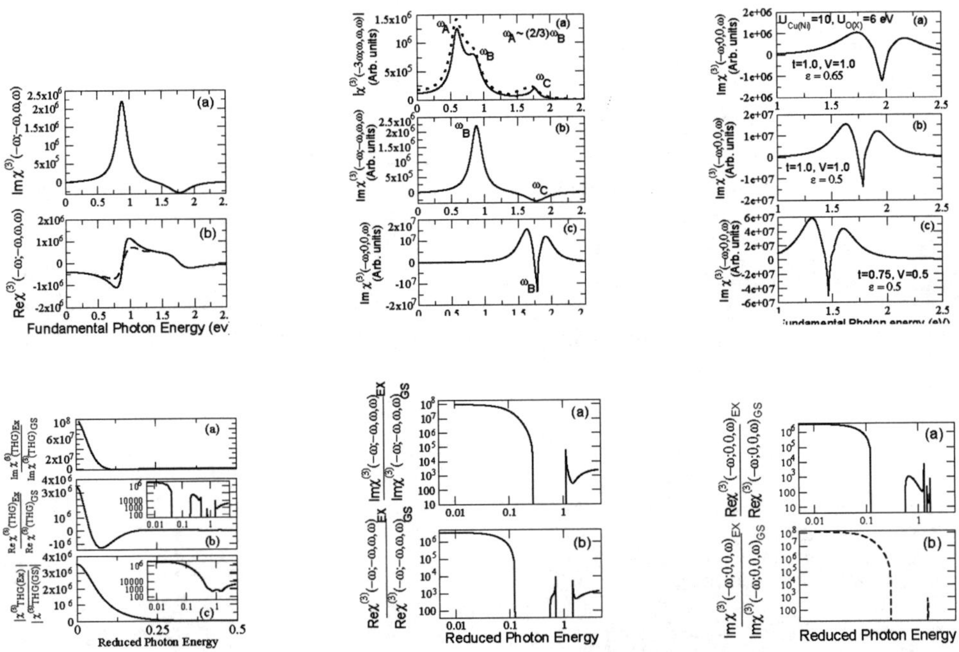

FIGURE 1. Theoretical predictions based on the exact calculation of finite size extended two band one dimensional Hubbard model. The upper panel (of figures have all same x-axis) left describes real and imaginary parts of the third order susceptibility that are related to TPA (a) and n_2 (b) respectively. The upper middle panel describes THG (a), TPA (b), EA (c). The dotted curve in (a) is for 8-site Cu-O system showing very small finite size effect. The upper right panel describe EA for various materials. The lower panel describes *excited state* nonlinear optical properties in comparision to the same measured from the ground state (for details see text).

Mott-Hubbard systems as well. Effect of lattice constant, charge transfer energy in the EA of isostructural one dimensional Mott-Hubbard insulators are shown in the figures of upper right panel. For same values of on-site coulomb repulsions, in the right chanel Fig **(a)** $\varepsilon = 0.65$, $V = t = 1$ eV ; in Fig **(b)** $\varepsilon = 0.5$ eV, $V = t = 1$ eV ; and in Fig **(c)** $\varepsilon = 0.5$ eV = V, $t = 0.75$ eV in order to simulate situation from Sr_2CuO_3 to $[Ni(chxn)_2Cl]Cl_2$ to $[Ni(chxn)_2Br]Br_2$.

Theoretical prediction of enhanced excited state nonlinearity in one dimensional Mott insulators are shown in the lower panel of figures. The left lower panel **(a)** represents ratio of imaginary part of the THG susceptibility from the two-photon allowed state (EX) to that from the ground state (GS). The later is the same as that presented in middle upper panel **(a)**. The panel **(b)** represents the ratio of the real part of the THG evaluated from the excited state to the ground state. The panel **(c)** represents the ratio of the magnitudes of the THG as indicated in the axes labels. The inset figures are indicative of the same as that of the main figure but in logarithmic scale. Excited state enhancement of imaginary **(a)** and real **(b)** parts of the third order susceptibility with respect to the ground state relevant for TPA in logarithmic scale are presented in the lower middle panel. Excited state enhancement of real **(a)** and imaginary **(b)** parts of the third order susceptibility with respect to the ground state relevant for EA in logarithmic scale are presented in the lower right panel.

Thus one dimensional strongly correlated materials are predicted to have several orders-of-magnitude larger *excited state* optical non-linearities in comparison to that from the ground state. One dimensional strongly correlated materials like Sr_2CuO_3, halogen bridged *Ni* compounds etc. have excited one- and two-photon allowed states almost energetically degenerate which leads to order(s)-of-magnitude larger dipole coupling between them in comparison to that between the ground and optical state. This causes several orders-of-magnitude enhancement in the optical non-linearities obtained from the first two-photon state in the wavelength region suitable for terahertz communications.

REFERENCES

1. P. W. Anderson, Nature, **235** (1987) 1196 ; S. R. Julian *et al.*, J. Phys. - Condens. Matt. **8** 9675-9688 (1996).
2. T. Ogasawara, M. Ashida, N. Motoyama, H. Eisaki, S. Uchida, Y. T okura, H. Ghosh, A. Shukla, S. Mazumdar and M. Kuwata-Gonokami, Phys. Rev. Lett. **85**, 2204 (2000).
3. H. Kishida *et al.*, Nature (London) **405**, 929 (2000).
4. Y. Mizuno, K. Tsutsui, T. Tohyama and S. Maekawa, Phys. Rev. B **62**, R4769-R4773 (2000).
5. G. P. Zhang, Phys. Rev. Lett. **86**, 2086 (2000).
6. C. Kim, J Elect. Spect. **117**, 503 (2001). S. Maekawa and T. Tohyama, Rep. Prog. Phys. **64**, 383 (2001).
7. Akira Takahashi, Hiroki Gomi and Masaki Aihara, Phys. Rev. B **69** 075116 (2004).
8. I. A. Zaliznyak *et al.*, Phys. Rev. Lett. **93**, 087202 (2004).
9. M. Ashida *et al.*, Applied Phys. Lett. **78**, 2831 (2001).
10. S. Iwai *et al.*, M. Ono *et al.*, Phys. Rev. Lett. **91**, 057401 (2003).
11. A. S. Moskvin *et al.*, Phys. Rev. Lett. **91**, 037001 (2003).
12. M. Ashida,T. Ogasawara, N. Motoyama, H. Eisaki, S. Uchida, Y. Taguchi, Y. Tokura, M. Kuwata-Gonokami, H. Ghosh, A. Shukla, and S. Mazumdar, Int. J. Mod. Phys B **15**, 3628 (2001).
13. A. Schülzgen *et al.*, Phys. Rev. Lett. **86**, 3164 - 3167 (2001). For femtosecond studies see, K. Matsuda *et al.*, Phys. Rev. B **50**, 4097 - 4101 (1994);R. A. Kaindl *et al.*, Science **287**, 470 - 473 (2000); For magnetic excitations see, J. D. Perkins *et al.*, Phys. Rev. Lett. **71**, 1621 - 1624 (1993); J. Lorenzana and G. A. Sawatzky, Phys. Rev. Lett. **74**, 1867 - 1870 (1995); N. Motoyama, H. Eisaki, and S. Uchida, Phys. Rev. Lett. **76**, 3212 - 3215 (1996); H. Suzuura *et al.*, Phys. Rev. Lett. **76**, 2579 - 2582 (1996).
14. M. Takahashi *et al.*, Phys. Rev. B **66** 125102. (2002). M. Ashida *et al.*, Europhys Lett. **58** 455 (2002).
15. H. Kishida, M. Ono *et al.*, Phys. Rev. Lett. **87**, 177401 (2001).
16. A. Maeda, M. Ono *et al.*, Phys. Rev. B **70**, 125117 (2004).
17. M. Ono, K. Miura, A. Maeda *et al.*, Phys. Rev. B 70, 085101 (2004).
18. Q. L. Zhou *et al.*, Phys. Rev. A **43**, 1673 (1991); D. C Rodenberger *et al.*, Nature **359**, 309 (1992);J. R. Heflin *et al.*, Y. M. Cai, and A. F. Garito, Phys. Rev. A **45**, R4233 - R4236 (1992) ; D. C. Rodenberger *et al.*, Phys. Rev. A **51**, 3234 - 3245 (1995).
19. Lin X. Chen and D. Philip Laibale, Chem. Phys. Lett. **270**, 255 (1997).
20. S. R. Marder *et al.*, Science **261**, 186 – 189 (1993); S. R. Marder *et al*, Science **263**, 511-514 (1994); S. R. Marder *et al.*, Science **265**, 632-635 (1994); S. R. Marder *et al.*,, Science **276**, 1233 (1997).
21. Haranath Ghosh, Europhys Lett. **75** (3) 468- 474 (2006); Haranath Ghosh, Phys. Rev. B **75** (2007).

Polaron in the t-J Models with Three-site Terms: the $SU(2)$ and the Ising Cases

Krzysztof Wohlfeld

Marian Smoluchowski Institute of Physics, Jagellonian University, Reymonta 4, PL-30059 Kraków, Poland

Abstract. We compare the role of the three-site terms added to the t-J models in the $SU(2)$ and the Ising cases in the extremely low doping regime, i.e. when a single hole added to the strongly interacting half-filled system becomes a polaron. We show that in the realistic Ising case the three-site terms play a vital role in the polaron movement and should never be neglected unlike in the $SU(2)$ case.

Keywords: polaron, t-J model, three-site terms, $SU(2)$, Ising
PACS: 75.10.Hk, 75.10.Jm, 75.30.Ds, 75.30.Et

INTRODUCTION

The Hubbard model is regarded as *the* model which can describe many of the spectacular phenomena observed in transition metal oxides (TMO) – such as e.g. Mott insulating ground states, antiferromagnetism, and perhaps even high temperature superconductivity [1]. Usually, however, it is the t-J model which is used to explain these phenomena [1]. This model: (*i*) is easier to solve than the Hubbard model [2], (*ii*) can be easily reduced to the Heisenberg model for the half-filled doping and hence naturally predicts the antiferromagnetic ground state of undoped TMO, (*iii*) is believed to give qualitatively similar predictions as the Hubbard model for it merely follows from a perturbative expansion of the Hubbard model in the physical regime of strong interactions [3].

A priori, one can have some doubts concerning the stated above $1:1$ correspondence between these two models. Actually, in a rigorous perturbative expansion of the Hubbard model one obtains the so-called three-site terms, in addition to the kinetic and magnetic terms in the t-J model [3]. These terms are often neglected in the small doping regime (close to the Mott insulating ground state) when they are much smaller than the two t-J model terms and only yield a small longer range hopping. Indeed, it was shown in Ref. [4] that the inclusion of the three-site terms in this standard $SU(2)$ symmetric case of the t-J models does not change qualitatively the low-energy physics of the system.

However, a different situation can arise when the Ising limit of the t-J model is taken and the $SU(2)$ symmetry is broken, which is quite often done for computational purposes in e.g. more elaborate t-J models [5]. Could the three-site terms play an important role and entirely change the solutions in the Ising limit? In this paper we answer this question by comparing the role of the three-site terms added to the t-J models in the $SU(2)$ and the Ising cases, cf. [6]. We perform calculations in the extremely low doping regime namely for the case of a single hole added to a half-filled ordered ground state of the models, i.e. when a single hole couples to the excitations of the ordered state in a polaronic way.

CP1014, *Lectures on the Physics of Strongly Correlated Systems XII: Twelfth Training Course,* edited by A. Avella and F. Mancini

t-J MODELS WITH THREE-SITE TERMS

In order to see how the three-site terms arise in the perturbation theory and why they are often neglected we introduce (a generalized version of) the Hubbard model:

$$H = -t \sum_{\substack{i,j(i),\\ \sigma,\sigma'}} z_{ij}^{\sigma\sigma'} c_{i\sigma}^{\dagger} c_{j\sigma'} + U \sum_{i} n_{i1} n_{i2}, \tag{1}$$

where we defined $n_{i\sigma} = c_{i\sigma}^{\dagger} c_{i\sigma}$, $\sigma = 1,2$ – a quantum number with two values (not necessarily a spin quantum number), and $z_{ij}^{\sigma\sigma'}$ is a hopping element between state σ on site i and state σ' on site j. Hence, this model allows for e.g. σ-dependent values of the hopping elements. As already mentioned, usually in realistic systems the hopping t is much smaller than the on-site interaction U so that we can safely perform a perturbative expansion in the kinetic terms [3]. We obtain the effective Hamiltonian:

$$H_{\text{eff}} = -t \sum_{\substack{i,j(i),\\ \sigma,\sigma'}} \tilde{t}_{ij}^{\sigma\sigma'} \tilde{c}_{i\sigma}^{\dagger} \tilde{c}_{j\sigma'} - \frac{1}{4} J \sum_{\substack{i,\sigma_1,\ldots,4,\\ j(i),j'(i)}} \left(z_{j'i}^{\sigma_4\sigma_3} z_{ij}^{\sigma_2\sigma_1} \tilde{c}_{j'\sigma_4}^{\dagger} c_{i\sigma_3} n_{i\bar{\sigma}_3} n_{i\bar{\sigma}_2} c_{i\sigma_2}^{\dagger} \tilde{c}_{j\sigma_1} \right) \tag{2}$$

$$= \underbrace{-t \sum_{\substack{i,j(i),\\ \sigma,\sigma'}} (\ldots) - \frac{1}{4} J \sum_{\substack{i,\sigma_1,\ldots,4,\\ j(i)=j'(i)}} (\ldots)}_{H_{t-J}} \underbrace{- \frac{1}{4} J \sum_{\substack{i,\sigma_1,\ldots,4,\\ j(i)\neq j'(i)}} (\ldots)}_{H_{3s}}, \tag{3}$$

where $\tilde{c}_{i\sigma}^{\dagger} = c_{i\sigma}^{\dagger}(1 - n_{i\bar{\sigma}})$ and $J = 4t^2/U$.

The first two terms in Eq. (3) form a generalized version of the t-J model (H_{t-J}) whereas the last one is the three-site term (H_{3s}). The physical interpretation of these terms is the following: (*i*) the first term describes the hopping of electrons with no double occupancies allowed and this term contributes to the total energy of the system as $\propto t\delta$, where δ is the number of doped holes (by definition we have no holes in the half-filled case), (*ii*) the second term $\propto J(1-\delta)^2$ describes the electron which makes a virtual hopping t from site j to an occupied site i which costs energy U and then comes back to site j, (*iii*) the third term $\propto J\delta$ describes the situation when the electron does not return to the same site j but goes to the site j' provided this site is unoccupied (hence in the half-filled case this term vanishes). Since in the low doping regime $\delta \ll 1$ and $J < t$, then naturally the last term has the smallest contribution to the total energy of the system and this is the reason why (as explained in the introduction) it is often neglected.

Let us also note the reasons for introducing the generalized version of the Hubbard and t-J models. Firstly, it enables us to treat the Ising and the $SU(2)$ cases on equal footing (see below). Secondly, it is possible to obtain the Ising limit from this model in such a way that it coincides with the model describing the correlated electrons in the systems with partial t_{2g} orbital degeneracy (with active yz, zx orbitals in the ferromagnetic plane) as e.g. in the d^1 system of Sr_2VO_4 [7]. This means that this specific Ising limit would not only correspond to a standard Ising spin approximation *but* would also constitute a realistic physical model [6].

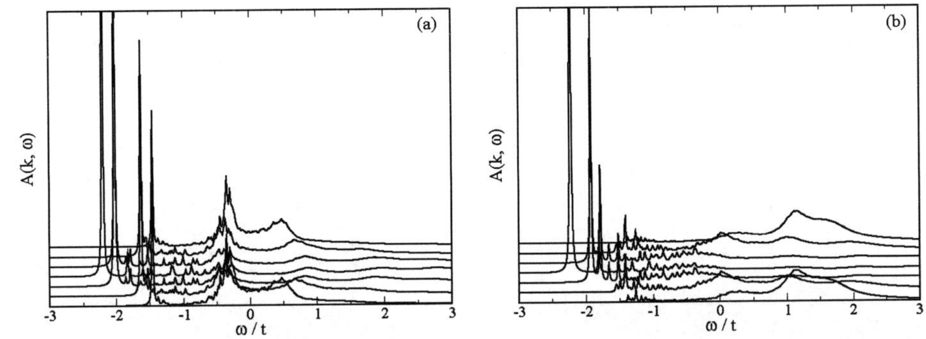

FIGURE 1. Spectral density $A(\mathbf{k},\omega)$ of the model Eq. (5) with $J = 0.4t$ along the $\Gamma - \mathrm{M}$ (from bottom to top) direction of the 2D Brillouin zone without (with) the first two terms of the three-site terms Eq. (6) shown on the left (right) panel, respectively.

POLARON IN THE $SU(2)$ CASE

In the $SU(2)$ symmetric case the quantum numbers $\sigma = \uparrow, \downarrow$ describe physical spins. Then

$$z^{\sigma\sigma'}_{\langle ij\rangle} = \begin{pmatrix} 1 & 0 \\ 0 & 1 \end{pmatrix}, \tag{4}$$

where the hopping is restricted to the nearest neighbour sites $\langle ij\rangle$. Substituting Eq. (4) to Eq. (3) we obtain for the *t-J* part

$$H_{t-J} = -t\sum_{\langle ij\rangle}(\tilde{c}^{\dagger}_{i\uparrow}\tilde{c}_{j\uparrow} + \tilde{c}^{\dagger}_{i\downarrow}\tilde{c}_{j\downarrow} + h.c.) + J\sum_{\langle ij\rangle}(\mathbf{S}_i\mathbf{S}_j - \frac{1}{4}\tilde{n}_i\tilde{n}_j), \tag{5}$$

where $S^+_i = \tilde{c}^{\dagger}_{i\uparrow}\tilde{c}_{i\downarrow}$, $S^-_i = \tilde{c}^{\dagger}_{i\downarrow}\tilde{c}_{i\uparrow}$, $S^z_i = \frac{1}{2}(\tilde{n}_{i\uparrow} - \tilde{n}_{i\downarrow})$. The three-site terms part reads

$$H_{3s} = -\frac{1}{4}J\sum_{\{j'ij\}}(\tilde{c}^{\dagger}_{j'\uparrow}\tilde{n}_{i\downarrow}\tilde{c}_{j\uparrow} + \tilde{c}^{\dagger}_{j'\downarrow}\tilde{n}_{i\uparrow}\tilde{c}_{j\downarrow} - \tilde{c}^{\dagger}_{j'\uparrow}\tilde{c}^{\dagger}_{i\downarrow}\tilde{c}_{i\uparrow}\tilde{c}_{j\downarrow} - \tilde{c}^{\dagger}_{j'\downarrow}\tilde{c}^{\dagger}_{i\uparrow}\tilde{c}_{i\downarrow}\tilde{c}_{j\uparrow}), \tag{6}$$

where we sum over all possible configurations of three adjacent sites $\{j'ij\}$ (with i denoting the middle site).

In the half-filled case it is easily seen that the ground state of the model is a quantum antiferromagnet (AF). The calculation of the ground state of the model with one additional hole in the AF is more complicated and one needs to use the self-consistent Born approximation (SCBA) [8]. The hole spectral density $A(\mathbf{k},\omega)$ obtained in this way for the two-dimensional (2D) lattice is shown in Fig. 1(a) [Fig. 1(b)] without [with] three-site terms included, respectively. In both figures we see a finite though strongly renormalized dispersion of the lowest peak. This sugests a formation of a *mobile polaron* as a hole moves by dressing up with spin excitations and acquires a large but finite effective mass. The inclusion of the three-site terms Eq. (6) merely strongly increases the spectral weight of the incoherent part for particular $\mathbf{k}$ values in the Brillouin zone, cf. Fig. 1(b).

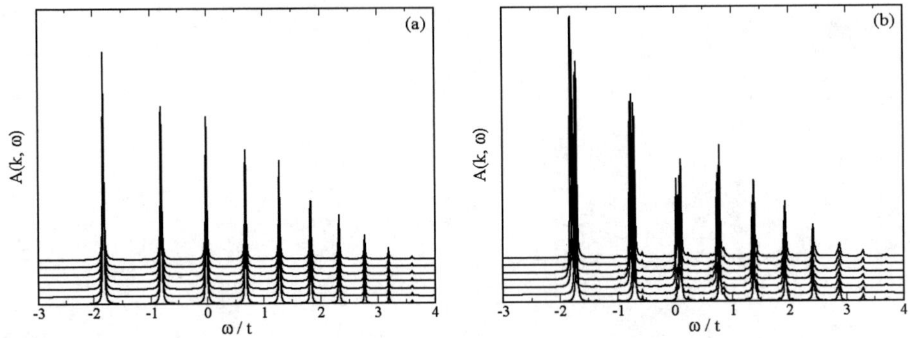

FIGURE 2. Spectral density $A(\mathbf{k},\omega)$ of the model Eq. (8) with $J = 0.4t$ along the $\Gamma - \mathrm{M}$ direction (from bottom to top) of the 2D Brillouin zone without (with) the first two terms of the three-site terms Eq. (9) shown on the left (right) panel, respectively.

POLARON IN THE ISING CASE

To obtain the Ising limit from the model Eq. (3) which at the same time serves as the realistic model for correlated electrons with t_{2g} degeneracy (see last paragraph of the second section) we assume $\sigma = yz, zx$ and

$$z^{\sigma\sigma'}_{\langle ij\rangle||\hat{x}} = \begin{pmatrix} 0 & 0 \\ 0 & 1 \end{pmatrix} \quad \text{and} \quad z^{\sigma\sigma'}_{\langle ij\rangle||\hat{y}} = \begin{pmatrix} 1 & 0 \\ 0 & 0 \end{pmatrix}, \tag{7}$$

where the hopping is between the nearest neighbour sites along the $\hat{x}$ and $\hat{y}$ directions depending on the value of σ [6]. Substituting Eq. (7) to Eq. (3) we obtain for the t-J part

$$H_{t-J} = -t \sum_{\langle ij\rangle||\hat{x}} (\tilde{c}^{\dagger}_{ib}\tilde{c}_{jb} + h.c.) - t \sum_{\langle ij\rangle||\hat{y}} (\tilde{c}^{\dagger}_{ia}\tilde{c}_{ja} + h.c.) + J \sum_{\langle i,j\rangle} (T^z_i T^z_j - \frac{1}{4}\tilde{n}_i\tilde{n}_j), \tag{8}$$

where $T^z_i = \frac{1}{2}(\tilde{n}_{ia} - \tilde{n}_{ib})$. The three-site terms part reads

$$H_{3s} = -\frac{1}{4}J \sum_{\{j'ij\}||\hat{x}} \tilde{c}^{\dagger}_{j'b}\tilde{n}_{ia}\tilde{c}_{jb} - \frac{1}{4}J \sum_{\{j'ij\}||\hat{y}} \tilde{c}^{\dagger}_{j'a}\tilde{n}_{ib}\tilde{c}_{ja}$$
$$+\frac{1}{4}J \sum_{\{j'ij\}||[\hat{y},\hat{x}]} \tilde{c}^{\dagger}_{j'a}\tilde{c}^{\dagger}_{ib}\tilde{c}_{ia}\tilde{c}_{jb} + \frac{1}{4}J \sum_{\{j'ij\}||[\hat{x},\hat{y}]} \tilde{c}^{\dagger}_{j'b}\tilde{c}^{\dagger}_{ia}\tilde{c}_{ib}\tilde{c}_{ja}, \tag{9}$$

where the three-site hop can be either straight (along $\hat{x}$ or $\hat{y}$ direction) or along the corner [e.g. first hop along the $\hat{x}$ direction and then along the $\hat{y}$ one, cf. the last term of Eq. (9)].

In the half-filled case the ground state of the model is a *classical* state with alternating orbitals occupied on each site (Ising AF in the spin language). The ground state of the model with one additional hole in the Ising AF can again be calculated using the SCBA. The hole spectral densities $A(\mathbf{k},\omega)$ for the 2D case are shown in Fig. 2 with and without three-site terms included. We see that in the case *without* the three-site terms the polaron is *immobile* [8] whereas *adding* the three-site terms yields a small *dispersion*.

CONCLUSIONS AND FINAL DISCUSSION

In conclusion, we studied a problem of a single hole doped into the half-filled ground state of two different cases of the t-J model. On one hand, in the $SU(2)$ case we noticed a quite big quantitative difference in the polaron behaviour between the models with and without the three-site terms. However, the polaron was always mobile and the inclusion of the three-site terms did not change qualitatively the low-energy physics of the system, cf. [4]. On the other hand, in the Ising case the polaron became mobile only *after* including the three-site terms in the model.

This means that for a t-J model in the Ising limit to become more realistic one should include the three-site terms even in the extremely low doping regime. In particular, one should also include these terms for the models describing the t_{2g} systems such as e.g. Sr_2VO_4 where the Ising limit arises not as an approximation to the $SU(2)$ case but as a genuine t-J model.

As a *postscriptum* one can discuss the reason why the three-site terms play any role in the Ising limit of the t-J model despite the fact that they are very small in the low doping regime (as stated in the introduction and explictly shown in the second section). We suggest the following answer: if there is a strong competition between two processes in the system (e.g. the magnetic interaction and the hole hopping in the t-J model) then the neglected third term (e.g. the three-site hopping) can decide about the ground state of the system. This is because it is the difference between these two larger terms in the Hamiltonian which matters and with which the magnitude of the neglected term should be compared.

ACKNOWLEDGMENTS

I would like to thank the organising committee of the course for their financial support. I thank Andrzej M. Oleś, Maria Daghofer and Peter Horsch for the extremely fruitful discussion during the common work on this subject. I am also particularly grateful to Andrzej M. Oleś for his invaluable help and ideas. This work was supported in part by the Foundation for Polish Science (FNP) and by the Polish Ministry of Science under grant No. N202 068 32/1481.

REFERENCES

1. M. Imada, A. Fujimori, and Y. Tokura, *Rev. Mod. Phys.* **70**, 1039 (1998).
2. The Hilbert space associated with the t-J Hamiltonian is smaller than for the Hubbard one: for N lattice sites the dimension of the Hilbert space for the half-filled case is $\left[\frac{N!}{(N/2)!(N/2)!}\right]^2$ for the Hubbard and 2^N for the t-J model.
3. K. A. Chao, J. Spałek, and A. M. Oleś, *Phys. Rev. B* **18**, 3453 (1978).
4. J. Bała, A. M. Oleś, and J. Zaanen, *Phys. Rev. B* **52**, 4597 (1995).
5. T. A. Costi and A. Liebsch, *Phys. Rev. Lett.* **99**, 236404 (2007).
6. M. Daghofer, K. Wohlfeld, A. M. Oleś, E. Arrigoni, P. Horsch, *Phys. Rev. Lett.* **100**, 066403 (2008).
7. A. Nozaki, H. Yoshikawa, T. Wada, H. Yamauchi, and S. Tanaka, *Phys. Rev. B* **43**, 181 (1991); Y. Imai, I. Solovyev, and M. Imada, *Phys. Rev. Lett* **95**, 176405 (2005).
8. G. Martinez and P. Horsch, *Phys. Rev. B* **44**, 317 (1991).

AUTHOR INDEX